Geometry An Introduction

by Günther Ewald

Foreword by Horst Struve

Geometry An Introduction

by Günther Ewald

Foreword by Horst Struve, University of Cologne

Translation of Foreword by Victor Pambuccian

Editing by Marvin Jay Greenberg

First Published in 1971

Copyright © 1971 by Günther Ewald

This Printing in September, 2013
by Ishi Press in New York and Tokyo

with a new introduction by Horst Struve

Copyright © 2013 by Horst Struve

ISBN 4-87187-718-3
978-4-87187-718-3

Ishi Press International
1664 Davidson Avenue, Suite 1B
Bronx NY 10453-7877
USA

1-917-507-7226

Printed in the United States of America

FOREWORD
HORST STRUVE, UNIVERSITY OF COLOGNE

Geometry was considered until modern times to be a model science. To be developed *more geometrico* was a seal of quality for any endeavor, whether mathematical or not. In the 17th century, for example, Spinoza set up his *Ethics* in a *more geometrico* manner, to emphasize the perfection, certainty, and clarity of his *pronouncements*. Geometry achieved this status on the heels of Euclid's Elements, in which, for the first time, a theory was built up in an axiomatic-deductive manner. Euclid started with obvious axioms - he called them "common notions" and "postulates" -, statements whose validity raised no doubts in the reader's mind. His propositions followed deductively from those axioms, so that the truth of the axioms was passed on to the propositions by means of purely logical proofs. In this sense, Euclid's geometry consisted of "eternal truths". Given its prominence, Euclid's Elements was also used as a textbook until the 20th Century.

Today geometry has lost the central importance it had during earlier centuries, but it still is an important area of mathematics, and is truly fundamental for mathematics from a variety of points of view. The "Introduction to Geometry" by Ewald tries to address some of these points of view, whose significance will be examined in what follows from a historical perspective.

Looking back at the history of geometry, one finds that the first significant development going beyond Euclid's geometry was the discovery and later the systematic development of projective geometry at the beginning of the 19th century. It was motivated by problems of descriptive geometry, coming from the perspective representation of three-dimensional bodies, problems that had already been addressed by artists of the Renaissance. Monge, Poncelet, Steiner, Plücker, and their contemporaries realized that one can extend the Euclidean plane by adding elements "at infinity", points and a straight line, to get a consistent new structure, a *projective plane*. It turned out that geometry was more than Euclid's theory, with which one can eminently describe physical space. The space of visual perception itself can be described mathematically. Moreover, projective geometry displayed a characteristic property that Euclidean geometry lacked. It satisfies the so-called duality principle, which states that the dual of any valid statement of projective geometry is also valid. One obtains that dual statement (in the planar case) by interchanging the words "point" and "line" (while leaving the word "incidence" unchanged). The nature of the geometric objects, whether they be points or lines, seemed at once to matter less than the relations among them.

Another discovery undermined the special status of Euclidean geometry. Euclid's axioms came with a claim of being beyond doubt, as well as recognizable as true by anyone with common sense. For many mathematicians throughout history, Euclid's fifth (and last) postulate, which was later called the Parallel Postulate, did not fulfill this requirement. Even the syntactic formulation that Euclid had chosen to avoid the notion of a straight line extended indefinitely was very complicated. Reformulations of that axiom as more reasonably sounding statements, such as "Given a line a and a point P not on a, there exists a unique line through P that does not intersect a", were seen as unconvincing (arguing that such reformulations make statements about straight lines extended beyond known bounds, and as such are not verifiable). This led to attempts to prove the Parallel Postulate from the other axioms, to allow for its removal from the list of obvious axioms. Famous are the attempts by Lambert and Saccheri, to derive a contradiction from the assumption that the Fifth Postulate does not hold. In hindsight, one may say that they developed the beginnings of a non-Euclidean geometry. It was Lobachevsky and Bolyai who first saw the possibility of building a consistent theory with the negation of the Fifth Postulate in the first half of the 19th century. Its consistency was only postulated by them. It became accepted by the mathematical community only after the discovery of models within the well-known Euclidean geometry (Beltrami, Klein, Poincare). If Euclidean geometry is consistent, which no one doubted, then so must the new structures be. What was all the more remarkable was that the same structure had several models in which the "points" and "lines" were different objects (lines could be chords of a circle or circular arcs or plane sections of quadrics, i. e. substructures of the circle geometry developed by Möbius). This was a stark departure from Euclid's geometry, where points and lines were univocally determined through definitions. For the new structures - should they be still called geometry? - the relations among the objects were apparently more important than the nature of those objects.

In his Foundations of Geometry from 1899 Hilbert answered the question asked earlier by presenting - on the example of Euclidean geometry - a new conception of mathematics, the formalist one. For Hilbert, as for Euclid, a mathematical theory is built in the axiomatic-deductive deductive manner. The axioms themselves, however, are for Hilbert no longer self-evident statements which structure a given realm of phenomena, but rather they define a structure whose objects are left undefined. Since the axioms do not specify the nature of the geometrical objects, but only the relations in which they stand to each other, the fundamental notions are only "implicitly" defined. The realization that this is sufficient for the formulation of a mathematical theory had been prepared by the discovery of the projective and non-Euclidean geometries. For Hilbert, the theorems of a theory are not true or false in the sense of the correct description of a given domain of phenomena, but only in terms of the presence or the absence of a derivation from the axiom system. If mathematics is thus understood, then the above structures are indeed geometries, namely non-Euclidean geometries.

The historical development of geometry led to the modern conception of geometry, which in turn led to the modern conception of mathematics. This is without a doubt a major cultural achievement of geometry that is worth emphasizing. All modern mathematical theories are currently set up *more geometrico*. A side effect of its own pioneering role and of the success of the axiomatic method throughout mathematics was the fact that geometry lost its special position within mathematics.

Hilbert builds Euclidean geometry in a step-wise manner. There are five groups of axioms, regarding incidence, order, congruence, parallelism and continuity. Since mathematical structures do not provide descriptions of realms living outside of mathematics, but are rather defined by axioms, one may ask questions regarding the scope of the axioms as well as regarding the structures that do not satisfy all of the axioms but only subsets thereof. Connections between geometry and algebra were gradually discovered, e. g. the simple relationship between three-dimensional projective geometry and (skew) fields. This was the beginning of a new area of mathematics, the so-called foundations of geometry, to whose development contributed not only geometers but also logicians, such as Tarski. One of the insights that arose not long after Hilbert's work, was that it is possible to build geometry without notions of order or continuity. An essential tool in this direction was the calculus of reflections, an idea that owes much to Hjelmslev. Bachmann has later deepened the study of reflection geometry in a systematic way and coined the concept of a metric plane. Metric planes are a class of structures that capture the core of the orthogonality properties common to the Euclidean and the classical non-Euclidean planes, the hyperbolic and the elliptic planes.[1] Bachmann provided axiomatizations both in terms of geometric objects, and in group theoretical terms. All Hilbert planes, i. e. all models of the plane axioms of Hilbert's axiom system, without the parallel axiom and the continuity axioms, turn out to be metric planes. Metric planes can be embedded in projective-metric planes, and thus can also be described analytically, i. e. in terms of coordinates. Reflection geometry emphasizes the interplay between geometry and group theory, the latter being a fundamental concept in many areas of mathematics.

The terms used by Hilbert for the axiomatization of Euclidean geometry, incidence, congruence, and order, indicate that even good old Euclidean geometry is a complex, multi-layered theory. Geometry is a discipline accurately described by Ewald as follows: "That man's mind is a creative mind is becoming beautifully apparent in geometry."

There are several books titled "Introduction to Geometry." The way in which the words "geometry" and "introduction" are interpreted depends on that book's author. Upon closer inspection, Günter Ewald's interpretation can be seen to be informed - both in substance and methodologically - by the major lines of development sketched above. English language books with the title "Introduction to Geometry" consist, following Coxeter's classic, of a wide variety of geometric problems, which offer a colorful, kaleidoscopic view of geometry.

These problems are dealt with on the basis of the school geometry knowledge of the reader. The "Introduction" by Ewald is of a different nature, and occupies a singular place in the English language literature. Ewald's book treats a central topic of geometry, the theory of metric planes in Bachmann's sense. It makes this theory accessible to readers of English, in a systematic manner, *more geometrico*, through an axiomatic-deductive approach.

Ewald defines metric planes in the sense of Bachmann both in geometric terms and in group theoretical language, and shows following Bachmann that metric planes can be embedded in projective-metric planes. For Euclidean and hyperbolic metric planes Ewald presents a stepwise construction (seen from the vantage point of the coordinate field, all the way to the field of real numbers) by adding order and continuity axioms. To make the book self-contained, the author first treats projective planes and models of non-Euclidean geometries. Models of hyperbolic and elliptic geometry are also treated as substructures of a circle geometry, the Möbius geometry. This geometry is also introduced axiomatically by using an axiom system of van der Waerden.

The publisher should be commended for the re-issue of this textbook, for the view of geometry presented therein cannot be found in any other English language textbook, and that view deserves to become better known to future generations of aspiring mathematicians.[2]

Translated from German by Victor Pambuccian.

[1] That metric planes can be axiomatized in terms of the notion of orthogonality alone was shown in V. Pambuccian, Orthogonality as single primitive notion for metric planes. With an appendix by Horst and Rolf Struve, Beiträge zur Algebra und Geometrie 48 (2007), 399-409.

[2] For a survey of more recent developments of the foundations of geometry, see the review by V. Pambuccian of Hilbert's work on the foundations of geometry in *Philosophia Mathematica* (III), 21 (2013), 255-277.

Preface

The indications are increasing that mathematicians once more are looking to geometry as a discipline holding a key position in mathematics. Whether or not this is a temporary trend will depend upon the way geometry is presented to those entering the field and the intellectual stimulation it provides. The neglect in the recent past of this subject that was of supreme importance in Euclid's time would seem to stem largely from the fact that geometry has been considered old-fashioned, hardly useful other than as a framework for puzzles with which to pass one's leisure hours. Although nothing should be said against those cute dissection problems, in which even the layman can see logic and intuitive ideas working in harmony, geometry is a serious business. From a general point of view it can be said to lie in the heart of modern mathematics.

The pattern for a thoroughly new approach to many areas of mathematics—the axiomatic approach offered in geometry—was set at the turn of the century when Hilbert's *The Foundations of Geometry* was published. Bourbaki's encyclopedic work, published in 24 volumes in 1960, is a direct consequence of this trend.

The axiomatic approach not only makes geometry a central part of mathematics. It also provides the basis for understanding many fundamental concepts in algebra, calculus, and even mathematical physics. One simple idea that contributes a great deal to the importance of geometry today is the idea of mapping—which brings new life to the static Euclidean approach. Combining mappings leads one to the idea of a group, thus building a bridge to the fundamentals of algebra. These combinations also help one to understand the idea of symmetry that permeates geometry with a particular flavor of beauty and fascination.

The group-theoretic viewpoint was first expressed in 1872 by Felix Klein in his *Erlanger Programm*. Although the ideas Klein expressed stimulated much research, it has taken almost a hundred years for his Programm to take its intended place—namely, in the teaching of geometry. Of course, we must not ignore the rigor of Euclid's arguments, nor the fact that for over two thousand years his *Elements* have had a stabilizing effect on teaching throughout the world. However, the dynamic approach of teaching geometry using mappings is just as rigorous, and we do Euclid as much honor by developing his ideas in this way as by slavishly following his *Elements*. Therefore, we use Klein's *Erlanger Programm* as a thread running through our work.

Although there are almost no prerequisites, the pace of the text and the amount of material covered suggest its use for mathematics majors of the upperclass or graduate level. The entire book would require a full year of study; however, the organization of the material makes it easy to use it for shorter courses, as outlined below.

Suggestions for further reading for all the chapters are given at the end of the book. Such reading is not necessary for understanding the text, but rather shows how the bibliography may be used to pursue a special topic more deeply.

Course Outlines

Three possibilities are given below for shortened courses based on the use of this book.

1. Choose only the sections shown in boldface type in the table.
2. Choose all boldfaced sections of Chapters 1–5; omit Chapter 6 and use all of Chapter 7.
3. Choose a union of the first and second possibilities. Add any additional sections or appendices that you wish, noting however that the three arrows in the table indicate sections that are logically linked.

Chapter 1   **1–14**, 15, **16–18**

Chapter 2   **1–10**, 11, **12**, 13–16

Chapter 3   **1–10**, 11–12, **13–14**, 15–16

Chapter 4   **1–12**, 13–14, 15–16

Chapter 5   **1–14**, 15, **16–18**

Chapter 6   **1–4**, 5–6, **7–11**

Chapter 7   1–12

In the planning stage of the book, F. Bachmann, L. M. Kelly, P. J. Kelly, and R. Lingenberg helped me with good suggestions. Corrections and additions were made by H. J. Arnold, R. Friedlein, D. Heitele, B. Kind, R. Kannerberg, G. Rahn, and K. A. Schmitt. Part of the drawing was done by Miss Mikela Ruth. J. Spira worked out the index. Mrs. H. Schmitt, Mrs. E. Rahn, and Mrs. M. Grosse-Dartmann did the typing. I wish to express my deep gratitude to all who have helped me.

Günter Ewald

Contents

---

viii

Contents

Chapter Three.   Affine Planes and Numbers

Chapter Four.   Affine Space and Vector Space

Chapter Five.   Projective Geometry

Chapter Six.   Inversive Geometry

Chapter Seven.   **Foundation of Euclidean and Non-Euclidean Geometry**

Appendix

Geometry:
An Introduction

Chapter One

Metric Planes and Motions

1.1 Points and Lines

We all have an intuitive idea of what a *straight line* is. We see and feel straight edges everywhere; we use rulers, drive along straight highways, span threads between our hands, see raindrops fall, and we know that rays of light move on straight lines from stars to our eyes. A *point* we might consider to be an "infinitely small spot." There are points in which the ceiling and the walls of our room meet; there are centers of rotating disks, dots of ink, periods after sentences, and intersections of lines drawn on paper. All of these we relate to the idea of a point.

Yet if asked "What is the mathematical or geometrical concept of a line? Which is that of a point?" we may have no answer. It is true that we are familiar with points and lines, but it is hard for us to say what they really are. Concerning lines, there seems to be an answer in the statement that a line is the shortest connection between two points. Apart from the fact that this definition refers only to line segments, it assumes that we know already what *length* is and what is meant by a *minimal property.* In short, we reduce a simple concept to a number of more or less complicated ones.

Let us try another answer: A line is that which remains fixed under a planar reflection. Indeed, later on this relationship will occupy us very often. To use it as a definition, however, would require some knowledge of reflections; so it would merely replace our problem with another one. In whichever way we proceed, there is a shift of the problem, not an answer to it. How, then, do we solve this dilemma?

Euclid (about 300 B.C.), one of the founders of synthetic geometry, wrote the following definitions:

A point is that which has no parts.
A line is length without width.

An interesting phenomenon lies in the fact that Euclid never made use of these rather mysterious definitions. What he really uses in his famous

---

Lake near Muir Pass, Sierra Nevada, California. Photograph by Ansel Adams.

Elements when proving geometrical theorems are "relations" between points and lines, as, for example,

"to be situated on."

A point either is situated on a given line or is not situated on that line. This relation follows certain rules, called *postulates*. One of these (in a modern version) states:

Given two different points *P*, *Q*, there is one and only one line *a* such that *P* and *Q* are both situated on *a*.

Euclid draws conclusions only from such relations, not from his definitions of points and lines. To illustrate this, let us show: "Given two different lines *a*, *b*, there is at most one point *P* situated on both *a* and *b*." Proof: Assume, on the contrary, that *Q* is different from *P* but is also situated on *a* and *b*. From the above postulate it follows that *a* = *b*, whereas *a* and *b* are supposed to be different. This contradiction proves our theorem.

In this proof one does not need to know what points and lines are. The same is true for all theorems about points and lines (and, analogously, for theorems about other mathematical objects). This fact shows that the dilemma mentioned above is unimportant. It does not matter what our intuitive concepts of points and lines are, whether we consider them to be dots and strokes of chalk or things "without parts" and "without width," respectively; if we handle all relations between geometrical objects correctly, we are in a position to build up geometry.

Does this mean that geometry is independent of our intuitive experience of nature and the technical world? We could state postulates about points and lines arbitrarily and call the totality of conclusions from these postulates "geometry." But this geometry would be of little value. *The source for our postulates, and also our guide for discovering geometrical theorems, is geometrical intuition that stems from experience. Geometrical theorems, however, are logical conclusions from these postulates.*

Concerning lines, for example, our mind has the ability to see a common feature in the following experiments:

(a) Span a thread between your hands (Fig. 1.1).
(b) Fold a piece of paper (Fig. 1.1).
(c) Arrange three pieces of cardboard with a little hole in each in such way that you can look through the holes (Fig. 1.2).

Figure 1.1

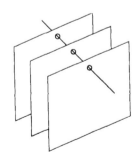

Figure 1.2

The common feature is "straightness"—of the thread, of the edge of a folded piece of paper, of a ray of light. We "see" a straight line in all three experiments. Let us express some of these observations on straightness in precise mathematical terms. We assume two sorts of objects to be given, "points" and "lines." We denote points by capital letters, P, Q, ..., and lines by small letters a, b, ... For any pair consisting of a point P and a line a, a relation "to be situated on" is assumed to be either true or false. Using symbols, we write

$$P \text{ I } a, \quad \text{instead of "} P \text{ is situated on } a\text{,"}$$

and

$$P \not{\text{I}} a, \quad \text{instead of "} P \text{ is not situated on } a\text{."}$$

So, given any pair P, a, one and only one of the relations P I a, P $\not{\text{I}}$ a can be given. There is nothing like "being partly situated on" or "almost situated on." We assume a clear-cut logical alternative in our pure mathematical treatment of the relations between points and lines.

Experiment (a) suggests the following postulate, or *axiom*, as we shall call it.

Axiom I.1 *Given two different points, there is a uniquely determined line on which both points are situated.*

In symbolic language, this axiom would read: If P is different from Q, there exists one and only one line a such that P I a and Q I a. For reasons of style we shall use alternative expressions for "the point P is situated on the line a"; for example,

> "a passes through P,"
> "P is a point of a,"
> "P is on a,"
> "P and a are incident."

The I in Axiom I.1 stems from the word incident. We call Axiom I.1 and Axiom I.2, below, *incidence axioms*.

Experiment (b) suggests the axiom "Two different planes that have a point in common meet in a line." This axiom will be explored in Chapter 4.

Since geometry is to consist only of conclusions from axioms, we cannot even take for granted that there exists any point. In fact, Axiom I.1 is trivially satisfied if no point at all exists. Therefore, we postulate two more axioms.

Axiom I.2 *On each line there are at least three different points.*

Axiom I.3 *There exist three points such that no line contains (passes through) all three.*

Intuitively we would expect that infinitely many points exist on every line. We could have postulated this instead of Axiom I.2. However, as we shall see in Section 1.5, Axiom I.2 is interesting from a logical standpoint.

In geometry texts, lines are often considered to be sets of points. Using Axiom I.2, we can characterize lines in that way also: To every line there is assigned uniquely the set of all points lying on the line. Two different points determine such a set (because of Axiom I.1). When we consider lines as sets of points, the incidence relation $P \mathrel{\mathrm{I}} a$ becomes the set-theoretic relation

$$P \in a.$$

It should be noted that without the postulation of Axiom I.2, there could be lines without any points on them. Such lines could not be characterized as sets of points.

Concerning the selection of axioms, the following should be kept in mind. There is no prescribed way in which geometrical properties must be taken as axioms. One can deduce theorems from axioms and then choose these theorems as axioms. If the original axioms are obtained as theorems deduced from the new axioms, we say that the two systems of axioms are equivalent (for an example, see Exercise 2). There are reasons of simplicity, practicality, and aesthetic taste that underlie the selection of axioms. The fact that there is a considerable degree of arbitrariness involved does not hurt the rigor, power, and beauty of axiomatic geometry.

Exercises

1. Show that every point is situated on at least three lines.
2. Prove that Axioms I.1–I.3 can be replaced by I.2, I.3, and the following axioms. I.1′: Any two points are joined by at least one line. I.1″: Two different lines intersect in at most one point.
3. Express Axioms I.1–I.3 in set-theoretic terms, considering lines as point sets.
4. Show that if any two lines intersect, all lines have the same number of points.
5. Show that if the number of points is infinite, the number of lines is also infinite, and conversely.
6. Does the statement of Exercise 5 remain true if Axiom I.3 is skipped?

1.2   Perpendiculars

The man-made world contains perpendicularity as one of its basic elements of form. Man feels at home in rectangular shaped homes; architecture of all ages illustrates this fact. This is so not only for practical reasons—there is a deep psychological basis underneath that tradition. The word perpendicular stems from a phenomenon in nature that has to do with the concepts of "horizontal" and "vertical," which, in turn, originate from gravity. Every plant and every human being standing upright creates a perpendicular in relation to a little flat piece of earth surface.

Although perpendicularity is often defined by use of the phrase "angle of 90 degrees," it refers to a very elementary phenomenon of our world. We may try to catch this phenomenon in a few axioms. But first, let us compare the following experiments, which are illustrated in Figure 1.3.

(a) Attach a weight to a thread and hold it above a flat surface of water (or a flat table).
(b) Fold a piece of paper twice, as in the figure, to obtain a right angle.

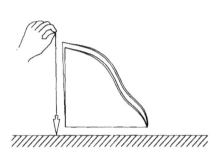

Figure 1.3

Holding the paper angle of (b) as it is illustrated shows a common feature in these experiments, namely that of "perpendicularity." Again, we find a common geometrical feature in two rather different experiments.

As in the case of the incidence relation between points and lines, we turn some of the basic observations on perpendicularity into mathematical axioms. There is a "perpendicular" relation defined for lines such that for any two lines a, b, either a is perpendicular to b or a is not perpendicular to b. In symbols we write

$$a \perp b, \quad \text{instead of "}a \text{ is perpendicular to } b\text{"}$$

and

$$a \not\perp b, \quad \text{instead of "}a \text{ is not perpendicular to } b\text{."}$$

In experiment (b), we can interchange the "lines" (by turning the paper) without loosing perpendicularity. This fact suggests the following axiom

Axiom P.1 *If a is perpendicular to b, then b is perpendicular to a.*

From the point of view of experiment (a), Axiom P.1 is not trivial. However, it is experiment (b) that frees perpendicularity from the position of the respective pair of lines in space. The possibility of dropping and erecting perpendiculars is expressed in the following axiom.

Axiom P.2 *If P is an arbitrary point and a is any line, then there exists a line b through P such that $a \perp b$. If P is on a, then b is uniquely determined.*
(See Fig. 1.4.)

The reason we do not postulate b to be uniquely determined when P is not on a will become apparent in Section 7 (see also Exercise 2 below). Lines in

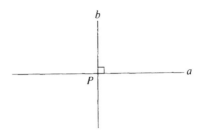

Figure 1.4

space can be perpendicular without intersecting. Since in this chapter we restrict our attention to plane geometry, we may postulate the following axiom.

Axiom P.3 *If $a \perp b$, then a and b meet in a single point.*

We say that a and b are perpendicular at this point.

Through axioms on perpendiculars, metric properties have entered our discussion without numerical data (such as "90 degrees"). This is typical for synthetic geometry.

Exercises

1. Can a line be perpendicular to itself?
2. Consider the idealized surface of the earth—that is, consider the surface of a sphere S. As "points" of a geometry, choose the points of S; as "lines," choose the great circles of S. Call two lines perpendicular if one taken as a "meridian" passes through the "poles" of the other taken as an "equator." Investigate by intuitive arguments (not rigorous proofs) which of the Axioms I.1–I.3 and P.1–P.3 are satisfied.
3. If in Exercise 2 we consider every pair of "opposite" points of S (antipodal points) as *one* point of an axiomatic geometry, which of the Axioms I.1–I.3, P.1–P.3 are true? (Again, use only intuitive arguments.)
4. Do Axioms I.1–I.3 and P.1–P.3 exclude the possibility that three lines exist, each of which is perpendicular to the other two? (Compare Exercise 3.)
5. Let the points inside a circular disk be the points of an axiomatic geometry; and let the chords of the circle be lines. Call two lines perpendicular if they intersect (inside the circular disk) and are perpendicular in the ordinary sense. Investigate by using intuitive arguments (not rigorous proofs) which of the Axioms I.1–I.3 and P.1–P.3 are satisfied.

6. Change Exercise 5 by calling any two lines perpendicular if the ordinary lines carrying the respective chords are perpendicular. Which axioms are now valid?

1.3 *Reflections*

One of our most elementary experiences with abstract forms in nature and in the man-made world involves symmetry, in particular, axial symmetry, or reflection. There are butterflies, men, and cars in front view; there are reflections in lakes, double doors, and carpet patterns. We can create those reflections by using mirrors or a folding process (see Exercise 5 at the end of this section). For the time being, we are discussing only plane geometry;

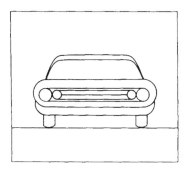

Figure 1.5

Figure 1.6

so we look at our world as a two-dimensional picture, as, for example, a photograph or drawing. Symmetries, like those in Figures 1.5 and 1.6, can be looked at as plane symmetries, though they represent spatial symmetries. By a reflection, we shall always mean a plane axial symmetry.

Reflections play a vital role in this book. The word reflection is to be understood as an operation called *mapping*, since our approach to geometry is a dynamic one. How can we crystallize the intuitive concept of a reflection in rigorous mathematical terms? A reflection preserves form, in particular the two basic form elements we met in the preceding sections—straightness and perpendicularity. We use them for defining reflections.

Before giving the definition, we remind ourselves what a mapping is. Let $\mathscr{S}$ and $\mathscr{T}$ be any two sets of objects, and let f be an assignment that assigns to every element x of $\mathscr{S}$ a unique element $f(x)$ of $\mathscr{T}$. Usually f is called a *function*; we speak also of a *mapping*, or *map*, in order to emphasize its geometrical character. The *argument*, x, can also be put on the left side:

$$(x)f.$$

It is useful for our purposes to do so. In the case of geometrical maps, we shall stick to this notation throughout the book. f is called *one-to-one*, in short *1-1*, if $(x)f = (y)f$ implies $x = y$. In case every element of $\mathscr{T}$ is an image under f, we say that f is *onto* $\mathcal{T}$. $\mathcal{S}$ is called the **domain** and $\mathcal{T}$ is called the **range** of the mapping f. The element $(x)f$ is said to be the **image** of x under f.

Now let $\mathcal{S} = \mathcal{T} = \mathcal{P}$, where $\mathcal{P}$ is the set of all points. Let a be a line. A **reflection** R_a (in a) is defined to be a one-to-one mapping of $\mathcal{P}$ onto itself with the following properties:

(1) Lines are mapped onto lines.
(2) Perpendicularity is preserved.
(3) Every point of a remains fixed.
(4) Not all points remain fixed.

By (1), R_a assigns to every point of an arbitrary line b a point of a line b'. We write

$$b' = (b)R_a.$$

Property (4) can also be expressed by saying that R_a is not the **identity map** I, that is, the mapping that maps every point onto itself; a is called the **axis** of the reflection R_a. A one-to-one mapping of $\mathcal{P}$ onto itself satisfying (1) is generally called a **collineation**. The collineation is said to be **orthogonal** if (2) holds. So a reflection is an orthogonal collineation with the additional properties (3) and (4). We postulate the following axioms on motions:

Axiom M.1 *Every line is the axis of one and only one reflection.*

Axiom M.2 *If R_a maps the point P onto the point P', it also maps P' onto P.*

A mapping that satisfies (4) and Axiom M.2 is generally called **involutoric**. So a reflection R_a is an involutoric orthogonal collineation leaving all points of a fixed.

Axiom M.1 provides a fundamental one-to-one correspondence between lines and reflections,

$$\boxed{\text{line } a \longleftrightarrow \text{reflection } R_a.}$$

This correspondence will enable us in many instances to shift problems on lines to those on reflections, and conversely.

Exercises

1. Give examples of sets $\mathcal{S}$, $\mathcal{T}$ and of mappings f of the following kinds: (a) f is onto $\mathcal{T}$ but not one to one, (b) f is one to one but not onto $\mathcal{T}$, (c) f is neither onto $\mathcal{T}$ nor one to one.

2. Prove that any reflection induces a one-to-one mapping of the set of lines onto itself.

3. Prove that a reflection R_a maps every line b perpendicular to a onto itself.

4. By using only the notation of points, lines, perpendiculars, and so forth, introduced so far, define "angular bisector of lines," "perpendicular bisector," and "middle parallel."

5. Create an ink-spot symmetry by putting some ink (or color) on paper, folding it, and then unfolding it.

6. Show that the mapping introduced for the definition of a reflection can be proved to be a mapping onto the set of points if it is only assumed to be *into* the set of points.

1.4   *Rotations and Translations*

If we turn an angular segment of paper over each edge successively, the position that the paper takes on both sides also can be obtained by

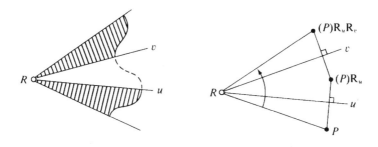

Figure 1.7   **Figure 1.8**

rotating the paper segment about its vertex R (Fig. 1.7). In Figure 1.8, the image of a point P under the successive application of reflections R_u and R_v is shown, where R is on u and v. Reflections and rotations are thus closely related to each other. Our experiment suggests this postulate: Every rotation can be replaced by successive applications of two reflections. However, rotations have not been defined yet, so the postulate cannot yet be made.

We help ourselves simply by *defining* a **rotation** as the result of two reflections, R_u, R_v, applied successively, where u and v pass through a point R, the *center* of the rotation. Although carrying out a rotation is a more elementary operation than carrying out two successive reflections, our definition is very convenient, since it takes advantage of properties of reflections that have already been introduced. We postulate one axiom about rotations that provides a further link between reflections and rotations:

Axiom M.3   *If a' is the image of a under a rotation, with center R on a, then there is a reflection R_b, with b passing through R, that maps a onto a' pointwise in the same way that the given rotation does (Fig. 1.9).*

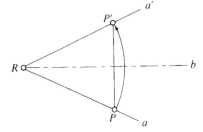

Figure 1.9

Now take a strip of paper with two parallel edges and turn it over, first one edge, then the other. The result is the same as that of a parallel displacement. This suggests that we define the concept of a parallel displacement analogously to that of a rotation. Since we do not at this stage want to talk about parallels, we shall use another name, that of a **translation**.

Let u, v be two nonintersecting lines perpendicular to a line t. Then the result of successive application of R_u and R_v is called a *translation* (*along* t) (Fig. 1.10).

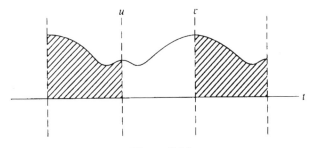

Figure 1.10

Axiom M.4 *If a' is the image of a under a translation along a perpendicular line t to a, then there exists a reflection R_b, with b perpendicular to t, that maps a onto a' pointwise in the same way that the given translation does* (Fig. 1.11).

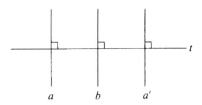

Figure 1.11

From Figures 1.10 and 1.11 and from the strong analogy between Axioms M.3 and M.4, we recognize that rotations and translations have many things in common. This will become clearer as we proceed to investigate them.

The successive application of a finite number of reflections, rotations, and translations is called a ***motion***. From the definitions of rotation and translation we have: A motion is obtained by successive application of a finite number of reflections. If the number of these reflections is even, we call the motion a ***proper motion***. In this sense, rotations and translations are proper motions.

A system of points and lines satisfying Axioms I.1–I.3, P.1–P.3, and M.1–M.4, is called a ***metric plane***.

Exercises

1. Show that the identity map I is a rotation.

2. If $(P)R_a = Q$, find a translation mapping P onto Q.

3. Let an equilateral triangle be given in the ordinary plane geometry. Express all motions mapping this triangle onto itself in terms of reflections.

4. Replace the triangle in Exercise 3 by an arbitrary regular n-gon.

5. Find in Exercise 4 the least number of reflections such that all other motions of the n-gon onto itself can be expressed as products of these reflections.

6. Introduce reflections in the example of Exercise 2 of Section 1.2 by ordinary spatial reflections of the surface of the sphere S in a plane through the center of S. Show that Axioms M.1, M.2, and M.3 are satisfied (no rigorous proofs). Why is Axiom M.4 also true?

1.5   The Problem of Parallels

Before we continue our investigation of metric planes (in Section 1.8), we shall study some key problems concerning the logical nature of geometrical axioms. Although "mathematical" geometry consists only of logical conclusions from axioms, we can observe two principles from what we have done so far:

1. Axioms must be motivated by "physical geometry," that is, by our intuitive experience of the world surrounding us.
2. New geometrical theorems and their proofs are often suggested by intuitive ideas or physical experience. Of course, a theorem is not established unless it is logically deduced from the axioms.

Let us look somewhat closer at the physical background of our development of synthetic geometry. We may describe the axioms introduced in the preceding sections in terms of the following operations:

Draw a line (segment) joining two points.

Drop (or erect) a perpendicular from a point onto a line.

Intersect perpendicular lines.

Reflect in a given line.

Rotate about a given point.

Translate along a given line.

These operations may be thought of as being carried out by a gardener, provided his garden is big enough. What is meant by *big enough*? If two perpendicular lines are drawn on the ground, their point of intersection may not belong to the garden. However, if all points and lines we start with are given in a sufficiently small, inner part of the garden, we need not leave the garden when achieving the above operations. For example, if the garden is the size of a football field, a square 20×20 yards in the center of the field will do for the inner part. In this sense, the axioms of a metric plane refer to the world "nearby." This is not accidentally so but is intentional, and it shows the result of a careful analysis of basic geometrical facts, initiated by the two-thousand-year-old problem of parallels.

What are parallel lines? We may simply say: Two lines in a plane are parallel if they do not intersect. However, is the decision on whether two lines do or do not intersect a matter of physical geometry? To intersect perpendicular lines in a garden is fine. Yet other lines drawn on the ground might intersect somewhere in the galaxy or beyond. So geometrical axioms that have to do with parallels no longer refer to the world nearby. In fact, the following arguments show how intuition may fool us concerning questions about parallels.

First assertion: Parallel lines intersect.

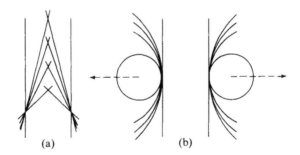

(a)   (b)

Figure 1.12

In order to show this, let two intersecting lines be given. We turn the lines continuously until they are parallel (see Fig. 1.12(a)). During these turns, the point of intersection moves farther and farther away and attains a position "at infinity," when the lines are parallel. So parallel lines do intersect.

Second assertion: Parallel lines do not intersect.

For a "proof," let two circles be given, neither of which is contained in the other or has a point of intersection with the other. We hold the two points of the circles that are nearest to each other fixed and "blow up" the circles in such a way that their centers move away from each other in opposite directions (Fig. 1.12(b)). Eventually the two circles become parallel lines. On the way, no point on one circle has come closer to a point on the other than the distance of the two points held fixed. So the two parallel lines do not intersect "at infinity."

Another difficulty of applying intuition when we have to deal with parallels is that the idea of a thread stretched between our hands or a folded piece of paper loses its meaning for long lines. What about rays of light? We might say that they provide us with lines of arbitrarily large extension. However, our visual experience of straightness is already based on the information coming to our eyes by rays of light. So what if these rays fool us by being curved in some physical way? As a matter of fact, here we touch on one of the crucial problems of Einstein's theory of relativity (see Appendix A).

In summarizing, we state that the two principles observed at the beginning of this section can be relaxed where parallels are concerned. Since intuition fails to give us a precise instruction on how to handle parallelism, we are free to choose various kinds of axioms. One is the classical *axiom of parallels* that follows.

Axiom Euc *Given a line a and a point P not on a, there exists a unique line through P that does not intersect a.*

A metric plane in which Axiom Euc holds is called a **Euclidean (metric) plane**. Two alternatives to the Euclidean axiom of parallels are of special interest:

Axiom Ell *Any two lines intersect.*

A metric plane with Ell as an additional axiom is said to be an **elliptic (metric) plane**.

Axiom Hyp *Given a line a and a point P not on a, there exist precisely two different lines through P having neither a point nor a perpendicular in common with a.*

A metric plane satisfying Hyp is denoted as a **hyperbolic (metric) plane.** So we have the division of metric planes into Euclidean, elliptic, and hyperbolic ones (there are still other alternatives that do not interest us here), diagrammed below.

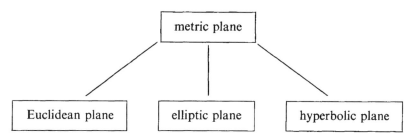

Clearly, those geometrical theorems that involve one of the Axioms Euc, Ell, or Hyp hold only in one type of plane. In the other types, each would contradict one of the axioms. In this way, we have arrived at three different kinds of geometries, each presenting a mathematical world of its own. Elliptic and hyperbolic planes are both called **non-Euclidean planes**.

The fact that there exists at all an alternative to Euclidean geometry was discovered a little more than a hundred years ago. The first discoverer was Gauss, who did not publish his results. Then Bolyai and Lobachevski independently discovered the hyperbolic plane. The discovery was made as a result of the old problem of deducing the axiom of parallels (Euc) from elementary incidence axioms, such as I.1 and I.2. It took a long time to see that it *cannot* be deduced from those elementary axioms; in our terminology, a metric plane may be Euclidean *or* elliptic *or* hyperbolic, and the axiom of parallels can be satisfied only in the Euclidean case.

Exercises

1. Prove the following theorem for Euclidean planes: If two different lines *a*, *b* are both perpendicular to different lines *c*, *d*, then they are parallel.

2. Show that Axiom Euc can be replaced by the following axiom: If two different lines *a*, *b* possess a common perpendicular, then any line perpendicular to *a* is also perpendicular to *b*.

3. Show that in a Euclidean metric plane a line has only one perpendicular passing through a given point.

4. Show that an elliptic plane contains at least seven points.

5. Show that in a hyperbolic plane at least four lines pass through every point.

6. (a) Show that the example of Exercise 3, page 8, represents an elliptic plane.
   (b) Given a line *a* in this plane, for which points *P* is the perpendicular through *P* onto *a* uniquely determined?

1.6   Models of Metric Planes

As we saw in Section 1.1, points and lines are not absolute objects in some philosophical sense; we may imagine them in different ways. Mathematically they are elements of abstract sets for which certain relations are defined satisfying certain axioms. This means, in particular, that we may choose other familiar objects, like circles, and consider them concrete realizations of abstract lines.

As an example, we restate that which we considered in Exercises 3 of Section 1.2, 6 of Section 1.4, and 6 of Section 1.5: Let the surface of a sphere, Σ, be given. Since we do not presume any knowledge of spatial geometry at this stage, we apply only intuitive arguments, by using, say, terms of geography. Σ may be thought of as the idealized surface of the earth. We call any great circle of Σ a line of an abstract geometry. Points are not the points of Σ, but pairs of antipodal points on Σ (that is, points of largest

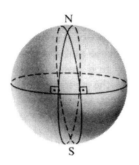

Figure 1.13

distance, like north pole and south pole). If a point of Σ lies on a great circle *a*, its antipodal point lies also on *a*, and so an incidence relation between points and lines is defined. Perpendicularity can be introduced as follows: If a great circle *a* is considered as an "equator," any "meridian" is perpendicular to *a*. A reflection in a "line" *a* is defined by the ordinary spatial reflection in the plane which intersects Σ in *a*. Such a spatial reflection maps Σ onto itself and preserves incidence and perpendicularity.

It can be shown without great difficulty that the system of points and lines thus introduced satisfies all axioms of a metric plane. (See the exercises mentioned above for explanations, though not rigorous proofs.) Clearly, *this is an example of an elliptic metric plane*, any two great circles of Σ intersecting in a pair of antipodal points.

Another example of a metric plane is found by using semicircles. Consider

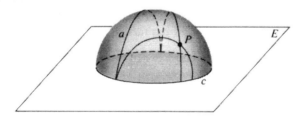

Figure 1.14

a semisphere *T* without its boundary circle *c*, and denote the plane which contains *c* by *E*. All semicircles on *T* that are perpendicular to *c* (and thus to *E*) are called "lines" of a geometry. (See Fig. 1.14.) "Points" are now the points of *T*. Incidence is the ordinary incidence of points and semicircles.

Two lines are called perpendicular at a point Q if they are perpendicular in the ordinary sense ("perpendicular tangents at Q"). A reflection may intuitively be described as follows: On E choose a point outside of c and consider it a source of light. We imagine T to be made of glass. A ray of light penetrates T in at most two points. If P is one of the two points, we denote the other by P', and we say P' is the image of P and P the image of P' under a mapping. Those points of T in which a ray only touches T can be shown to form a "line." We consider them fixed points of the mapping and call the mapping a reflection.

It is possible to show that the system of "points" and "lines" obtained in this way is a metric plane. (A proof is given in the last section of Chapter 7.) If a is a line of this metric plane and P a point not on a, any other line b through P either intersects a in a point or has a common boundary point with a on c (not belonging to the metric plane) or it has, as can be shown, a common perpendicular with a. In the second case, a and b have neither a point nor a perpendicular in common. As is seen from Figure 1.14, there are precisely two "lines" through P of that sort. So *we have found an example of a hyperbolic metric plane.*

A concrete realization of an abstract system of axioms, as we have in the above two examples, is called a *model* of this system of axioms. So the example of a hyperbolic metric plane is a model of the system of Axioms I.1–I.3, P.1–P.3, M.1–M.4, and Hyp. Similarly, the example of an elliptic plane at the beginning of this section is a model of the system of Axioms I.1–I.3, P.1–P.3, M.1–M.4, and Ell. Exercise 2 following this section provides a second model in the hyperbolic case, showing that there may be many models to a given set of axioms. In fact, our analysis of the question of parallels in the preceding section has shown that there are quite different models to the system of Axioms I.1–I.3, P.1–P.3, and M.1–M.4, namely those with different properties concerning parallels.

Using the above model of elliptic and hyperbolic metric planes, we may comment on the question of "physical geometry" as follows: In a sufficiently small spot of the surface of a large sphere, no physical difference can be made between a great circle, as in the elliptic model; a semicircle, as in the hyperbolic model; and a Euclidean line. So the extension of the "small world" of points and lines surrounding us may be thought of as being Euclidean, elliptic, or hyperbolic. As in Section 1.5, we conclude that physical observations leave open the question of parallels.

Exercises

1. Explain by intuitive arguments the validity of the axiom of a hyperbolic plane in the model represented by Figure 1.14.

2. Project the points of the hyperbolic plane in the model onto the plane E by lines perpendicular to E. Describe the model of a hyperbolic plane obtained in this way.

3. Discuss the difference between the model of a hyperbolic plane in Exercise 2 and the system of points and lines introduced in Exercise 5 of Section 1.2.

4. In the model of a hyperbolic plane, choose the points of the "boundary circle" *c* as additional "points." Any "line" is thus enlarged by two points; however, no new lines are introduced. Does this system of points and lines also represent a metric plane?

5. Choose, on the surface of a sphere *S*, all circles through a given point *N* as "lines" of an axiomatic geometry—"points" are all points of *S* except *N*. Show by intuitive arguments that this system of points and lines can be considered as a Euclidean metric plane. (Take the existence of reflections for granted; details will be found in Chapter 6.)

6. Find a general definition for "model of a system of axioms."

1.7   Consistency

A metric plane cannot be both Euclidean and hyperbolic or both elliptic and hyperbolic or both Euclidean and elliptic. Otherwise there would be a logical contradiction involved in the geometry. How about the axioms of a metric plane themselves? Can we be sure that there are no contradictions involved secretly? Axioms Euc and Ell are directly seen to contradict each other. However, it may be that a contradiction not directly recognized in the axioms of a metric plane (without Axioms Euc, Hyp, or Ell) will occur in conclusions drawn from the axioms of a geometry. To make this concept clear, let us take as a system of axioms that of a Euclidean metric plane and postulate as an additional axiom:

(A) There exist three lines, each two of which are perpendicular.
There is no obvious reason why this new axiom should contradict the axioms of a Euclidean metric plane. However, as we shall see in the following section (after a number of logical steps), (A) implies Axiom Ell and, therefore, contradicts Axiom Euc.

How can we ascertain, then, that no conclusions drawn from the axioms of a metric plane lead to contradictory theorems? The answer is that we construct a model of a metric plane consisting of only nine points and twelve lines. This model enables us to test, by a number of logical steps, whether or not all metric-plane axioms do hold simultaneously.

One may ask: Why not take one of the models of Section 1.6, the nine-point model being a rather artificial one? The answer is that the models of Section 1.6, though they deal with familiar points and circles, involve infinitely many geometrical objects; so there is no way to test the validity of, for example, Axiom I.1 individually for any pair of points. The use of algebraic equations would not help. We would have to make sure first that all algebraic theorems involved are free of contradictions.

The nine-point model consists of nine objects arbitrarily chosen as "points."

We select twelve triplets of these objects as "lines" in a way that is indicated schematically in Figure 1.15 (the lines drawn are only to indicate which three "points" form a line). Which lines are considered perpendicular can be

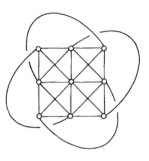

Figure 1.15

seen from the figure; the auxiliary lines drawn are perpendicular in the Euclidean sense at the common point of perpendicular lines. This is, of course, easily made precise by collecting all pairs of perpendicular lines explicitly.

A "reflection" in a "line" a is obtained simply by interchanging, on any perpendicular of a, the two points that do not lie on a. It is now easily seen that all axioms of a metric plane are true (see Exercises 1–4 below).

A system of axioms that is free of contradictions is said to be **consistent**. So we have shown the consistency for the axioms of a metric plane. We are lucky, though, that there exists a model with a finite number of elements and that this number is small enough to allow a quick test. In general, consistency proofs are much more complicated.

Exercises

1. Describe the nine-point geometry by using a 9×12 rectangular array in which each column represents a line a, each row a point P. If $P \mathrel{\text{I}} a$, then put a one where row and column meet, otherwise a zero.

2. Show that in the nine-point geometry all axioms of a metric plane are satisfied. (*Hint*: Take advantage of the symmetries in Figure 1.15 to shorten your proof.)

3. Is the nine-point metric plane elliptic, hyperbolic, or Euclidean?

4. Show that the nine-point geometry provides a "minimal model" of a metric plane, that is, that a metric plane has at least nine points and twelve lines.

5. Let a line in a Euclidean metric plane have only finitely many, say $m \geq 3$, points. Show that
   (a) all lines have equally many points,
   (b) through every point there pass exactly $m + 1$ lines,
   (c) the total number of lines is $m(m + 1)$,
   (d) the total number of points is m^2.

6. Show that Axioms I.1–I.3 and Ell are consistent by finding a geometry with seven points, each line having three points.

1.8   *Groups of Plane Motions*

In Section 1.4 we defined a motion as the successive application of a finite number of reflections. We now investigate more thoroughly what is involved when we speak of "successive application."

If M and M′ are any two motions of a metric plane and we wish to say that motion M is followed by motion M′, we write

(motion M)(motion M′),

or simply

MM′.

So MM′ means successive application of M and M′. By the functional notation (see Section 1.3), MM′ can alternatively be defined by the equation

$$[(x)\mathrm{M}]\mathrm{M}' = (x)\mathrm{MM}',$$

which is to hold for all points x of the plane.

In general, MM′ will not be the same as M′M, as is seen, for example, by choosing M = R_a, M′ = R_b, where a and b are neither identical nor perpendicular. The associative law, however, is always satisfied. Thus,

$$(\mathrm{MM}')\mathrm{M}'' = \mathrm{M}(\mathrm{M}'\mathrm{M}'').$$

This is true for simple logical reasons. If we apply any three actions A, B, C successively, it does not matter whether we consider A and B or B and C to be a whole. Let, for example, A = pouring coffee into a cup, B = putting cream into the cup, C = drinking what is in the cup; then A(BC) and (AB)C yield the same result: we get coffee with cream in our stomachs. Yet BC and CB are different; in the first case, we drink coffee with cream, in the second case, coffee without cream.

Because of the associative law, we may omit brackets. If we still do use them, it is to emphasize a subterm of a term. The definition of a motion M can now be restated as follows:

$$\mathrm{M} = R_{a_1} R_{a_2} \ldots R_{a_k}, \quad k \text{ any positive integer.}$$

A *proper motion* is given by a term

$$\mathrm{M} = R_{a_1} \ldots R_{a_{2l}},$$

that is, by an even number of reflections.

The "identity map" I is defined by

$$(X)\mathrm{I} = X$$

for all points X of the plane.

Alternatively, I can be defined by

$$\mathrm{MI} = \mathrm{IM} = \mathrm{M};$$

that is, if I "stands still" before or after applying M, the result will always be the same. Axiom M.2 (Section 1.3) may be restated as follows:

$$\mathrm{R}_a \mathrm{R}_a = \mathrm{I}.$$

This property of a reflection will be used very often. If $\mathrm{M} = \mathrm{R}_{a_1}, \ldots, \mathrm{R}_{a_k}$ is any motion, then choosing $\mathrm{M}' = \mathrm{R}_{a_k}, \ldots, \mathrm{R}_{a_1}$, we obtain for MM',

$$
\begin{aligned}
\mathrm{MM}' &= (\mathrm{R}_{a_1} \ldots \mathrm{R}_{a_{k-1}} \mathrm{R}_{a_k})(\mathrm{R}_{a_k} \mathrm{R}_{a_{k-1}} \ldots \mathrm{R}_{a_1}) \\
&= \mathrm{R}_{a_1} \ldots \mathrm{R}_{a_{k-1}}(\mathrm{R}_{a_k} \mathrm{R}_{a_k})\mathrm{R}_{a_{k-1}} \ldots \mathrm{R}_{a_1} \\
&= \mathrm{R}_{a_1} \ldots \mathrm{R}_{a_{k-1}} \mathrm{I} \; \mathrm{R}_{a_{k-1}} \ldots \mathrm{R}_{a_1} \\
&\;\;\vdots \\
&= \mathrm{R}_{a_1} \mathrm{R}_{a_1} = \mathrm{I}.
\end{aligned}
$$

So, given any motion M, we find a motion M′ such that $\mathrm{MM}' = \mathrm{I}$. We denote M′ by M^{-1}, so that

$$\mathrm{MM}^{-1} = \mathrm{I}.$$

We are now ready to introduce one of the most fundamental discoveries of modern mathematics. Many families of geometrical operations—motions in our case—can be treated in some ways as if they were number systems. "Successive application" of two operations stands for "multiplication" or "addition" in the ordinary number system. I takes the place of 1 under multiplication or 0 under addition. Finally, M^{-1} replaces $1/\mathrm{M}$ under multiplication or $-\mathrm{M}$ under addition.

We now introduce the general concept of a ***group***.

Let $\mathscr{G}$ be a nonempty set of "things," $g, h, \ldots$, called ***elements of*** $\mathscr{G}$, and let an operation, "$\circ$" ("times"), be defined for each pair of elements of $\mathscr{G}$. Then $\mathscr{G}$ is said to be a ***group*** if the following rules are satisfied:

(G.1) *If g and h are in $\mathscr{G}$, then $g \circ h$ is in $\mathscr{G}$.*

(G.2) *For any g, h, l in $\mathscr{G}$, the associative law holds:*

$$g \circ (h \circ l) = (g \circ h) \circ l.$$

(G.3) *There exists an element I in $\mathscr{G}$ such that*

$$g \circ \mathrm{I} = \mathrm{I} \circ g = g \quad \text{for any } g \text{ in } \mathscr{G}.$$

(G.4) *For each g in $\mathscr{G}$ there exists an element g^{-1} in $\mathscr{G}$ such that*

$$g \circ g^{-1} = g^{-1} \circ g = \mathrm{I}.$$

g^{-1} is called the ***inverse*** or ***inverse element*** of g. From the above arguments we have the following theorem.

Theorem 1.8.1 *The set of all motions in a metric plane is a group under the operation of successive application.*

By analogy to the multiplication of numbers, MM' is often called the **product** of M and M'. Remember, however, that the "multiplying" of motions does not conform to the same laws that apply to the multiplication of numbers. The group is called **commutative**, or **Abelian**, if, in addition to the group axioms, the following rule is satisfied:

(G.5)   *For all g, h of $\mathscr{G}$ the commutative law holds; thus,*

$$g \circ h = h \circ g.$$

Whereas groups of ordinary numbers are commutative, the group of all motions in a metric plane is not commutative.

The following list of some groups shows that not only in geometry but also in algebra the concept of a group provides a roof under which different kinds of structures can be united.

	Elements of $\mathscr{G}$	Operation
1	All integers	Addition
2	All even integers	Addition
3	All rationals	Addition
4	All rationals $\neq 0$	Multiplication
5	All positive rationals	Multiplication
6	All plane motions	Successive application
7	All proper motions	Successive application
8	All motions having the same fixed point F	Successive application
9	One reflection R_a and I	Successive application

There is a degenerate case of a group, namely, I alone. We denote this group by {I}. A group is said to be **finite**, if it contains only finitely many elements. Examples of finite groups are {I} and 9 of the preceding list.

If $a_1, \ldots, a_m$ are the elements of a finite group $\mathscr{G}$, we denote this relationship by

$$\mathscr{G} = \{a_1, \ldots, a_m\}.$$

We present now some examples of abstract calculation in arbitrary groups by proving three theorems.

Theorem 1.8.2   *If g, h are elements of a group $\mathscr{G}$, then the equations*

$$g \circ x = h \qquad\qquad\qquad (a)$$

and

$$y \circ g = h \qquad\qquad\qquad (b)$$

have unique solutions in x or y, respectively.

Proof   By Rule (G.4), there is a g^{-1} in $\mathscr{G}$ such that $g^{-1} \circ g = $ I. Multiplying (a) "from the left" yields

$$g^{-1} \circ (g \circ x) = g^{-1} \circ h.$$

But from (G.2), (G.4), and (G.3),

$$g^{-1} \circ (g \circ x) = (g^{-1} \circ g) \circ x = \text{I} \circ x = x.$$

Hence $x = g^{-1} \circ h$. So, if there is a solution of (a), it can be only $g^{-1} \circ h$. But $g^{-1} \circ h$ is indeed a solution, since

$$g \circ (g^{-1} \circ h) = (g \circ g^{-1}) \circ h = I \circ h = h.$$

This proves part (a).

The same proof applies for part (b) if "multiplying from the left" is replaced by "multiplying from the right." ◆

It should be noted that (a) and (b) in Theorem 1.8.2 may have different solutions, since $\mathcal{G}$ need not be commutative. Choosing $g = h = I$ in 1.8.2, we obtain Theorem 1.8.3.

Theorem 1.8.3 *The identity element I of a group is uniquely determined.*

Also by letting $h = I$, we obtain the next theorem.

Theorem 1.8.4 *If g is any element of a group $\mathcal{G}$, then its inverse element g^{-1} is uniquely determined.*

The concept of a group supplies an elegant and powerful tool for the development of geometry that we are undertaking.

Exercises

1. Show explicitly that examples 2, 7, 8, and 9 of the list of groups in this section are indeed groups.
2. Which of the groups 1–9 are commutative?
3. Why do the following systems *not* represent groups?
   (a) All integers under multiplication.
   (b) All rationals under multiplication.
   (c) All odd integers and zero under addition.
   (d) All reflections of a metric plane under successive application.
4. Find motions g, h such that, in Theorem 1.8.2, x and y are different (no rigorous proof).
5. Show that all permutations of a set of objects (digits) form a group under successive application.
6. Show that in the sense of Section 1.7 the axioms of a group are consistent.

1.9   Subgroups

If $\mathcal{G}$ is a group and $\mathcal{H}$ a nonempty subset of elements of $\mathcal{G}$, then $\mathcal{H}$ is called a **subgroup** of $\mathcal{G}$ if it is a group under the group operation of $\mathcal{G}$.

So the group of all proper motions is a subgroup of the group of all plane motions. The group of integers under addition is a subgroup of the group of all rationals under addition. In order to find out whether a given subset $\mathscr{H}$ of a group $\mathscr{G}$ is a subgroup, we need not check all group axioms, as the following theorem shows.

Theorem 1.9.1 *A nonempty subset $\mathscr{H}$ of a group $\mathscr{G}$ is a subgroup of $\mathscr{G}$ if and only if it possesses the following two properties:*
(H.1) *If h and h' are in $\mathscr{H}$, then hh' is in $\mathscr{H}$.*
(H.2) *If h is in $\mathscr{H}$, then h^{-1} is in $\mathscr{H}$.*

Proof If $\mathscr{H}$ is a subgroup, then (H.1) and (H.2) are satisfied by definition of a group. Let, conversely, $\mathscr{H}$ satisfy (H.1) and (H.2). Then Rules (G.1) and (G.4) hold in $\mathscr{H}$. The associative law (G.2) holds, since $\mathscr{H}$ is a subset of $\mathscr{G}$. Since there exists at least one element h in $\mathscr{H}$ (by assumption, $\mathscr{H}$ is nonempty), we find by (H.1) and (H.2) that $hh^{-1} = I$ is in $\mathscr{H}$. So Rule (G.3) is satisfied and $\mathscr{H}$ is a subgroup of $\mathscr{G}$. ◆

The definition of a subgroup does not exclude the possibility $\mathscr{G} = \mathscr{H}$. So every group $\mathscr{G}$ has $\mathscr{G}$ itself and $\{I\}$ as "trivial" subgroups. It is often useful to indicate by a diagram how subgroups of a group are linked. If $\mathscr{H}$ is a subgroup of $\mathscr{G}$, we write

or something similar. An example of such a diagram (P is a fixed point of a metric plane) is shown below.

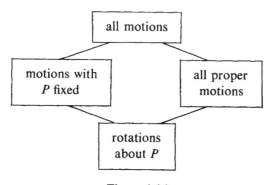

Figure 1.16

The group of all proper motions is evidently not a subgroup of the group of motions with a fixed P.

Another example: For any integer n, the multiples of n form a group under addition. The identity element is $0 = n \cdot 0$, the inverse of an element $n \cdot k$ is $-n \cdot k$. We denote this group by $\mathscr{Z}^{(n)}$. Clearly, if m is a multiple of n, the group $\mathscr{Z}^{(m)}$ is a subgroup of $\mathscr{Z}^{(n)}$.

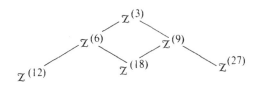

Figure 1.17

The diagram of subgroups in Figure 1.17 may be extended to an arbitrarily large diagram by choosing further multiples $3k$ of 3 in $\mathscr{Z}^{(3k)}$.

Exercises

1. In the list of groups presented in Section 1.8, find out which groups are subgroups of others.

2. Why is the set of odd integers not a subgroup of the group of all integers?

3. Find the diagram of all groups $\mathscr{Z}^{(n)}$ which possess $\mathscr{Z}^{(12)}$ as a subgroup.

4. Let a, b be lines such that $(R_a R_b)^2 = I$. Show that the set $\{I, R_a, R_b, R_a R_b\}$ is a group. Find the diagram of all subgroups.

5. Find the diagram of all subgroups of the group of all permutations of three objects (compare Exercise 5, Section 1.8).

6. Show that in Theorem 1.9.1, (H.1) and (H.2) can be replaced by the following single condition: If h and h' are in $\mathscr{H}$, then hh'^{-1} is in $\mathscr{H}$.

7. For two finite subsets $\mathscr{S}, \mathscr{T}$ of a group $\mathscr{G}$, let the set of all products st with s in $\mathscr{S}$ and t in $\mathscr{T}$ be denoted by $\mathscr{S}\mathscr{T}$. Show that $\mathscr{S}$ is a subgroup of $\mathscr{G}$ if and only if $\mathscr{S}\mathscr{S} = \mathscr{S}$.

1.10 Cyclic Groups

Let $g = R_a R_b$ be a rotation about a point R. If we repeat the same rotation over and over again, there are two possibilities: After a finite number of steps, either we arrive at the identity map or we always obtain new rotations. In the first case, consider a point $P \neq R$ and all its image points. Join successive points by a line. Then the figure of these points and lines is mapped onto itself by $g, gg, \ldots$ (Fig. 1.19). For convenience we write

$$g^2 \quad \text{instead of} \quad gg$$

and, more generally,

$$g^n \quad \text{instead of} \quad g^{n-1}g.$$

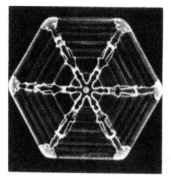

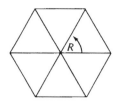

Figure 1.18   **Figure 1.19**

Defining further,

$$(g^{-1})^n = g^{-n} \quad \text{and} \quad g^0 = I,$$

from which we obtain the general "power rule" for groups:

$$g^m g^n = g^{m+n}; \quad m, n \text{ any (positive or negative or zero) integers.}$$

The two cases observed above for rotations can be analyzed for any group element g as follows:

(a) Let $g^k = I$, where $k > 0$, and let k be the smallest positive integer of this kind.   Then

Theorem 1.10.1   *The set $\mathscr{Z}_k = \{I, g, g^2, \ldots, g^{k-1}\}$ is a subgroup of $\mathscr{G}$.*

Proof   We have to show that (H.1) and (H.2) of Theorem 1.9.1 are satisfied.
(H.1): Let g^m, g^n be in $\mathscr{Z}_k$.   If $m + n < k$, then clearly $g^n g^m = g^{n+m}$ is in $\mathscr{Z}_k$.   If $m + n = ks + r$, where $0 \le r < k$, then from $g^{ks} = (g^k)^s = I^s = I$, we find:

$$g^m g^n = g^{ks+r} = g^{ks} g^r = g^r,$$

which is in $\mathscr{Z}_k$.
(H.2): If g^m is in $\mathscr{Z}_k$, then $g^{-m} = I\, g^{-m} = g^k\, g^{-m} = g^{k-m}$ is in $\mathscr{Z}_k$.
This proves the theorem.   ◆

$\mathscr{Z}_k$ is called a *cyclic group*.

(b) Let $g^h \neq I$ for every integer $h \neq 0$.   Then

$$\mathscr{Z} = \{\ldots, g^{-n}, \ldots, g^{-1}, I, g, \ldots, g^m, \ldots\}$$

forms a group. $\mathscr{Z}$ is called an *infinite cyclic* group.   It is essentially the same as the group of all integers under addition.
If for an arbitrary group $\mathscr{G}, g^k = I$ and $g^m \neq I$ for $0 < m < k$, we call k the *order* of g.   If all powers of g are $\neq I$, we say that g has *infinite order*. So, in the group of all motions, a reflection R_a always has order 2.

Exercises

1. Find the diagram of all subgroups of $\mathscr{Z}_6$.

2. Determine the orders of all elements of $\mathscr{Z}_{12}$.

3. Show that the order of an element of a finite group $\mathscr{G}$ divides the number of all elements of $\mathscr{G}$.

4. A finite group may be characterized by its "multiplication table," as, for example, $\mathscr{Z}_3$:

	I	g	g^2
I	I	g	g^2
g	g	g^2	I
g^2	g^2	I	g

Find the multiplication table of $\mathscr{Z}_n$.

5. Let $\mathscr{D}_3$ be the group of all (proper and improper) motions that map an equilateral triangle (in ordinary plane geometry) onto itself. Find the multiplication table.

6. In Exercise 5, replace the equilateral triangle by a square.

1.11   Fixed Points and Lines

If $(P)\mathrm{M} = P$ under a mapping M, then P is called a *fixed point* or an *invariant point* of M. If a line a is mapped onto itself by M in the sense that for all points X on a the image points $(X)\mathrm{M}$ are on a too, then a is called a *fixed line* or an *invariant line* of M. If, in particular, every point of a is a fixed point, we say that a is *pointwise fixed* under M. Correspondingly, we say that P is *linewise fixed* if each line passing through P is a fixed line. From these definitions some theorems immediately follow.

Theorem 1.11.1 *If P, Q are different fixed points under a motion M, then the line PQ is a fixed line of M. If a, b are different fixed lines through a point P, then P is a fixed point.*

By definition, a reflection R_a leaves all points of the line a fixed. Since R_a is an orthogonal collineation (see Section 1.3), it maps any line b perpendicular to a onto a line b' perpendicular to a. But a and b intersect in a point P (by Axiom P.3) and $(P)\mathrm{R}_a = P$. So, by Axiom P.2,

$$b' = b.$$

Theorem 1.11.2 states this more briefly.

Theorem 1.11.2 *Every perpendicular of a line a is a fixed line under the reflection R_a.*

A less trivial theorem follows.

Theorem 1.11.3 Fixed-line theorem. *The only fixed lines of a reflection R_a are a and all perpendiculars of a.*

Proof We show a slightly more general fact: *If an orthogonal collineation* M *leaves a line a pointwise fixed and possesses a line u that is neither equal nor perpendicular to a as a fixed line, then* M *is the identity map* I. Since $R_a \neq$ I, this implies our theorem.

Suppose a, u are two such lines. Let U be an arbitrary point on u. We drop a perpendicular v from U onto a. By the same argument that proved Theorem 1.11.2, we see that v remains fixed under M. So, by Theorem 1.11.1, U remains fixed under M, and hence u remains pointwise fixed. Let P be an arbitrary point. Drop a perpendicular c from P onto a (Axiom P.2). a and c intersect in a point C (Axiom P.3). Choose a point U on u that is neither on a nor on c (Axiom I.2). Drop a perpendicular c' from P onto the line d that joins U and C (Axiom P.2). By Theorem 1.11.1, d is a fixed line. Furthermore, d is not perpendicular to a since otherwise $d = c$ (Axiom P.2), and hence U is on c contrary to

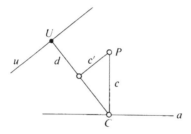

Figure 1.20

assumption. Therefore, d is also pointwise fixed under M. Then also c' is a fixed line. Finally, $c \neq c'$, since, otherwise, $d = a$ (Axiom P.2) and U on a would follow. So, by Theorem 1.11.1, P is a fixed point. Since P has been arbitrary, M = I. ◆

We draw two conclusions, stated in the next two theorems, from the fixed-line theorem:

Theorem 1.11.4 $a' = (a)R_b$ *is true if and only if*

$$R_b R_a R_b = R_{a'}.$$

Proof Since Theorem 1.11.4 is an "if and only if" statement, we have to proceed in two steps:

I. Let $a' = (a)R_b$. We assert: $R_bR_aR_b = R_{a'}$. If, in fact, X' is an arbitrary point on a', then $(X')R_b = X$ is on a and

$$(X')R_bR_aR_b = (X)R_aR_b = (X)R_b = X'.$$

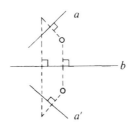

Figure 1.21

Therefore, $R_bR_aR_b$ leaves every point of a' fixed. Since the product of orthogonal collineations is again an orthogonal collineation, $R_bR_aR_b$ is such. So it satisfies Conditions 1–3 for a reflection (Section 1.3) and hence is either equal to I or is the reflection $R_{a'}$ (Condition (4) of Section 1.3 being satisfied). If $R_bR_aR_b$ were equal to I, then multiplying by R_b from left and right would yield $R_a = I$, contradicting the definition of a reflection. Thus $R_bR_aR_b = R_{a'}$.

II. Let $R_bR_aR_b = R_{a'}$. We assert: $a' = (a)R_b$. Let, in fact, $a'' = (a)R_b$. Then, by part I of this proof, $R_bR_aR_b = R_{a''}$, and so $R_{a''} = R_{a'}$, which implies $a'' = a'$. ◆

Theorem 1.11.5 $a \perp b$ *if and only if* $R_aR_b = R_bR_a$ *and* $a \neq b$.

Proof By Theorems 1.11.2 and 1.11.3, $a \perp b$ is true if and only if $a = (a)R_b$ and $a \neq b$. By Theorem 1.11.4, this also is equivalent to

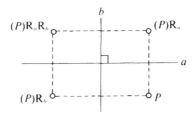

Figure 1.22

$$R_bR_aR_b = R_a, \quad a \neq b,$$

or

$$R_bR_a = R_aR_b, \quad a \neq b. \quad ◆$$

We know that $a \neq b$ is equivalent to $R_a \neq R_b$, which, in turn, is equivalent to $R_aR_b \neq I$.

Furthermore, $R_b R_a R_b = R_a$ can be written as

$$(R_a R_b)(R_a R_b) = (R_a R_b)^2 = I.$$

So Theorem 1.11.5 can be rephrased as follows (for the definition of "involutoric" see Section 1.3).

Theorem 1.11.6 *$a \perp b$ if and only if $R_a R_b$ is involutoric.*

Exercises

1. Illustrate Theorem 1.11.4 by constructing the image of a triangle (a) under $R_b R_a R_b$ and (b) under $R_{a'}$ (use ruler and right angle).

2. Show that for any motion M,

$$M^{-1} R_a M = R_{(a)M}.$$

3. Prove that in a Euclidean metric plane, any motion with three noncollinear fixed points is the identity map I.

4. In the model of an elliptic plane discussed in Section 1.6, find all fixed points and fixed lines under a reflection R_a (no rigorous proofs).

5. Assume that in a given metric plane there is no "polar triangle," that is, there are no three lines each two of which are perpendicular. Give a proof of Theorem 1.11.5 without using Theorems 1.11.3 and 1.11.4. (*Hint*: Reflect an arbitrary point X as indicated in Figure 1.22.)

6. Find a simple proof of Theorem 1.11.3 for Euclidean and hyperbolic metric planes.

1.12 Elliptic Planes

In Section 1.6 we discussed a model of an elliptic plane, that is, a metric plane in which any two lines intersect. We considered great circles of an ordinary sphere to be "lines," and we called pairs of antipodal points of the sphere "points." To every great circle a taken as an "equator" of the sphere there corresponds a "north pole" and a "south pole." The pair of these antipodal points, taken as a point of the metric plane, we call a "pole" P of a. Lines through P (that is, meridians relative to a) are perpendicular to a. This motivates the definitions that follow.

In any metric plane, a point P is called a *pole* of a line a if there exist two different lines u, v passing through P and perpendicular to a. The line a is called the *polar line* of P (see Fig. 1.23). One of the main results of this

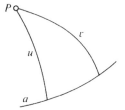

Figure 1.23

section is the fact that either every line possesses a pole or no line possesses a pole, the former case being that of an elliptic plane (see Theorem 1.12.8 below).

Theorem 1.12.1 *If P is a pole of a, then P is not on a.*

Proof If this were not so, there would be two different perpendiculars of a passing through P, though P is on a.   This contradicts Axiom P.2. ◆

Theorem 1.12.2 *If P is a pole of a and if $b \perp a$, then P ı b.*

Proof Let A be the intersection of a and b.   Since perpendiculars of a remain fixed under R_a, we find that

$$(P)R_a = P.$$

Furthermore, by Theorem 1.11.1,

$$(A)R_a = A.$$

Using Theorem 1.11.1 again, we find that the line PA remains fixed under R_a.   From Theorem 1.11.3, we conclude that $PA \perp a$ and so, by Axiom P.2, $PA = b$.   Therefore, P ı b. ◆

Theorem 1.12.3 *If P is a pole of a and if b ı P, then $b \perp a$.*

Proof Let Q ı b and $Q \neq P$.   Drop a perpendicular b' from Q onto a. By Theorem 1.12.2, P ı b'; hence, by Axiom I.1, $b = b'$. ◆

Theorem 1.12.4 *Every line has at most one pole.*

Proof Let P, P' be two poles of a and let u, v be different perpendiculars of a through P.   Then, by Theorem 1.12.2, P' ı u and P' ı v; hence $P' = P$.
◆

Theorem 1.12.5 *Every point has at most one polar line.*
The proof of 1.12.5 is left to the reader (see also Exercise 1 of this section).

Theorem 1.12.6 *If in a metric plane there exists one line that possesses a pole, then every line possesses a pole and every point a polar line.*

Proof Let P be the pole of a and let Q be an arbitrary point different from P. (See Fig. 1.24.)   If Q ı a, then $PQ \perp a$ (by Theorem 1.12.3), and the perpendicular b of PQ in P is also perpendicular to a.   Hence, b is the polar line of Q.

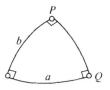

Figure 1.24

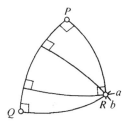

Figure 1.25

Let Q not lie on a. The line PQ and its perpendicular in P both are perpendicular to a. Hence, PQ possesses a pole R. If $b \mathbin{\text{\scriptsize I}} R$ and $b \perp QR$, then b is the polar line of Q (Fig. 1.25).

It can be shown similarly that every line possesses a pole (Exercise 2). ◆

Theorem 1.12.7 *If in a metric plane there exists a line that possesses a pole, then the plane is elliptic; that is, each two lines intersect.*

Proof Let a, b be any lines and let P, Q be their poles, respectively (existence by Theorem 1.12.6). If $a \neq b$, then $P \neq Q$ (by Theorem 1.12.5), and the line PQ is perpendicular to a and b (by Theorem 1.12.3). Hence, a and b intersect in the pole of PQ (by Theorem 1.12.2). ◆

The results obtained so far in this section can be summarized in the theorem that follows.

Theorem 1.12.8 *In a metric plane the following conditions are equivalent (each implies all the others):*
  (a) *The plane is elliptic; that is, any two lines intersect.*
  (b) *Every line has a pole.*
  (c) *Every point has a polar line.*
  (d) *There exists a line that has a pole.*
  (e) *There exists a polar triangle.*

To carry out the proof of Theorem 1.12.8, it is sufficient to verify the following chain of implications:

$$\text{(a)} \Rightarrow \text{(e)} \Rightarrow \text{(d)} \Rightarrow \text{(c)} \Rightarrow \text{(b)} \Rightarrow \text{(a)}.$$

This can easily be done by using the above theorems.

We are now able to find a counterpart of the fixed-line theorem, 1.11.3.

Theorem 1.12.9 Fixed-point theorem *The only fixed points of a reflection* R_a *are* (I) *the points of* a, *if the plane is not elliptic,* (II) *the points of* a *and the pole of* a, *if the plane is elliptic.*

Proof If P is fixed under R_a and P is not on a, then joining P by two different points of a provides two fixed lines u, v of R_a. By the fixed-line theorem 1.11.3, $u \perp a$ and $v \perp a$. Hence P is a pole of a and, by Theorem 1.12.8, the plane is elliptic. ◆

Exercises

1. Prove Theorem 1.12.5.
2. Prove the second part of Theorem 1.12.6 (using the first part).
3. Carry out the proof of Theorem 1.12.8.
4. Prove that in an elliptic plane, every rotation has a fixed line.
5. Prove that a metric plane is elliptic if and only if there exists a motion not equal to I possessing three noncollinear fixed points.
6. In an elliptic plane, let P lie on a. How many polar triangles exist having P as a vertex and a as a side?

1.13 Theorems on Three Reflections

An important role in our analysis of motions is played by the following theorem.

Theorem 1.13.1 First theorem on three reflections *If R is on a, b, c, then there exists a line d passing through R such that*

$$R_a R_b R_c = R_d.$$

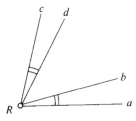

Figure 1.26

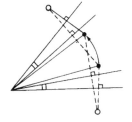

Figure 1.27

Proof $R_b R_c$ is a rotation about R. We may assume $b \neq c$ (otherwise $R_a R_c R_c = R_a$ and there is nothing to prove). We let $a' = (a)R_b R_c$. Apply Axiom M.3: There exists a reflection R_d such that

$$(P)R_d = (P)R_b R_c \quad \text{for all } P \text{ I } a.$$

Hence

$$(P)R_b R_c R_d = P \quad \text{for all } P \text{ I } a.$$

So, by definition of a reflection, it is only left to show that $R_bR_cR_d \neq I$;
then

$$R_bR_cR_d = R_a \quad \text{and} \quad R_aR_bR_c = R_d.$$

Suppose $R_bR_cR_d = I$; then $R_cR_d = R_b$, and hence $R_cR_d = R_dR_c$, which, by Theorem 1.11.5, means $c \perp d$. Similarly, it follows that $b \perp d$. Since $R \perp b, c, d$, we conclude from Axiom P.2 that $c = b$, contradicting the assumption that $c \neq b$. ◆

A direct consequence of Theorem 1.13.1 is the following statement about rotations:

Theorem 1.13.2 *If $g = R_uR_v$ is a rotation about R, then, given any line v' passing through R, we can find a line u' passing through R such that*

$$g = R_{u'}R_{v'}.$$

In ordinary plane geometry, this corollary amounts to the following: A rotation is determined by its center and the angle of rotation.

Another application of the first theorem on three reflections leads to the four theorems that follow.

Theorem 1.13.3 *The set of all rotations about a given point R is a subgroup of the group of all motions.*

Proof According to Theorem 1.9.1, it suffices to verify (H.1) and (H.2).
  (H.1): Let S, S' be two rotations about R. If $S' = R_uR_v$ then S can be expressed as $S = R_wR_u$ (according to Theorem 1.13.2). Hence

$$SS' = R_wR_uR_uR_v = R_wR_v,$$

which is again a rotation about R.
  (H.2): If $S = R_wR_u$, then $S^{-1} = R_uR_w$, since

$$SS^{-1} = R_wR_uR_uR_w = R_wR_w = I.$$

S^{-1} is also a rotation about R. ◆

The following three theorems are strictly analogous to Theorems 1.13.1–1.13.3; they refer to translations instead of rotations.

Theorem 1.13.4 Second theorem on three reflections *If t is perpendicular to a, b, c, then there exists a line d perpendicular to t such that*

$$R_aR_bR_c = R_d.$$

Proof Let $b \neq c$ (otherwise $R_aR_bR_b = R_a$). If b and c intersect, then the plane is elliptic and Theorem 1.13.4 reduces to Theorem 1.13.1. Let, therefore, the metric plane be not elliptic, so that b and c do not intersect (Fig. 1.28). Using Axiom M.4 instead of Axiom M.3, we show, as in the proof of Theorem 1.13.1, that

$$(P)R_bR_cR_d = P \quad \text{for all } P \perp a.$$

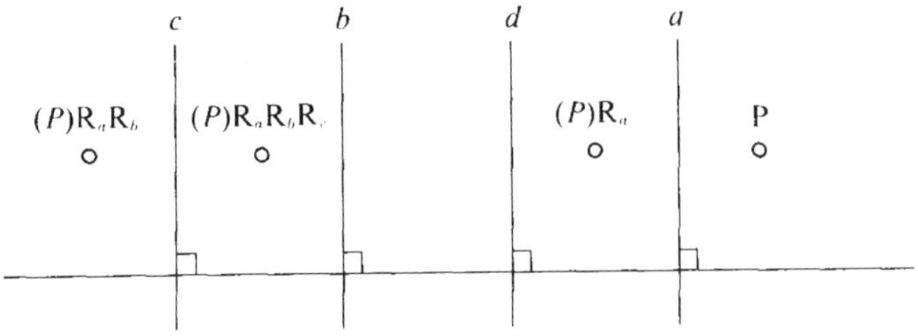

$(P)R_aR_b$ $(P)R_aR_bR_c$ $(P)R_a$ P
   o   o   o   o

c   b   d   a

Figure 1.28

If $R_bR_cR_d$ were equal to I, then c would be perpendicular to d, and t, c, d would form a polar triangle, contradicting our assumption that the metric plane is not elliptic (see Theorem 1.12.8).

So $R_bR_cR_d = R_a$ and $R_aR_bR_c = R_d$.   ◆

Theorem 1.13.5   *If* $T = R_uR_v$ *is a translation along* t, *then, given any line* v' *perpendicular to* t, *we can find a line* u' *perpendicular to* t *such that*

$$T = R_{u'}R_{v'}.$$

This is a direct consequence of Theorem 1.13.4.

The proof of Theorem 1.13.3 applies directly to the following theorem.

Theorem 1.13.6   *The set of all translations along a line* t *is a subgroup of the group of all motions.*

It is to be emphasized, however, that in an elliptic plane this group consists only of the identity element, since we defined R_uR_v to be a translation only if $u, v \perp t$; and u, v do not intersect unless $u = v$.

We conclude this section by proving a theorem on "triangles" which, in a striking manner, illuminates the power of group-theoretic methods in geometry. Without speaking of "angles," it contains the theorem on angular bisectors of a triangle, well-known in Euclidean geometry as well as in spherical geometry. We define a **triangle** *ABC*, or, in short, $\triangle ABC$, to be a set of three noncollinear points called the vertices of the triangle. *AB*, *BC*, and *CA* are called side lines of $\triangle ABC$. The line *AB* is said to be *opposite C* and *C* is said to be *opposite AB*, and so forth.

Theorem 1.13.7   *If* a, b, c *are the side lines of a triangle,* $\triangle ABC$, *and if* u, v *are lines meeting in a point* M *and satisfying* $(a)R_u = b$ *and* $(b)R_v = c$,

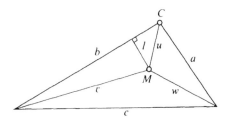

Figure 1.29

then there exists a line w passing through M such that $(c)R_w = a$ (see Fig. 1.29).

Proof Let l ɪ M and $l \perp b$. By Theorem 1.13.1, there is a line w such that $R_u R_l R_v = R_w$. Then

$$(a)R_w = (a)R_u R_l R_v = (b)R_l R_v = (b)R_v = c. \qquad \blacklozenge$$

Exercises

1. Illustrate Theorem 1.13.1 by constructing the image of a triangle under R_d and under $R_a R_b R_c$ in ordinary plane geometry (use ruler and compass).

2. *Theorem on perpendicular bisectors.* If $\triangle ABC$ is any triangle, if $(A)R_u = B$, $(B)R_{\bar{v}} = C$, and if u, v intersect in a point M, then there exists a line w through M such that $(C)R_w = A$.

3. Show that in the proof of Theorem 1.13.7 the line l is uniquely determined.

4. Let P lie on the lines $a_1, a_2, \ldots, a_{2k+1}$. Show that $R_{a_1} R_{a_2} \ldots R_{a_{2k+1}}$ is a reflection.

5. Prove that the group of all rotations about a point P is commutative.

6. Prove that the group of all translations along a line t is commutative.

1.14   Half-Turns

A counterpart to a reflection is a half-turn. In ordinary plane geometry it can be characterized as either a rotation of 180 degrees about a point R or as a mapping under which R is the midpoint of any segment formed by a point and its image point. A figure of the plane is said to have a center R if it is mapped onto itself by a half-turn about R (see Fig. 1.30).

Figure 1.30

In a metric plane we define a **half-turn** H_P to be a rotation R_uR_v, where u and v are perpendicular at P (Fig. 1.31). H_P may also be called a **reflection in P**. However, in general, we will reserve the term "reflection" for reflections in lines. In order to realize that H_P does depend only on P, we need the following theorem.

Theorem 1.14.1 *If H_P is a half-turn, then for any two lines u', v' ɪ P satisfying $u' \perp v'$ it follows that $H_P = R_{u'}R_{v'}$.*

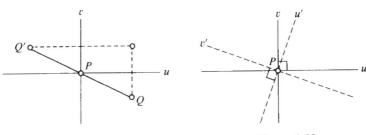

Figure 1.31   **Figure 1.32**

Proof Let $H_P = R_uR_v$, where $u \perp v$. By Theorem 1.13.2, we conclude that there exists a line u'' such that $R_uR_v = R_{u''}R_{v'}$ and P ɪ u''. Since $R_uR_v = R_vR_u$ (by Theorem 1.11.5), we find also that $R_{u''}R_{v'} = R_{v'}R_{u''}$; hence $u'' \perp v'$. By Axiom P.2, this implies that $u'' = u'$. ◆

Theorem 1.14.2 *Points and half-turns are in one-to-one correspondence:*

$$\boxed{P \longleftrightarrow H_P}$$

This follows immediately from Theorem 1.14.1.

Theorem 1.14.3 *In an elliptic plane, half-turns and reflections coincide. In a nonelliptic plane, no half-turn coincides with a reflection.*

Considering Theorem 1.12.8, the proof of Theorem 1.14.3 is contained in the following statement.

Theorem 1.14.4 $H_P = R_a$ *if and only if P is a pole of a.*

Proof I. Let $H_P = R_a$. Setting $H_P = R_uR_v$, we conclude $R_uR_v = R_a$ and $R_aR_u = R_v$; hence, by Theorem 1.11.5, $a \perp u$. Similarly, $a \perp v$. So a, u, v form a polar triangle and P is a pole of a.

II. Let P be a pole of a, and let A be an arbitrary point on a. Setting $H_P = R_uR_v$, where $u = PA$, we conclude from Theorem 1.12.3 that u, v, a

form a polar triangle; hence A is a pole of v. By the fixed-point theorem, 1.12.9,

$$(A)H_P = (A)R_u R_v = (A)R_v = A.$$

So every point of a is a fixed point under the orthogonal collineation H_P. Since $H_P \neq I$, we conclude from the definition of a reflection that

$$H_P = R_a.$$

Theorem 1.14.5   *Every line passing through P is a fixed line under H_P.*

Proof   Let $P \perp a$. By Theorem 1.14.1, $H_P = R_a R_b$ where $b \perp a$, $b \perp P$. Now $(a)H_P = (a)R_a R_b = (a)R_b = a$. ◆

Theorem 1.14.6   *The only fixed lines of H_P are*
  (I) *all lines passing through P together with the polar line of P if the plane is elliptic.*
  (II) *all lines through P if the plane is nonelliptic.*

Proof   If a remains fixed under H_P and if P is not on a, then, by Theorem 1.11.1, all points of a remain fixed; hence, $H_P = R_a$. Theorem 1.14.6 now follows from Theorems 1.14.4 and 1.12.8. ◆

Theorem 1.14.6 readily implies the next theorem.

Theorem 1.14.7   *The only fixed points of H_P are*
  (I) *P and all points on the polar line of P if the plane is elliptic.*
  (II) *P alone if the plane is not elliptic.*

Figure 1.33 illustrates the following theorem.

Theorem 1.14.8 $H_{P'} = R_a H_P R_a$ *if and only if* $P' = (P)R_a$.

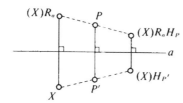

Figure 1.33

Proof   I. Let $H_{P'} = R_a H_P R_a$. We set $H_P = R_u R_v$ (where $u \perp v$) and $u' = (u)R_a$; $v' = (v)R_a$. Then

$$(u')H_{P'} = (u')R_a H_P R_a = (u)R_a R_a H_P R_a = (u)R_u R_v R_a = (u)R_v R_a$$
$$= (u)R_a = u'.$$

Hence u' either passes through P' or it is the polar line of P' (by Theorem 1.14.6). Since there are at least three lines passing through each point (see

Exercise 1, Section 1.1), we can assume that u', v' are not polar lines of P. Then $P' \perp u'$, v'; hence $P' = (P)R_a$.

II. Let $P' = (P)R_a$ and $P' \neq P$. The line $b = PP'$ is perpendicular to a. Let $c \perp P$; $c \perp b$; $c' \perp P'$; $c' \perp b$. Then $c' = (c)R_a$ and

$$R_a H_P R_a = R_a R_c R_b R_a = R_a R_c R_a R_b = R_{c'} R_b$$

(by Theorem 1.11.4). ◆

Exercises

1. Illustrate Theorem 1.14.8 by drawing the image of a triangle (a) under $H_{P'}$ and (b) under $R_a H_P R_a$.

2. If a is mapped onto a' by a reflection R_b and if a and a' intersect in a single point P, show that there is a reflection different from R_b that also maps a onto a'.

3. Show that all half-turns are involutoric.

4. Show that the product of two d'fferent half-turns H_P, H_Q is a translation along the line PQ.

5. Replace the product of a half-turn and a reflection by the product of a translation along a line and tl e reflection at this line.

6. Prove that the set of all translations along a given line t and all half-turns H_P with P on t is a group.

7. Show that the group in Exercise 6 is commutative.

8. Let a, b, c, have common perpendiculars, and let $A \perp a$; $B \perp b$; $C \perp c$. Show that $R_a R_b R_c H_A H_B H_C$ is involutoric.

1.15   *Group-Theoretic Calculation with Points and Lines*

We call an element α of a group *involutoric* if it is of order 2, that is, if $\alpha^2 = I$ and $\alpha \neq I$ (see also Section 1.3). So far we have found two kinds of involutoric motions : reflections and half-turns; compare Axiom M.2 and Section 1.14. (In Chapter 7, we show that these are the only involutoric motions). Theorem 1.11.6 states that $a \perp b$ if and only if $R_a R_b$ is involutoric. This characterizes perpendicularity in group-theoretic terms. A similar characterization can be obtained for incidence.

Theorem 1.15.1 $P \perp a$ *if and only if* $H_P R_a$ *is involutoric.*

Proof   By Theorems 1.12.9 and 1.14.4, $P \perp a$ if and only if $(P)R_a = P$, $H_P \neq R_a$. This, by Theorem 1.14.8, is equivalent to $H_P = R_a H_P R_a$, $H_P \neq R_a$ or $(H_P R_a)^2 = I$, $H_P R_a \neq I$. ◆

These characterizations of incidence and perpendicularity enable us to translate all geometric considerations about metric planes into group-theoretic ones, namely, in the group of all motions. The following table shows in detail how this is done:

line	$a \leftrightarrow R_a$
point	$P \leftrightarrow H_P$
incidence	$P \mathrel{I} a \leftrightarrow H_P R_a$ involutoric
perpendicularity	$a \perp b \leftrightarrow R_a R_b$ involutoric
reflection	$P' = (P)R_a \leftrightarrow H_{P'} = R_a H_P R_a$

[1.15]

Metric Planes and Motions

If we identify every line a with the reflection on R_a and every point P with H_P, then we are able to "multiply" points and lines. Two lines are perpendicular if their product is of order 2, and so forth. So we have represented our geometry in algebraic terms (without coordinates)!

Instead of "AB involutoric," we write simply "$A \mid B$." If AB is not involutoric, we set $A \nmid B$. Then $a \perp b$ is equivalent to $R_a \mid R_b$, and $P \mathrel{I} a$ is equivalent to $H_P \mid R_a$. Let us restate some of the theorems we have obtained so far:

(1) To any H_P, H_Q there is an R_a such that $H_P \mid R_a$, $H_Q \mid R_a$.

(2) H_P, $H_Q \mid R_a$, R_b implies $H_P = H_Q$ or $R_a = R_b$.

(3) If R_a, R_b, $R_c \mid H_P$, then there exists a d such that

$$R_a R_b R_c = R_d.$$

(4) If R_a, R_b, $R_c \mid R_t$, then there exists a d such that

$$R_a R_b R_c = R_d.$$

(5) There are R_a, R_b, R_c such that

$$R_a \mid R_b, \quad R_a \nmid R_c, \quad R_b \nmid R_c, \quad R_c \nmid R_a R_b.$$

Theorems (1)–(5) not only express some properties of metric planes in group-theoretic terms. They even characterize metric planes in the following sense: Given a group $\mathscr{M}$ and a subset $\mathscr{R}$ of $\mathscr{M}$, the elements of $\mathscr{R}$ may be called reflections, and the elements of $\mathscr{M}$, generally motions. We assume $\mathscr{M}$ to be generated by $\mathscr{R}$; that is, every element of $\mathscr{M}$ is a product of finitely many elements of $\mathscr{R}$. Let, furthermore, the elements of $\mathscr{R}$ all be involutoric (have order 2) and let $M^{-1}R_a M$ be in $\mathscr{R}$ if R_a is in $\mathscr{R}$ and M in $\mathscr{M}$ (we say "$\mathscr{R}$ remains invariant under all **inner automorphisms**: $x \to M^{-1}xM$ of $\mathscr{R}$"). If R_a, $R_b \in \mathscr{R}$ and $R_a R_b$ is involutoric, we also call $R_a R_b$ a half-turn. If Theorems (1)–(5) are satisfied by the reflections and half-turns thus defined, *one can construct a geometry by assigning lines to reflections and points to half-turns and defining incidence and perpendicularity according to the above table.* In this way, one obtains a *metric plane possessing the original group $\mathscr{M}$ as its group of all motions.* We do not carry out the proof here, since this fact is not essential for our further development of geometry. *It should, however, be kept in mind, that one can in this way give a purely group-theoretic foundation of plane Euclidean and non-Euclidean geometry.*

Exercises

1. Given a group $\mathcal{G}$ and a subset $\mathcal{R}$ of $\mathcal{G}$ having the properties mentioned in the text of this section, define points and lines as explained above.   Show that the incidence Axiom I.1 is satisfied for this geometry.

2. Extend Exercise 1 to show that Axioms P.1 and P.2 are satisfied.

3. State Axioms Euc, Hyp, and Ell using reflections and the symbol "|."

4. Show that the set of inner automorphisms of a group is also a group under "successive application."   It is called the *group of inner automorphism* of the given group.

5. Find all inner automorphisms of a commutative group.

6. Prove that if all the elements except I of a group $\mathcal{G}$ are involutoric, the group is commutative.

1.16   Pencils of Lines

The set of all lines passing through a given point P is called a *pencil* of lines or, more specifically, a *pencil of the first kind* (see Figure 1.34).

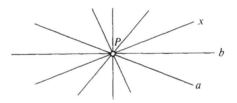

Figure 1.34

Let a, b be different lines of the pencil.   We give the following characterization for a line x to lie in the pencil:

Theorem 1.16.1   *A line x belongs to the pencil of all lines through P if and only if*
(1) $R_a R_b R_x$ *is a reflection.*

Proof   This theorem is a restatement of the first theorem on three reflections together with its converse.   Since the first theorem on three reflections was proved in Section 1.13, we need only show that if (1) is true, then x passes through P.

Suppose this is not so. Then we drop a perpendicular b' from P onto x. By the first theorem on three reflections, $R_a R_b R_{b'}$ is a reflection which may be denoted by $R_{a'}$. Furthermore, by assumption, $R_a R_b R_x = R_d$ for some d. Let x and b' intersect in P'. Then from

$$R_{a'} R_{b'} R_x = R_a R_b R_x = R_d,$$

we have (see Section 1.14)

$$R_{a'} R_d = R_{b'} R_x = H_{P'},$$

and hence, by Theorem 1.15.1, P' is on a'. Since we assume $P' \neq P$, we conclude from Axiom I.1 that $b' = a'$ and hence, from $R_a R_{b'} = R_a R_b$, that $a = b$, contrary to the assumption that $a \neq b$. ◆

By a ***pencil of the second kind*** we mean the set of all lines perpendicular to a given line p (see Figure 1.35). Let a, b again be two different lines of the pencil.

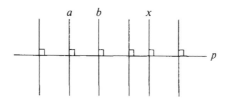

Figure 1.35

Theorem 1.16.2 *A line x belongs to the pencil of all lines perpendicular to a given line p if and only if (1) is true.*

Proof This is a restatement of the second theorem on three reflections together with its converse (Section 1.13). We need only show that (1) implies $x \perp p$. We can again set $R_a R_b R_x = R_d$. Let P be a point of x. In case $R_p = H_P$, that is, P is a pole of p, clearly $x \perp p$ (see Section 1.12). So let $R_p \neq H_P$. Then from Theorem 1.12.4 it follows that there is a unique perpendicular x' of p through P. Suppose $x' \neq x$; then let (by the second theorem on three reflections)

$$R_a R_b R_{x'} = R_{d'},$$

where d' is perpendicular to p. This implies

$$R_d R_x = R_{d'} R_{x'}.$$

If $x \neq x'$ Theorem 1.16.1 tells us that P is on d'; hence $d' = x'$ and, from $R_a R_b = R_{d'} R_{x'}$, we have $a = b$. This is a contradiction. ◆

If the given metric plane is elliptic, the first and the second kinds of pencils will coincide (compare Theorem 1.12.8). If the plane is not elliptic, no pencil of the first kind will also be of the second kind, and conversely.

In a Euclidean plane, it is readily seen that two lines either intersect or have common perpendiculars.

In a hyperbolic plane, however, two lines a, b may neither intersect nor have a common perpendicular; that is, they may be hyperbolic parallels. So they are not simultaneously in a pencil of the first or the second kind. In this case we *define* a pencil as the set of all x satisfying (1). We call it a *pencil of the third kind*. Then pencils of the first kind are in one-to-one

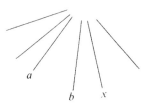

Figure 1.36

correspondence with the points. The pencils of the second and third kind we may intuitively consider to represent "ideal" points, or points that "cannot be reached." In Chapter 7, we enlarge any metric plane by calling all pencils "points" and identifying the pencils of the first kind with the original points. Naturally, any two of these generalized points are to be joined by a unique line. The following section will contribute to solving this join problem.

Exercises

1. Let a, b have a common point or a common perpendicular. Show that $R_aR_bR_x$ involutoric, $R_aR_bR_y$ involutoric, and $a \neq b$ imply $R_aR_xR_y$ involutoric (transitivity theorem).

2. Let $e \neq f$, but let them have a point in common. Show by using Theorem 1.16.1 that if $R_eR_fR_a$, $R_eR_fR_b$, $R_eR_fR_c$ are involutoric, then $R_aR_bR_c$ is involutoric.

3. Let u, v, w lie in a pencil, let a not lie in this pencil, and let $(a)R_uR_vR_w = a'$. Show that if a, w, and $(a)R_uR_v$ lie in a pencil, then $a = a'$.

4. Show by using Exercise 3 that if a, b, c are not in a pencil, if $(a)R_u = b$, $(b)R_v = c$, if u, v, w are in a pencil, and if c, w, a are in a pencil, then $(c)R_w = a$.

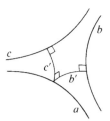

Figure 1.37

5. Show that two hyperbolic parallels determine a unique pencil of the third kind in which they both lie.

6. Let each pair of the lines a, b, c be in a pencil of the third kind (that is, be hyperbolic parallels) and let these pencils be different. From a point on a, drop perpendiculars b', c' onto b, c, respectively (Fig. 1.37). Show that

$$b' \perp c'.$$

1.17   A Theorem of Hjelmslev

We begin with a lemma.

Lemma 1.17.1 *Let A, C be different points and let v be the line joining them. $H_A R_b H_C$ is a reflection if and only if $b \perp v$. (See Fig. 1.38.)*

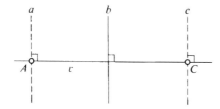

Figure 1.38

Proof I. Let $H_A R_b H_C$ be a reflection R_d. We have to show $b \perp v$. Let $H_A = R_v R_a$, $H_C = R_v R_c$ (see Section 1.14). Then

$$H_A R_b H_C = R_v R_a R_b R_v R_c = R_d;$$

hence

$$R_b R_v = R_a R_v R_d R_c = R_v R_a R_d R_c$$

or

$$R_v R_b R_v = R_a R_d R_c.$$

By Theorem 1.11.4 $R_v R_b R_v$ is a reflection R_w. So we can write

$$R_a R_d R_c = R_w \quad \text{or} \quad R_a R_c(R_c R_d R_c) = R_w.$$

Setting $R_c R_d R_c = R_{d'}$, we have, by Theorem 1.16.2, $d' \perp v$ and also $d \perp v$ (using $R_c = R_d R_c R_{d'}$). Now $w \perp v$, by the second theorem on three reflections. Finally, from

$$R_v R_b R_v = R_w \quad \text{and} \quad R_v R_w = R_w R_v,$$

we have

$$R_b = R_v R_w R_v = R_v R_v R_w = R_w;$$

hence $b = w$ and thus $b \perp v$.

II. Let $b \perp v$. Letting $H_A = R_a R_v$ and $H_c = R_v R_c$, we obtain

$$H_A R_b H_C = R_a R_v R_b R_v R_c = R_a R_b R_v R_v R_c = R_a R_b R_c.$$

By the second theorem on three reflections, this is a reflection. ◆

The following theorem will play an important role for pencils of lines.

Theorem 1.17.2 Theorem of Hjelmslev Let $R_a R_{a'} = H_A$ and $R_c R_{c'} = H_C$, where $A \neq C$. Furthermore, let $R_a R_b R_c = R_d$ and let v join A and C. Then $R_{a'} R_b R_{c'}$ is a reflection if and only if $d \perp v$.

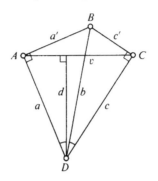

Figure 1.39

Proof

$$R_{a'} R_b R_{c'} = R_{a'} R_a R_a R_b R_c R_c R_{c'} = H_A R_a R_b R_c H_C = H_A R_d H_C$$

is, by Lemma 1.17.1, a reflection if and only if $d \perp v$. ◆

It should be noted that neither a and c nor a' and c' need to intersect. In ordinary Euclidean geometry, Theorem 1.17.2 could (for A, B, C, not on a line) be restated as follows: Let $ABCD$ be a quadrangle with right angles at A and C and let d pass through D. The angle between AD and d is the same as the angle between CD and BD if and only if $AC \perp d$ (see Fig. 1.39).

Our main application of Hjelmslev's Theorem is stated in the following theorem.

Theorem 1.17.3 Join Theorem A pencil of the first kind and a pencil of second or third kind have one and only one line in common.

Proof Drop perpendiculars a, c, d from D onto a', c', AC, respectively. Theorem 1.17.2 implies that a line b joins B and the pencil determined by a', c' if and only if $d \perp v$. Since d is uniquely determined, so is b. ◆

Exercises

1. Show that any four of the following five equations imply the fifth equation:

$$R_a R_{a'} = H_A;$$

$$R_c R_{c'} = H_C;$$
$$R_a R_b R_c = R_d;$$
$$R_{a'} R_b R_{c'} = R_{d'};$$
$$H_A R_d H_C = R_{d'}.$$

2. Prove, as a degeneration of Hjelmslev's theorem, the following theorem: Let a, a', b not lie in a pencil and let $a \perp a'$; $H_A = R_a R_{a'}$ (that is, given a "right triangle"). If b ("hypotenuse") is reflected in a and a', then the images $d = (b)R_a$, $d' = (b)R_{a'}$ satisfy

$$(d)H_A = d'.$$

3. Let a, b, c pass through P. Construct, in ordinary plane geometry, the line d for which $R_d = R_a R_b R_c$.

4. Prove: If $R_a H_B R_c$ is a half-turn, there exists a line v such that $v \perp a$, c and B I v.

5. In a Euclidean metric plane, let ABC and $A'B'C'$ be triangles such that AB is parallel to $A'B'$ and AC is parallel to $A'C'$. Assume further that AA' is perpendicular to AB and BC is perpendicular to CC'. Let, finally, AA', BB', and CC' intersect in a point. Show that BC is parallel to $B'C'$ (see Fig. 1.40).

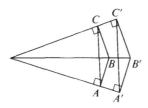

Figure 1.40

6. State and prove the converse of the theorem in Exercise 5. In other words, assume BC is parallel to $B'C'$ and prove that AA', BB', and CC' intersect in a point.

1.18   *Classification of Motions*

In this section we shall prove a fundamental result that gives us an insight into the structure of motions. We begin with two definitions.

A rotation has been defined as a product $R_a R_b$ where a, b intersect in a point. In an elliptic plane any such product is a rotation. In a Euclidean or hyperbolic plane, a and b may not intersect, but may have a common perpendicular t; in this case, $R_a R_b$ defines a translation along t. In hyperbolic planes, $R_a R_b$ may be neither a rotation nor a translation. If a and b are intersecting lines, then $R_a R_b$ is a rotation by definition. If a and b are any two lines, which may or may not intersect and which may or may not be identical, we shall call the product $R_a R_b$ a **quasi-rotation**. If $a = b$, then the product is the identity. We can regard this as analogous to the fact in ordinary Euclidean geometry that a rotation about a point through 360 degrees is the identity. Later we shall be able to say in what sense the product can be regarded as a rotation when a and b do not intersect.

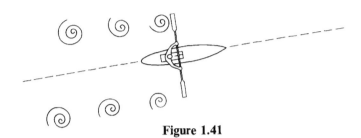

Figure 1.41

We define a **glide reflection** or a **paddling** (see Fig. 1.41) as the product of a translation along t and the reflection at t. Equivalently, we can say that a glide reflection is a product $R_a R_b R_t$, where $a \perp t$, $b \perp t$ (see Fig. 1.42). This again shows that a glide reflection can be looked upon as the product of a reflection R_a and a half-turn $H_W = R_b R_t$. Remember that a proper

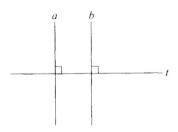

Figure 1.42

motion was defined as the product of an even number of reflections, and an improper motion as the product of an odd number of reflections. Our main result is stated in the theorem that follows.

Theorem 1.18.1 Classification Theorem *Every proper motion is a quasi-rotation. Every improper motion is a glide reflection.*

Before proving this theorem, we shall first prove two lemmas.

Lemma 1.18.2 *Every product $R_u R_v H_W$ can be replaced by a product $R_a R_b$.*

Proof We can assume $u \neq v$. Then u, v span a pencil and, by Axiom I.1 or Theorem 1.17.3, there is a line l belonging to this pencil and passing through W (Fig. 1.43). By definition of a pencil, $R_u R_v R_l$ is a reflection R_a.

Hence

$$R_u R_v = R_a R_l.$$

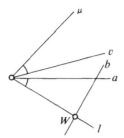

Figure 1.43

Furthermore, we can let

$$H_W = R_l R_b.$$

Now

$$R_u R_v H_W = R_a R_l R_l R_b = R_a R_b. \qquad \blacklozenge$$

Lemma 1.18.3 *Every product $R_u R_v R_w$ can be replaced by a product $R_a H_B$.*

Proof Again $u \neq v$ can be assumed. Let U be a point on u (Fig. 1.44) and

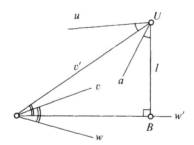

Figure 1.44

let v' join U and the pencil formed by v, w (Axiom I.1 or Theorem 1.17.3). $R_w R_v R_{v'}$ is a reflection $R_{w'}$; hence

$$R_v R_w = R_{v'} R_{w'}.$$

(If v, w intersect in a point, this means that we change the representation of the rotation $R_v R_w$). Let l be a perpendicular of w' through U. Letting

$$H_B = R_l R_{w'} \quad \text{and} \quad R_u R_{v'} R_l = R_a,$$

we find

$$R_u R_v R_w = R_u R_{v'} R_{w'} = R_u R_{v'} R_l R_l R_{w'} = R_a H_B. \qquad \blacklozenge$$

Proof of Theorem 1.18.1   By Lemmas 1.18.2 and 1.18.3, every product of four reflections can be reduced to a product of two reflections:

$$R_t R_u R_v R_w = R_t R_a H_B = R_c R_d.$$

Therefore, every product of an even number of reflections can be replaced by that of two reflections, that is, a quasi-rotation.   The product of an odd number of reflections reduces to a product of three and hence, by Lemma 1.18.3, to a glide reflection.   ◆

Since in an elliptic plane every reflection is a half-turn and conversely, we may, by Lemma 1.18.3, replace every product of three reflections by a product of two reflections.   So we find the following result.

Theorem 1.18.4   Classification of elliptic motions   *In an elliptic plane, every motion is a rotation.*

Another immediate conclusion from Theorem 1.18.1 follows.

Theorem 1.18.5   Classification of Euclidean motions   *In a Euclidean plane, every proper motion is either a translation or a rotation; every improper motion is a glide reflection.*

Finally, we show that an improper motion possessing a fixed point is a reflection.

Theorem 1.18.6   *If, in a metric plane, a proper motion possesses a fixed point, it is a rotation.   If in a nonelliptic plane an improper motion possesses a fixed point, it is a reflection.*

Proof   The first statement is clear from Theorem 1.18.1.   To prove the second part, let an arbitrary improper motion be expressed as a glide reflection $R_a R_b R_t$ (according to Theorem 1.18.1), where $a \perp t$ and $b \perp t$. Let, furthermore, P be a fixed point of $R_a R_b R_t$.   Then

$$(P)R_a R_b R_t = P.$$

By Theorem 1.13.5, we can let $R_a R_b = R_{a'} R_{b'}$ where $a' \perp t$, $b' \perp t$, and $P \perp a'$.

From $R_{b'} R_t = H_B$ for some point B on t, we now have

$$(P)H_B = P.$$

From Theorem 1.14.7, it follows that $P = B$; hence, by Axiom P.2, $a' = b'$, and thus $a = b$.   Therefore, $R_a R_b R_t = R_t$.   ◆

Exercises

1. Show that the square of an improper motion is a translation.

2. Show that any product $R_u R_v R_w$ (and hence every improper motion) can be replaced by a product $H_A R_b$.

3. Given a point R, show that any product $R_u R_v R_w$ (and hence every improper motion) can be replaced by the product of a reflection and a rotation about R. (*Hint*: Proceed analogously to the proof of Lemma 1.18.3, starting with a line w' connecting R and the pencil formed by v, w.)

4. Let $M \neq I$ be a motion. Suppose that for any three lines a, b, c the products $R_a M$, $R_b M$, $R_c M$ are involutoric. Prove that $R_a R_b R_c$ is involutoric.

5. Find all motions of a Euclidean metric plane that map every line onto itself or a parallel of itself.

6. Find all motions of a Euclidean metric plane that leave all lines of one pencil of parallel lines fixed.

7. Prove that the set of all translations and glide reflections in a line t is a commutative group.

8. Prove that the set of all quasi-rotations $R_x R_y$ with x, y in a pencil is a group.

Chapter Two

Order and Congruence

In this chapter we shall develop those concepts of plane geometry that involve "order" or "betweenness." By the introduction of order axioms, elliptic planes are automatically excluded (Theorem 2.2.4). So we are dealing mainly with the common basis of Euclidean and hyperbolic planes (sometimes called *absolute geometry*). We shall pursue the axiomatic treatment of these planes up to a foundation of *ordinary* Euclidean and hyperbolic planes in the sense that every line is essentially the same as the set of real numbers. Thus, we present a counterpart to Hilbert's *Foundation of Geometry*, except that we restrict our attention to plane geometry (an extension to spatial geometry follows, in Chapter 4).

The main difference between Hilbert's treatment and ours lies in the fact that Hilbert introduces axioms of congruence, whereas here the idea of congruence flows out of the more vivid idea of a motion.

After the problem of a foundation of ordinary plane geometry has been solved (Section 2.10), we shall go into separate treatments of Euclidean and hyperbolic planes, discussing mainly those properties of Euclidean planes that are direct consequences of congruence properties. Although this is a little out of date we spend one section on constructions by ruler and compass, mainly as a preparation for showing the impossibility of such concepts as angular trisection in the next chapter.

2.1   Free Mobility

Let a metric plane be given—that is, a system of "points" and "lines" satisfying Axioms I.1–I.3, P.1–P.3, M.1–M.4 introduced in the first four sections of Chapter 1. If a, a' are two lines, then Axioms M.3 and M.4 guarantee the existence of a reflection mapping a onto a'

only under the assumption that there already exists a rotation or a translation moving a onto a'. Such a rotation or translation, however, need not exist, as can be seen from the example of a nine-point plane (Euclidean) presented in Section 1.7. In our future discussion we would be hindered by a nonexistence of this kind. Therefore, we have "free mobility" guaranteed by the following axiom.

Axiom FM (*Free mobility*) *Any two lines possessing a common point or a common perpendicular can be mapped onto each other by a reflection.*

Later on (Section 2.9) we introduce an axiom of "continuity" that implies FM, so that eventually we may drop Axiom FM. As an immediate consequence of Axiom FM, we prove the following theorem.

Theorem 2.1.1 *Two lines that intersect can be mapped onto each other by a rotation. Two lines possessing a common perpendicular but no common point, can be mapped onto each other by a translation.*

Proof Let R_b map a onto a' (existence of R_b by Axiom FM). $R_b R_{a'}$ is a rotation if a and a' intersect. $R_b R_{a'}$ is a translation along t if $t \perp a$ and $t \perp a'$, in which case $t \perp b$, (see Section 1.11), and if a, a' do not intersect. Furthermore, we have in both cases

$$(a)R_b R_{a'} = (a')R_{a'} = a'. \qquad \blacklozenge$$

Theorem 2.1.2 *Any two different points can be mapped onto each other by a reflection and also by a half-turn. Both mappings are uniquely determined.*

Proof Let P, P' be the points, and let $u = PP'$. The perpendiculars of u through P, P', respectively, are denoted by a, a'. If $(a)R_b = a'$ (existence of R_b by Axiom FM), then clearly $b \perp u$ and hence

$$(P)R_b = P'.$$

Also, letting $H_R = R_u R_b$, we find

$$(P)H_R = (P)R_u R_b = (P)R_b = P'.$$

By Theorem 1.18.6, R_b is uniquely determined. $\blacklozenge$

R is called the **midpoint** of P and P'. It is evident that R is uniquely determined, so any two points possess a well-determined midpoint.

Theorem 2.1.3 *In a nonelliptic plane any two points P, P' determine a unique translation T along PP' mapping P onto P'.*

Proof Let R_b map P onto P' (see Theorem 2.1.2) and let a be perpendicular to PP' at P. Then either $a = b$ or a and b do not intersect, the plane not being elliptic. Hence $T = R_a R_b$ maps P onto P'.

Assume that T' is another translation along PP' mapping P onto P'. By Theorem 1.13.5, T' can be expressed as

$$T' = R_a R_{b'}.$$

From $(P)R_b = P'$ and $(P)R_a R_{b'} = P'$, we conclude (by Theorem 2.1.2) that $b = b'$. Therefore, $T = T'$. $\blacklozenge$

Exercises

1. Show that two lines a, a' intersecting in a point can be mapped onto each other by precisely two reflections.

2. Show that any two lines can be mapped onto each other by a motion.

3. Show that any two points can be mapped onto each other by a rotation that is not a half-turn.

4. Define parallelograms and diagonals of parallelograms in a Euclidean metric plane. Assume Axiom FM to be true. Show that the two diagonals intersect in the common midpoint of two pairs of vertices.

5. Define squares in a Euclidean metric plane. Again, assume Axiom FM to be true. Show that any two given points are common vertices of exactly three squares.

6. Find, in the nine-point plane introduced in Section 1.7, all pairs of lines that cannot be mapped onto each other by reflections.

2.2   Axioms of Order

There is a large gap in Euclid's original foundation of geometry. He took for granted those facts dealing with "betweenness." This gap has been filled in during the second half of the nineteenth century, mainly by Moritz Pasch. In this section we follow, in general, the ideas of Pasch (as is done in Hilbert's *Foundation of Geometry*, too).

All theorems on points and lines in Chapter 1 have been based on two fundamental relations:

"incidence"   and   "perpendicularity."

The first relates point and line; the second relates line and line. Now we introduce a third relation,

"betweenness,"

which relates triplets of points. We shall consider a subset of the set of all ordered triplets of different collinear points. If (A, B, C) is such a triplet, we say

"B lies between A and C,"

or we write it

$[ABC]$.

The following axioms will hold:

Axiom O.1 $[ABC]$ *implies* $[CBA]$ (Fig. 2.1).

Figure 2.1

Figure 2.2

Axiom O.2 *If A, C are different points, there exist points B, D such that [ABC] and [BCD] (Fig. 2.2).*

Axiom O.3 *If A, B, C are collinear, then [ABC] or [BAC] or [ACB].*

A **line segment** (AB) we define as the set of all points X such that [AXB] We call A, B the **end points** of (AB) (they do not belong to (AB)). Another way of expressing [ABC] is to say "B is on (AC)." A **closed line segment** [AB] is the set of all points of (AB) together with the end points A and B.

Axiom O.4 (**Pasch's axiom**). *Let A, B, C be noncollinear, and let a not pass through any of these points. If a intersects (AB), it also intersects either (AC) or (BC) (see Fig. 2.3).*

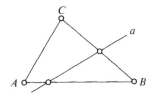

Figure 2.3

The following axiom provides an important link between motion and congruence (the latter being defined in Section 2.6).

Axiom O.5 *For any reflection R$_a$, [ABC] implies*

$$[(A)R_a(B)R_a(C)R_a].$$

In other words, the relation "between" remains unchanged under reflections, and hence under all motions. A metric plane with an order relation satisfying Axioms O.1–O.5 is called an **ordered metric plane**. We now draw some conclusions from Axioms O.1–O.5.

Theorem 2.2.1 *If P′ = (P)R$_a$ and P ≠ P′, then a intersects (PP′) (Fig. 2.4).*

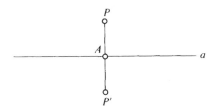

Figure 2.4

Proof By Axiom P.3, *a* and *PP'* meet in a point *A* (Fig. 2.4). Suppose [*PAP'*] is not true. Then, by Axiom O.3, either [*APP'*] or [*AP'P*]. Since $(A)R_a = A$ and $(P)R_a = P'$, we conclude from [*APP'*] that

$$[(A)R_a(P)R_a(P')R_a]$$

and hence [*AP'P*], which is a contradiction. Analogously, [*AP'P*] is impossible. ◆

Theorem 2.2.2 [*ABD*], [*ACD*], *and* $B \neq C$ *imply either* [*ABC*] *or* [*ACB*] (Fig. 2.5).

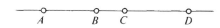

Figure 2.5

Proof Let *U* be a point not on *AB*, and let *V* lie on (*UB*). (See Fig. 2.6.) Since *A* is not on (*BD*) nor on (*CD*), we obtain, by applying Axiom O.4 to *B*, *D*, *U* and *C*, *D*, *U*, respectively, that *AV* intersects (*UD*) and (*UC*). Therefore, by applying Axiom O.4 to *B*, *C*, *U*, we see that *AV* cannot intersect (*BC*). This excludes [*BAC*], so that, by Axiom O.3, either [*ABC*] or [*ACB*]. ◆

By an analogous argument, the following theorem is readily proved:

Theorem 2.2.3 [*ABC*] *and* [*ACD*] *together imply* [*ABD*].

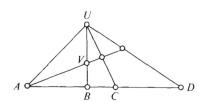

Figure 2.6

Theorem 2.2.4 Nonellipticity *Any ordered plane is nonelliptic; that is, no line possesses a pole.*

Proof Suppose *P* is a pole of *a*. (See Fig. 2.7.) Let *A* be on *a* and let *B* be a point such that [*ABP*] (existence by Axiom O.2).

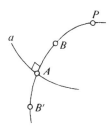

Figure 2.7

We let $B' = (B)R_a$. By Theorem 1.12.9, B and B' are different. Since $(A)R_a = A$ and $(P)R_a = P$, we obtain, by Axiom O.5: $[AB'P]$. This implies, by Theorem 2.2.2, either $[ABB']$ or $[AB'B]$. But $[ABB']$ implies, by Axiom O.5, $[AB'B]$ and the converse. This contradicts Theorem 2.2.3. ◆

Exercises

1. Show that on any line there are infinitely many points.

2. Show that the midpoint of two points lies between these points.

3. Carry out in detail the proof of Theorem 2.2.3.

4. Show that $[ABD]$, $[ACD]$, and $B \neq C$ imply either $[ABC]$ or $[CBD]$.

5. Let $P_1 P_2 \ldots P_h P_1$ be a closed "polygonal train" consisting of $P_1, \ldots, P_h$ and the line segments $(P_1 P_2), (P_2 P_3), \ldots, (P_{h-1} P_h), (P_h P_1)$. Show that if a line meets a closed polygonal train in one of its line segments $(P_i P_{i+1})$, it meets the polygonal train in at least two points (possibly coinciding).

6. Show, without rigorous proofs, that the example of a hyperbolic metric plane in Section 1.6 is an ordered metric plane.

2.3 Equivalence Relations

We leave for a moment the development of plane geometry and introduce the fundamental mathematical notion of equivalence, which will be needed in the following section. Let us call two human beings A and B equivalent if they have the same nationality—so that we can study a few properties of this relation "equivalent." A trivial property is the following:

(a) *A is equivalent to itself.*

This would not be so for a relation like "being taller" or "having less money than." These examples also would not have the following property:

(b) *If A is equivalent to B, then B is equivalent to A.*

However, in all of our three examples:

(c) *If A is equivalent to B and B is equivalent to C, then A is equivalent to C.*

Using the symbol "$\sim$" as an abbreviation for equivalent, we may restate (a)–(c) as follows:

(a) $A \sim A$.
(b) $A \sim B$ implies $B \sim A$.
(c) $A \sim B$ and $B \sim C$ imply $A \sim C$.

Generally we call a relation between the elements of a set an *equivalence relation* if it obeys (a)–(c). Hereby a relation may be defined abstractly as a subset of the set of all pairs of the elements of the given set. In the above example, the pairs would be those consisting of two people with the same nationality. Other examples are the following ones:

The ordinary equality in logic, in particular the equality of natural numbers, clearly satisfies (a)–(c). In fact, (a)–(c) can be used as axioms in the foundation of logic.

In number theory, the congruence relation $a \equiv b \pmod{m}$, defined by "$a - b$ is an integer multiple of m," is an equivalence relation, since $a - a = 0m$; $a - b = rm$ implies $b - a = (-r)m$; $a - b = rm$ and $b - c = sm$ imply $a - c = (r + s)m$.

A third example is obtained by splitting an arbitrary set into disjoint subsets and calling two elements equivalent if they belong to one and the same of these subsets. This example is, however, more than an example. We now show that every equivalence relation can be characterized by subsets of a set. Let an equivalence relation on a set $\mathscr{S}$ be given. We assign to every element x of $\mathscr{S}$ the subset $\mathscr{U}_x$ of $\mathscr{S}$, defined by all elements y satisfying $x \sim y$. If z is any element of $\mathscr{U}_x$, that is, if $x \sim z$, then, by (b), $z \sim x$ and so x is an element of $\mathscr{U}_z$. In other words, $\mathscr{U}_x = \mathscr{U}_z$ if and only if $x \sim z$. Therefore, the subsets $\mathscr{U}_x$ thus defined depend only on the relation "$\sim$," not on special elements they contain. We call these subsets *distinguished subsets*.

Every element x belongs to a distinguished subset, since, by (a), $x \sim x$.

Suppose two distinguished subsets $\mathscr{U}_x$, $\mathscr{U}_y$ have a point z in common. Then $x \sim z$ and $y \sim z$; hence, by (b) and (c), $x \sim y$. This implies, as we have seen, $\mathscr{U}_x = \mathscr{U}_y$. So two distinguished subsets are either identical or disjoint.

In summary, we have found that if an equivalence relation on a set $\mathscr{S}$ is given, then $\mathscr{S}$ can be split into nonempty disjoint subsets, called *equivalence classes*, such that two elements are equivalent if and only if they belong to the same equivalence class.

In the introductory example, the equivalence classes are the nationalities. The equivalence classes of a number-theoretic relation $a \equiv b \pmod{m}$ are sets, $\{\ldots, a - 2m, a - m, a, a + m, a + 2m, \ldots\}$. If, for example, $m = 2$, there are precisely two equivalence classes, namely the odd integers and the even integers.

Equivalence relations provide a very helpful tool in many mathematical problems. The following section presents one of these problems.

Exercises

1. Call two elements x, y of an arbitrarily given group $\mathscr{G}$ equivalent if there exists an element u of $\mathscr{G}$ such that

$$y = u^{-1}xu.$$

(a) Show that this is an equivalence relation.

(b) What can be said about equivalence classes if $\mathscr{G}$ is commutative?

2. Let $\mathscr{H}$ be a subgroup of a commutative group $\mathscr{G}$. Set $a \sim b$ if ab^{-1} belongs to $\mathscr{H}$. Show that "$\sim$" is an equivalence relation.

3. In an arbitrary set, call two elements equivalent if and only if they coincide. Show that an equivalence relation is thus defined. Determine the equivalence classes.

4. Give examples of relations that satisfy two of the properties of equivalence but are not equivalence relations.

5. In a Euclidean metric plane, call two lines parallel if they do not intersect or if they coincide. Show that "parallel" is an equivalence relation and find the equivalence classes.

6. In a metric plane, let a, b be two intersecting lines. Call two lines x, y equivalent if $R_a R_b R_x$ and $R_a R_b R_y$ are reflections. Show that an equivalence relation is thus obtained. Find the equivalence classes. (See Section 1.16.)

7. Show that the properties of equivalence (b) and (c) do not imply (a). (*Hint*: Use Exercise 4.)

8. Why is the following "proof" that properties (b) and (c) imply property (a) false? "$a \sim b$ implies, by (b), $b \sim a$; hence, by (c), $a \sim a$."

2.4   *Rays and Oriented Lines*

We introduce now a geometrical concept that is very helpful in building up the theory of order and congruence, namely, the concept of a ray.

Let a be an arbitrary line and let A, P, Q be points on a. If $[PAQ]$, we say that P, Q are **on different sides** of A (relative to a). If, however, $[APQ]$ or $[AQP]$ or $P = Q \neq A$, then P and Q are said to be **on the same side** of A (relative to a). See Figure 2.8. The set of all points on one side of A

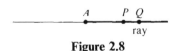

$$A \qquad P \ Q$$
ray

Figure 2.8

relative to a is called an **open ray**. A is called the **end point** of the ray. A ray together with its end point is called a **closed ray** or, briefly, a **ray**. We expect that every point A of a is an end point of exactly two rays. This must be a conclusion of the order axioms we have introduced, and we now state it as a theorem.

Theorem 2.4.1   *If A is on a, then "to be on the same side of A" is an equivalence relation on the set of all points of a not equal to A, subdividing this set into two open rays as equivalence classes.*

Proof We denote the relation "to be on the same side of A" by "$\sim$" and assert that it is an equivalence relation. By definition, $P \sim P$ for any $P \neq A$ on a, and also $P \sim Q$ implies $Q \sim P$. So let $P \sim Q$, $Q \sim R$, say $[APQ]$ and $[ARQ]$. By Theorem 2.2.2, $[APR]$ or $[ARP]$ or $P = R$; that is, $P \sim R$. Theorem 2.2.3 implies that there are precisely two equivalence classes. ◆

We denote rays by letters $r, s, \ldots$. If we wish to say that A is the end point of r, we denote the ray by the pair A, r. The line a on which r lies is called the **carrier** of the ray. The two rays in which a is subdivided by its point A are called **opposite rays**. We say also "opposite sides," instead of "different sides."

If a motion maps a ray r onto a ray s, it maps the end point A of r automatically onto the end point of s, since, otherwise, points on different sides of the image of A would have inverse images on the same side of A, contradicting the conclusion made after Axiom O.5.

A fundamental fact about the "mobility" of motions is expressed in the following two theorems.

Theorem 2.4.2 *Any proper motion leaving a ray A, r fixed (that is, mapping A onto A and r onto r) is the identity map.*

Proof By Theorem 1.18.6, a proper motion leaving a point A fixed is a rotation S about A. Let a carry A, r, and let $S = R_a R_b$ (see Section 1.4). From $(a)R_a R_b = a$, we have $(a)R_b = a$; hence either $b \perp a$ or $b = a$ (by the fixed-line theorem, 1.11.3). In other words, either $S = H_A$ or $S = I$. In the first case, we conclude from Theorem 2.2.1 that S maps A, r onto its opposite ray, contrary to our assumption. Therefore, $S = I$. ◆

Theorem 2.4.3 *Two rays can be mapped onto each other by one and only one proper motion.*

Proof Let A, r and B, s be the given rays. Let a carry A, r, and let b carry B, s (Fig. 2.9). We map A onto B by a translation T (see Theorem 2.1.3) and then map $a' = (a)$T onto b by a rotation S. The image of A, r under

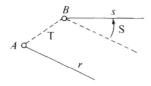

Figure 2.9

the product TS is either B, s or the opposite ray of B, s. In the latter case, SH_B maps A, r onto B, s. To show uniqueness, assume that two proper motions M, M′ map A, r onto B, s. Then MM'^{-1} is a proper motion leaving A, r fixed. By Theorem 2.4.2, $MM'^{-1} = I$; hence $M = M'$. ◆

We shall rephrase Theorem 2.4.3 after introducing the following terminology, which is used widely nowadays in algebra and geometry: Because of Theorem 2.4.3, the group $\mathscr{G}$ of all motions can be considered to interchange, or "permute," the elements of the set of all rays. So $\mathscr{G}$ "operates" not only on the set of all points or on the set of all lines, but also on other sets (as another example, take the set of all "half-planes" to be introduced in Section 2.5).

Let, in general, a group $\mathscr{G}$ of transformations "operate" on a set $\mathscr{M}$; that is, let it consist of one-to-one mappings of $\mathscr{M}$ onto itself. If any element of $\mathscr{M}$ can be mapped onto any other element of $\mathscr{M}$, we call $\mathscr{G}$ *transitive* on $\mathscr{M}$. We also say that $\mathscr{G}$ *operates transitively* on $\mathscr{M}$. If there is one and only one mapping of $\mathscr{G}$ that maps any prescribed element on any other prescribed element of $\mathscr{M}$, we say that $\mathscr{G}$ is *simply transitive* on $\mathscr{M}$, or *operates simply transitively on $\mathscr{M}$*. Now we can express Theorem 2.4.3 as follows:

Theorem 2.4.4 Ray transitivity theorem *The group of all proper motions is simply transitive on the set of all rays.*

As a direct consequence of Theorem 2.4.3 or 2.4.4, we state the following theorem.

Theorem 2.4.5 *Given two pairs P, P' and Q, Q' of different points, there is at most one proper motion that maps P onto P' and Q onto Q'.*

Now we consider a subgroup of the group of all proper motions, namely, that of all translations along a given line a (compare Theorem 1.13.6). a remains fixed under the translations of this group. In particular, rays carried by a are permuted. Let A, r and B, s be two of those rays. By Theorem 2.1.3, there exists a unique translation mapping A onto B. It maps A, r onto either B, s or the opposite ray of B, s. So the group of translations along a subdivides the set of all rays carried by a into two subsets. On each of these subsets it operates transitively.

The pair consisting of a and one of the subsets of rays is called an *oriented line*. So every ray carried by a determines an *orientation* of a defined by the subset to which the ray belongs.

Exercises

1. Show that if A, B belong to a ray r, all points of (AB) belong to r.

2. Why does a ray possess only one end point?

3. Prove that every line segment (AB) contains infinitely many points.

4. Show that (a) any proper motion leaving two different points fixed is the identity map, and (b) any motion leaving three noncollinear points fixed is the identity map.

5. Prove that if a motion M interchanges two rays, then $M^2 = I$.

6. In a Euclidean ordered metric plane, define betweenness for parallel lines. Find the notions that correspond to "ray" and "line segment."

7. Let a group $\mathscr{G}$ operate on a set $\mathscr{S}$. Call two elements of $\mathscr{S}$ equivalent if they can be mapped onto each other by an element of $\mathscr{G}$. Show this relation to be an equivalence relation. How many equivalence classes are there if $\mathscr{G}$ operates transitively on $\mathscr{S}$?

2.5 Half-Planes and Orientation of the Plane

Let us consider a line a and let us call two points P, Q not on a, *on the same side of a* if $P = Q$ or if (PQ) does not contain a point of a (see Fig. 2.10).

Theorem 2.5.1 *"To be on the same side of a" is an equivalence relation for the set of all points not on a. It splits this set into two nonempty disjoint subsets, called "sides" of a.*

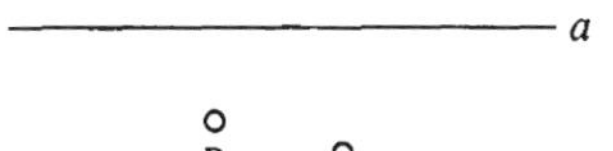

———————————————— a

∘   ∘
P Q

Figure 2.10

Proof Let "$\sim$" stand for "on the same side of a," and denote by $\mathscr{P}_a$ the set of all points not on a. Clearly, $P \sim P$ for any P in $\mathscr{P}_a$. Furthermore, from $(PQ) = (QP)$ (compare Axiom O.1), $P \sim Q$ implies $Q \sim P$. Finally, $P \sim Q$ and $Q \sim R$ imply $P \sim R$, since, otherwise, (PR) would contain a

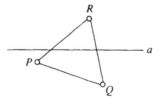

Figure 2.11

point of a, contradicting Axiom O.4 or, if P, Q, R, are collinear, Theorem 2.2.2. We now show that there are precisely two equivalence classes. Clearly there is a point P not on a (Fig. 2.11). If A lies on a, then there is a point Q such that A is between P and Q (Axiom O.2). Hence, there are at least two equivalence classes, which we call *sides* of a. For any other point R of $\mathscr{P}_a$, either (PR) or (QR) contains a point of a (Axiom O.4), so R is on one of the two sides of a. ◆

By a **half-plane** we mean the points of a line a together with all points on one side of a; a is called the **boundary** of the half-plane. So every line is a boundary of precisely two half-planes.

Theorem 2.5.2 *A reflection* R_a *interchanges the sides of* a.

Proof By Theorems 2.2.1 and 2.4.1, for any P of $\mathscr{P}_a$, we conclude that P and $P' = (P)R_a$ are on different sides of a. Since, by Axiom O.5, betweenness is preserved under reflections, Theorem 2.5.2 follows. ◆

Theorem 2.5.3 *Translations along a line* a *leave the sides of* a *unchanged.*

Proof Let P be not on a and let b be a perpendicular of a not passing through P (Fig. 2.12). We set $P' = (P)R_b$. If we can show that P and

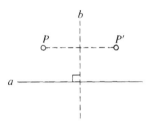

Figure 2.12

P' are on the same side of a, we are done, since any translation along a is a product of two such reflections. Assume that the line PP' intersects a in a point A. Since PP' is perpendicular to b, A would be a pole of b and hence the plane would be elliptic. This contradicts Theorem 2.2.4. So there is no point of a on (PP') and hence P, P' are on the same side of a.

By definition, the following theorem is clear.

Theorem 2.5.4 *If* a *and* b *intersect in a point* P, *then the rays into which* b *is divided by* P *are each contained in opposite sides of* a.

Consider now a half-plane α and a ray A, r which is carried by the boundary line a of α (Fig. 2.13). We call the triplet A, r, α a **flag** and now prove the following theorem.

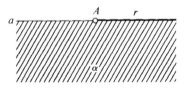

Figure 2.13

Theorem 2.5.5 *The group of all motions is simply transitive on the set of all flags.*

Proof Let A, r, α and B, s, β be two given flags where a is the boundary of α and b is the boundary of β. By Theorem 2.4.4, there is a uniquely determined proper motion M that maps the ray A, r onto the ray B, s. If M' is an improper motion mapping A, r onto B, s, then $M^{-1}M'R_b$ is a proper motion leaving B, s fixed. Hence, by Theorem 2.4.2,

$$M^{-1}M'R_b = I \quad \text{or} \quad M' = MR_b.$$

Either M or MR_b, but not both, will map A, r, α onto B, s, β. ◆

Theorem 2.5.5 implies the following one.

Theorem 2.5.6 *If we call two flags equivalent if they can be mapped onto each other by a proper motion, we obtain an equivalence relation with two equivalence classes.*

Every such equivalence class is called an **orientation** of the plane. The given plane together with an orientation is called an **oriented plane**.

Exercises

1. Define the interior of a triangle, $\triangle ABC$, by using half-planes and show that the interior of a triangle is **convex**, that is, it contains all points of (PQ) whenever it contains P and Q.

2. Define closed convex polygons and define the interior of convex polygons. Show that the interior of a convex polygon can be covered by finitely many triangles such that every point of the interior belongs to at most two triangles (proof by induction).

3. Show that if a half-plane α with boundary a is mapped onto itself by a reflection in a line b, then a is perpendicular to b.

4. Prove that if a ray does not intersect a line, it is contained in one side of this line.

5. Prove that if in a Euclidean or a hyperbolic ordered metric plane two different lines are parallel, each is contained in one side of the other.

6. In a Euclidean ordered metric plane, characterize half-planes by using the concepts introduced in Exercise 6 of Section 2.4.

2.6 Congruence

Two line segments (PQ) and $(P'Q')$ are said to be **congruent** if there is a motion M mapping P onto P' and Q onto Q'. We write

$$(PQ) \equiv (P'Q').$$

Clearly, (PQ) and $(P'Q')$ are congruent if any point of (PQ) is mapped under M onto a point of $(P'Q')$. More generally, we call any two point sets congruent if they can be mapped onto each other by a motion.

Theorem 2.6.1 *Given two different points A, B and a ray r' with end point A', there is one and only one point B' on r' such that $(AB) \equiv (A'B')$ (Fig. 2.14).*

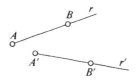

Figure 2.14

Proof Let A, r be the ray with end point A that contains B. Then there is a unique proper motion M mapping A, r onto A', r' (see Theorem 2.4.3). As is shown in the proof of Theorem 2.5.5, the only improper motion mapping A, r onto A', r' is $MR_{a'}$, where a' carries r'. So

$$B' = (B)M = (B)MR_{a'}$$

is uniquely determined and satisfies $(AB) \equiv (A'B')$. ◆

Theorem 2.6.2 *The relation "congruent" is an equivalence relation on the set of all line segments.*

Proof This follows readily from the properties of a group: Since I is in the group $\mathcal{M}$ of all motions,

$$(AB) \equiv (AB).$$

If $M \in \mathcal{M}$, then also $M^{-1} \in \mathcal{M}$. Therefore,

$$(AB) \equiv (CD) \quad \text{implies} \quad (CD) \equiv (AB).$$

If M maps (AB) onto (CD) and M' maps (CD) onto (EF), then MM' maps (AB) onto (EF). Hence,

$$(AB) \equiv (CD) \quad \text{and} \quad (CD) \equiv (EF) \quad \text{implies} \quad (AB) \equiv (EF). \qquad ◆$$

Let A, r and A, s be rays. Then the pair consisting of the rays r and s is called an **angle** (Fig. 2.15), which we denote by

$$\sphericalangle(r, s) \quad \text{or} \quad \sphericalangle(s, r);$$

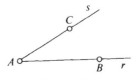

Figure 2.15

r and s are said to be *sides* of $\angle(r, s)$. The angle (r, s) is uniquely determined by the common end point A of r, s and by two points B, C on r, s, respectively. We let

$$\angle(r, s) = \angle BAC = \angle CAB.$$

If the carriers of r, s are different, we call $\angle(r, s)$ a *proper angle*. In the case that s, r are opposite, $\angle(r, s)$ is said to be *straight*. If a proper angle (r, s) is given, consider the open half-plane containing r in its interior and having s on its boundary, also the half-plane containing s and having r on its boundary. The intersection of these half-planes is called the *interior* of $\angle(r, s)$ (Fig. 2.16). The set of points neither in the interior of $\angle(r, s)$ nor on either A, r or A, s is called the *exterior* of $\angle(r, s)$. So the interior and the exterior are defined for all proper angles.

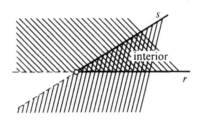

Figure 2.16

Two angles $\angle(r, s)$ and $\angle(r', s')$ are called *congruent* if there exists a motion mapping r onto r' and s onto s':

$$\angle(r, s) \equiv \angle(r', s'),$$

as in the case of line segments.

Theorem 2.6.3 *The relation "congruent" is an equivalence relation on the set of all angles.*

The next theorem readily follows from Theorem 2.5.5.

Theorem 2.6.4 *Given a proper angle $\angle(r, s)$ and a flag A', r', α', there is one and only one ray s' with end point A' and contained in α' such that $\angle(r, s) \equiv \angle(r', s')$. (Fig. 2.17.)*

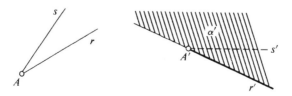

Figure 2.17

Exercises

1. Show that $(AB) \equiv (BC) \equiv (CA)$ implies $\sphericalangle CAB \equiv \sphericalangle ABC \equiv \sphericalangle BCA$.

2. Define right angles and show that any two right angles are congruent.

3. Show that any two squares are congruent if they are congruent in one side.

4. Let S be a rotation of order n (that is, $S^n = I$ and $S^k \neq I$ for $0 < k < n$). Let P not be fixed under S.
   (a) Show that $P, (P)S, (P)S^2, \ldots, (P)S^{n-1}$ are the vertices of a convex closed polygon (in the sense of Exercise 2, Section 2.5).
   (b) Find the group of all motions mapping the polygon onto itself.

5. Show that the group of Exercise 4(b) is generated by two elements (that is, every element of the group is expressible as a product with these two elements as only factors).

6. Find all subgroups of the group in Exercise 4(b) for $n = 5$.

2.7 Triangles

A set of three noncollinear points is called a **triangle**. If A, B, C are these points, we denote the triangle by

$$\triangle ABC,$$

or, briefly, by $\triangle$. A, B, C are called **vertices** of $\triangle$. The (closed) segments $[AB]$, $[BC]$, and $[AC]$ are said to be the **sides** of $\triangle$. These are to be distinguished from the lines AB, BC, and CD, which we call **side lines** of $\triangle$.

The vertex A is said to be **opposite** the side $[BC]$, etc. Every side is on the boundary of a half-plane containing the vertex opposite the side. The intersection of these half-planes is called the **interior** of $\triangle$ (Fig. 2.18).

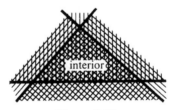

interior

Figure 2.18

Theorem 2.7.1 *If P is in the interior of $\triangle ABC$, the line AP intersects the line segment $[BC]$.*

Proof By Axiom O.2, there exists a point Q on $[AP]$ (Fig. 2.19). By Axiom O.4, the line BQ meets either $[AC]$ or $[PC]$ in a point R. Axiom O.4 now applied to $\triangle BRC$ yields the theorem. ◆

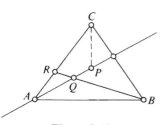

Figure 2.19

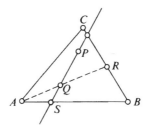

Figure 2.20

Theorem 2.7.2 *If P is in the interior of $\triangle ABC$, any line through P intersects two of the sides $[AB]$, $[BC]$, $[AC]$.*

Proof Let Q be a point $\neq P$ on the given line (Fig. 2.20). If Q is interior to $\triangle ABC$, then, by Theorem 2.7.1, AQ intersects BC in a point R. Applying Axiom O.4 to $\triangle ARC$ and $\triangle ABR$ shows Theorem 2.7.2 to be true. Suppose Q is not interior to $\triangle ABC$. Then for at least one of the lines AB, BC, AC, say AB, either Q is on AB or P, Q are on different sides of AB. In both cases there exists a point of intersection S of PQ and AB. If S is on $[AB]$, Axiom O.4 or Theorem 2.7.1 shows that PQ intersects $[BC]$ or $[AC]$, too. If S is not on $[AB]$, apply Theorem 2.7.1 to $\triangle SBC$ and then Axiom O.4 to $\triangle ABC$ in order to realize that the two points of intersection exist. ◆

Remember that, by our general definition of congruent sets (Section 2.6), two triangles are congruent if they can be mapped onto each other by a motion.

Theorem 2.7.3 *Two triangles $\triangle ABC$, $\triangle A'B'C'$ are congruent if and only if the vertices can be numbered in such a way that*

$$[AB] \equiv [A'B'], \qquad [BC] \equiv [B'C'], \qquad [CA] \equiv [C'A']$$

and

$$\sphericalangle BAC \equiv \sphericalangle B'A'C', \qquad \sphericalangle ABC \equiv \sphericalangle A'B'C', \qquad \sphericalangle BCA \equiv \sphericalangle B'C'A'.$$

Proof If $\triangle ABC \equiv \triangle A'B'C'$, there exists a motion mapping the points A, B, C in some way onto the points A', B', C'. By adjusting the notation, we can assume that A is mapped onto A', B is mapped onto B', and C is mapped onto C'. Then clearly all congruence relations of Theorem 2.7.3 are satisfied. If, conversely, the conditions of the theorem hold, then assume M to be the unique motion mapping the flag A, r, α onto the flag A', r', α', where B is on r, B' is on r', C is on α, and C' is on α' (existence by Theorem 2.5.5). From Theorem 2.6.4, we conclude that under M the angle $\sphericalangle BAC$ is mapped onto $\sphericalangle B'A'C'$. Now Theorem 2.6.1 implies that $(B)M = B'$ and $(C)M = C'$. ◆

The proof of Theorem 2.7.3 shows that not all congruence assumptions of the theorem are needed in order to guarantee $\triangle ABC \equiv \triangle A'B'C'$. Rather, one easily verifies the following well-known theorems on congruence stated in Theorem 2.7.4.

Theorem 2.7.4 *Two triangles $\triangle ABC$ and $\triangle A'B'C'$ (which may coincide) are congruent if and only if there is a one-to-one correspondence of vertices, defining a correspondence of sides and angles, respectively, such that one of the following conditions is true:*

(a) *Two sides in one triangle and the angle containing them are respectively congruent to the corresponding sides and angle of the other triangle.*

(b) *Two angles of one triangle and the side they both contain are respectively congruent to the corresponding angles and side of the other triangle.*

(c) *The three sides of one triangle are respectively congruent to the corresponding sides of the other triangle.*

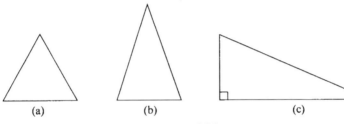

(a)   (b)   (c)

Figure 2.21

It is clear what is meant by a triangle's **angle bisectors** and a **perpendicular bisector** of a side of a triangle. Furthermore, equilateral and isosceles triangles can be defined in the ordinary way. Finally, in a right triangle, **legs** and the **hypotenuse** have the usual meaning. An angle formed by perpendicular rays is called a **right angle**. Since an ordered plane is not elliptic, a right triangle has a unique right angle and a unique hypotenuse. (See Fig. 2.21.)

Exercises

1. Prove that if a triangle can be mapped onto itself by three different motions, it is an equilateral triangle.

2. Prove that if there exist two different motions mapping a triangle onto another triangle, then both triangles are isosceles.

3. In $\triangle ABC$, let the midpoints of $[AC]$ and $[BC]$ be denoted by P and Q, respectively. Show that the lines AB and PQ have a common perpendicular. (*Hint*: Reflect at a line passing through the midpoint of $[AC]$ and perpendicular to AB.)

4. Prove Theorem 2.7.4.

5. Define a closed polygonal train as in Exercise 5, Section 2.2. Suppose all the line segments $(P_1P_2), \ldots, (P_{n-1}P_n), (P_nP_1)$ are congruent without the polygonal train being "regular," that is, without all $\sphericalangle P_nP_1P_2$, $\sphericalangle P_1P_2P_3$, $\ldots$, $\sphericalangle P_{n-1}P_nP_1$ being congruent. Let n be even. How many of these angles at most may be congruent?

6. Let a Euclidean ordered metric plane be given. Show that in any triangle $\triangle ABC$ "the sum of angles is two right angles," that is, there exist in a half-plane rays r, s, t, r' with a common end point such that r, r' are opposite and $\sphericalangle CAB \equiv \sphericalangle(r, s)$, $\sphericalangle ABC \equiv \sphericalangle(s, t)$, $\sphericalangle BCA \equiv \sphericalangle(t, r')$.

7. Show by an example that the theorem of Exercise 6 is not true for ordered hyperbolic metric planes (see Section 1.6; compare Exercise 6 of Section 2.2).

2.8 *Triangle Inequality*

So far we have not discussed the concept of "length" of a segment or "size" of an angle. We can, however, say whether or not two segments or two angles are equal in size, as when they are congruent. Correspondingly, we can say whether a line segment (PQ) is smaller than a line segment $(P'Q')$ by comparing the two: We define (PQ) to be *smaller* than $(P'Q')$ and $(P'Q')$ to be *larger* than (PQ) if there exists a point R' on $(P'Q')$ such that

$$(PQ) \equiv (P'R').$$

Also we define a proper angle $\sphericalangle(r, s)$ to be *smaller* than a proper angle $\sphericalangle(r', s')$; and we say $\sphericalangle(r', s')$ is *larger* than $\sphericalangle(r, s)$ if r can be mapped by a motion onto r' so that the image of s under this motion lies in the interior of $\sphericalangle(r', s')$. For improper angles $\sphericalangle(r', s')$, we extend these definitions as follows:

Any proper angle is said to be smaller than $\sphericalangle(r', s')$ if r', s' are opposite rays of a line, that is, if $\sphericalangle(r', s')$ is a *straight* angle.

So the concepts of "larger" and "smaller" have been introduced without defining a measure. In fact, these concepts are more primitive than the concept of a measure. This may be illustrated by two little boys quarreling over who is the taller one. In order to decide, they will stand back to back and compare their heights rather than both measure their lengths in inches.

An angle is called *acute* if it is smaller than a right angle, *obtuse* if it is larger than a right angle.

Theorem 2.8.1 *The nonright angles of a right triangle are acute angles.*

Proof Assume that in $\triangle ABC$ the right angle is at A. If $\sphericalangle ABC$ were obtuse, then the line perpendicular to AB at B would pass through the interior of $\triangle ABC$ and hence, by Theorem 2.7.1, meet $[AC]$ in a point P.

Then P would be a pole of AB, contrary to our assumption that the plane is not elliptic (see Theorem 2.2.4). ◆

Theorem 2.8.2 *Each leg of a right triangle is smaller than the hypotenuse.*

Proof If $\triangle ABC$ has its right angle at A, reflect A at the angular bisector of $\sphericalangle CBA$ (Fig. 2.22). The image A' of A under this reflection is a point on CB. We must show that C is not on the segment $[AB]$. If AA' meets the bisector of $\sphericalangle CBA$ in Q, then Theorem 2.8.1 applied to $\triangle ABQ$ shows $\sphericalangle QAB$ to be acute. This excludes $C = A'$ and also C on $(A'B)$, since, in the latter case, $\sphericalangle A'AB$ and hence $\sphericalangle QAB$ clearly would be obtuse, a contradiction. ◆

If P is between A and B, if $[AP]$ is smaller than $[CD]$, and if $[PB]$ is smaller than $[EF]$, then we say that $[AB]$ is **smaller than the combined size** of $[CD]$ and $[EF]$.

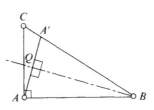

Figure 2.22

Now we prove an inequality that looks rather trivial but is of utmost importance for many branches of mathematics:

Theorem 2.8.3 Triangle inequality *Each side of a triangle is smaller than the combined size of the other two sides.*

Proof It is readily seen that the triangle cannot possess two obtuse angles (Exercise 4). So there is a side, say $[AB]$, whose two adjacent angles are acute. Drop a perpendicular from the opposite vertex, C, of $[AB]$ onto AB. This perpendicular meets (AB) in a point (see Exercise 3). Now Theorem 2.8.3 can be readily deduced from Theorem 2.8.2. ◆

Exercises

1. Let $\triangle ABC$ be a right triangle with right angle at A, and let P be the intersection of (AC) and the angular bisector of $\sphericalangle ABC$. Show that (AP) is smaller than (PC).

2. Let $\triangle ABC$ have an obtuse angle at A. Show that the foot of the perpendicular from C to the line AB lies outside (AB).

3. Show that in a triangle $\triangle ABC$ with acute angles at A and B, the foot of the perpendicular from C to AB lies in (AB).

4. Prove that a triangle has at most one obtuse angle.

5. Let $\triangle ABC$ and $\triangle AB'C'$ be right triangles with right angles at B, B', respectively, such that $[ABB']$ and $[ACC']$. Show that (BC) is smaller than $(B'C')$.

6. Extend the triangle inequality to convex polygons (see Exercise 2, Section 2.5) and prove this extension by induction.

7. Show that the relation "smaller than" of line segments and angles is preserved under motions.

2.9 *Continuity*

All axioms introduced so far do not guarantee that there are no "gaps" on a line as there are "gaps" in the set of all rational numbers ("irrationals") or gaps in the set of all algebraic numbers ("transcendental numbers" like π). So we complete our list of axioms by an additional one that makes any line "continuous."

Axiom C (*Dedekind*) *If the points of a line are divided into two nonempty disjoint subsets neither of which possesses a point between two points of the other, there exists a point on the line not lying between two points of either of the two subsets.*

We call a Euclidean or hyperbolic ordered metric plane satisfying Axiom C an **ordinary** Euclidean or hyperbolic plane, respectively. The partition of a line occurring in Axiom C is known under the name of a **Dedekind cut**. Dedekind has used those cuts for defining real numbers.

Let P be a point not on the line a. We join P to all points of a by rays emanating from P and call the system of these rays a **fan P, a**. (See Fig. 2.23.) The order relation on a carries over in a natural way to an order relation on the elements of the fan.

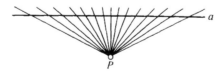

Figure 2.23

Theorem 2.9.1 *The rays of a fan satisfy Axiom C if the word "point" is replaced by "ray."*

Let P not lie on a, so that P and a determine a fan. By an **asymptote** of this fan, we mean a ray emanating from P that is neither between two rays of the fan nor between two rays emanating from P that are outside the fan.

Theorem 2.9.2 *Any fan possesses two and only two asymptotes.*

Proof Let P, a determine the fan and let r be an opposite ray of some ray of the fan, so that r does not lie in the fan (Fig. 2.24).   Join a point A

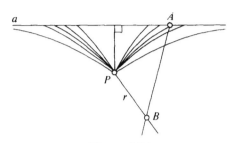

Figure 2.24

on a to a point B on r such that A, B, P are not collinear.   Then P and AB determine a second fan the elements of which are partitioned into those in the first fan and those not in the first fan.   By Theorem 2.9.1, there is a ray in the second fan that is not between any two rays of either one of the sets of the partition.   It is readily seen to be an asymptote of the first fan. Reflecting it at the perpendicular of a through P provides a second asymptote.   ◆

Theorem 2.9.3 (a) *In a hyperbolic plane the asymptotes of a fan P, a are carried by the two parallel lines to a through P.*

(b) *In a Euclidean plane the asymptotes of a fan P, a are opposite rays of the parallel to a through P.*

Proof (a) Remember that in a hyperbolic plane two different lines are called parallel if they neither intersect nor have a common perpendicular. Suppose Theorem 2.9.3 is false.   Then the given fan possesses an asymptote r carried by a line b such that a, b possess a common perpendicular c (Fig. 2.25).   Let c intersect a, b in A, B, respectively. B divides b into

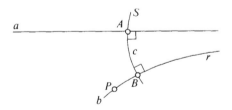

Figure 2.25

two rays, one of which clearly is an asymptote of the fan determined by B, a.   We call it r'.   The asymptote r contains, or is contained in r'. Let s be a ray emanating from B and containing A.   Now one of the parallels of a through B passes through the interior of $\measuredangle(r', s)$ (the two parallels are obtained from one another by reflecting in c); hence it contains a ray of the fan B, a, and, therefore, intersects a. This is a contradiction.

(b) is trivial. ◆

In this section, neither Axiom FM (free mobility, Section 2.1) nor conclusions drawn from the axiom have been applied. Thus, we avoid a circular argument as we now prove Axiom FM as a theorem and thus eliminate it as an axiom under the assumption of Axiom C.

Theorem 2.9.4 *Given two lines a, a' that intersect or have a common perpendicular, there exists a reflection mapping a onto a'.*

Proof Let a, a' intersect in P, and let b be a line joining a point $\neq P$ of a to a point $\neq P$ of a'. Let r, r' be the rays emanating from P (Fig. 2.26(a)) and carried by a, a', respectively, that intersect b. By the **reflection** at a ray, we mean the reflection at the carrier of this ray. We consider a Dedekind cut in the fan P, b: A fan ray s is called of the first kind if the image r^* of r under reflection at s is such that r' lies between r and r^*; otherwise we say it is of the second kind. Clearly no ray of the first kind lies between two rays of the second kind, and conversely. r' is of the first kind, r of the second kind. So, by Theorem 2.9.1, the fan P, b possesses a ray, the reflection at which maps r onto r' and hence a onto a'.

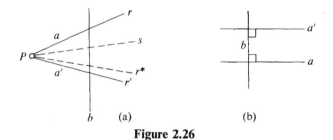

Figure 2.26

If a, a' possess a common perpendicular b (Fig. 2.26(b)), consider all perpendiculars of b. They are in one-to-one correspondence with the points on b obtained as points of intersection. b is mapped onto itself by any R_c where $c \perp b$. We introduce a Dedekind cut for the points on b corresponding to the Dedekind cut of the fan P, b above. We find a point Q on b such that for the perpendicular d of b through Q we have $(a)R_d = a'$. ◆

Exercises

1. With the aid of "betweenness," introduce a relation " $<$ " (less than) for points on a given line a. Restate Axioms O.1–O.3 and Axiom C using " $<$ ".

2. In an ordinary Euclidean plane, introduce "betweenness" for the lines of a pencil of parallel lines. Show that an analogue of Axiom C is satisfied.

3. Prove the following theorem, known as an "Archimedean postulate" (we state it in the terminology of Exercise 1): If A, B, X are points of a and if $[ABX]$, there exist points $X_1, \ldots, X_k$ on a such that

$$B = X_1 < X_2 < \cdots < X_{k-1} \leq X < X_k$$

and

$$(X_i X_{i+1}) \equiv (AB)$$

for $i = 1, \ldots, k - 1$ (that is, X can be "reached" in a walk of finitely many steps of constant step length). (*Hint*: Subdivide the points on the line into two subsets: those that can be "reached" from A and those that cannot. Apply Axiom C.)

4. State and prove the Archimedean postulate for angles.

5. Show that any ordinary Euclidean or hyperbolic plane can be covered by countably many triangles.

6. Show that Axiom C can be proved in general if we postulate it for only a single line.

2.10   *Foundation of Ordinary Plane Geometry*

By ordinary geometry we might intuitively mean a geometry involving all theorems on elementary geometry that can be found in books on this subject. If, however, we include non-Euclidean geometry, we cannot speak about *one* geometry, since there would be contradictory statements in it. Also if we restrict our attention to Euclidean planes, it is not very easy to say precisely what "ordinary" geometry is.

The two basic concepts of any axiomatic theory, completeness and categoricity, help clarify the problem. In order to explain what is meant by completeness of a system of axioms, let us for a moment anticipate the coordinatization of plane geometries introduced in the next chapter. If pairs of numbers (x, y) can be assigned to the points of an axiomatic geometry in such a way that a linear equation $ax + by + c = 0$ is assigned to every line. It may be that only rational numbers occur for x, y, a, b, c (that is, fractions m/n; m, n integers); we shall see that this may be so in a Euclidean metric plane. If we add to such a geometry all points (x, y) with x, y arbitrary real numbers (as $\sqrt{2}$, π, etc.) and all lines $ax + by + c = 0$ with a, b, c real, we also obtain a geometry satisfying, say, the axioms of a Euclidean metric plane. So we have augmented the geometry by new objects (points and lines) such that all previous relations (incidence, perpendicularity) remain unchanged and such that in the enlarged system the same axioms are true as in the original system. However, if all pairs (x, y) with x, y arbitrary real numbers are points and all linear equations $ax + by + c = 0$ with a, b, c real and a, b not both equal to 0 are lines, such an extension is impossible, as will be seen. We say generally that a system of axioms is **complete** if its objects cannot be augmented by new objects in such a way that all previous relations are carried over and such that all previous axioms remain true in the enlarged system. We shall show in this section (without using co-ordinates) that the system of axioms of a Euclidean metric plane is made complete if Axiom C is added to it.

When is a system of axioms called categoric? Let us look again at an example first. Consider two "planes in physical space." In both planes we may study the same "plane geometry." If this is done axiomatically, the two planes are "copies" of the same axiomatic plane. So they are the same and yet not the same. We say, in a precise manner, the two planes are **isomorphic**; that is, their points and lines are in one-to-one correspondence such that incidence and perpendicularity (and possibly other relations) are carried over. Clearly, "isomorphic" is an equivalence relation. So the equivalence classes of isomorphic planes represent an "abstract" or axiomatic geometry. (In Chapter 4, we shall explicitly demonstrate the isomorphism between two planes of a space.) Any element of the equivalence class is called a **model** of the system of axioms.

Now the question arises: Given a system of axioms, are any two models isomorphic? In general, no. As an example, consider the axioms of a metric plane. There are Euclidean metric planes and hyperbolic metric planes, both satisfying the axioms of a metric plane. But a Euclidean and a hyperbolic plane are never isomorphic. If any two systems of objects that fulfill the requirements of a certain system of axioms are isomorphic, we say that this system of axioms is **categoric**. So the system of axioms of a metric plane is not categoric.

We may now specify the intuitive idea of "ordinary" geometry as follows: An axiomatic geometry is called an **ordinary** geometry if the system of its axioms is complete and categoric. There is, of course, no logical necessity for such a definition. But we have in some way made precise, from an axiomatic standpoint, what we mean by ordinary.

Theorem 2.10.1 *A system of points and lines satisfying Axioms* I.1–I.3, P.1–P.3, M.1–M.4, O.1–O.5, *and* C *is complete*

(a) *if Axiom* Euc *is valid,*
(b) *if Axiom* Hyp *is valid.*

In case (a), we speak of an **ordinary Euclidean** plane, in case (b), of an **ordinary hyperbolic** plane. In Section 7.12, we show via coordinates that these systems of axioms are also categoric.

Proof of Theorem 2.10.1 We call the points and lines of the given plane **old** points and lines. Suppose Theorem 2.10.1 is false and we can introduce additional points and lines such that I.1–I.3, P.1–P.3, M.1–M.4, O.1–O.5, and (a) or (b) are still satisfied. We call the additional elements **new** points and lines. If there exists a new line, it contains a new point, since any two old points determine an old line joining them. Suppose U is a new point

(Fig. 2.27). We join it to an old point A. As is seen from Theorem 2.9.1, AU is an old line. Axiom C implies, furthermore, that U is not between any two old points of AU. Let P be an old point not on AU. The line PU is again, by Theorem 2.9.1, an old line.

Figure 2.27

If (a) is true, the line PU must be the old parallel of AU through P, since it does not intersect AU in an old point and since there exists only one such line. This is a contradiction, since the only candidate for a new parallel of AU through P is the old parallel (no new lines through P existing). Hence there exists no new point and thus no new line.

If (b) is valid, the only candidates for asymptotes of the fan P, AU are again the old ones. Since PU does not carry a new asymptote, it is, by Theorem 2.9.3, not an old parallel of AU; hence it possesses a common perpendicular with AU, contradicting the fact that our plane is not elliptic (see Theorem 2.2.4). So again there are neither new points nor new lines.

◆

Exercises

1. Define "categoric" in analogy to "complete" for any axiomatic theory (using "objects" and "relations").

2. Are the axioms of a group (a) complete, (b) categoric (the relation being "c is a product of a and b")?

3. Show that the axioms of a group together with the following axiom are complete but not categoric: "The group contains precisely four elements."

4. Show that the axioms of a group together with the following axiom are categoric but not complete: "The group has infinitely many elements and consists of all powers of a single group element."

5. Introduce an additional axiom to the axioms of a Euclidean metric plane such that the system of axioms is categoric and such that the nine-point plane is one of its models.

6. Given the axioms of a metric plane together with Axiom C and the following axiom: "Either Axiom Euc or Hyp is true." Is this system of axioms (a) complete, (b) categoric?

2.11   Hyperbolic Planes

As mentioned in the previous section, we shall show in Chapter 7
that the axioms of an ordinary Euclidean or an ordinary hyperbolic plane are
categoric.   Considering that the Klein model of a hyperbolic plane discussed
in Section 1.6 is an ordinary hyperbolic plane (details in Chapter 6), we draw
an interesting conclusion: Up to an isomorphism, the Klein model represents
the only ordinary hyperbolic plane.

Klein's model will be discussed more extensively in the framework of
inversive geometry (Chapter 6).   So in this section we can restrict ourselves
to some remarks about hyperbolic planes.   We point out a few geometrical
objects that do not occur in the Euclidean case: doubly and trebly asymptotic
triangles, horocycles, hypercycles.

Klein's discovery has taken quite a bit of mystery away from Bolyai-
Lobachevskian geometry.   It is true, there is no physical or philosophical
preference for either Euclidean or non-Euclidean geometry (see Section 1.6).
Not only can we find models of non-Euclidean planes in Euclidean space
(Klein's model), but we can also find models of Euclidean planes in non-
Euclidean geometry.   Basically, it is possible that intelligent beings on some
other planet might develop non-Euclidean geometry first and later on discover
Euclidean planes.   They might proceed in a manner opposite that of Gauss and
Bolyai-Lobachevski.   However, Euclidean geometry can be considered some-
what simpler in its structure than other geometries. So the chances are that any-
one who discovers axiomatic geometry will discover the Euclidean one first!

We wish to investigate parallelism in hyperbolic planes a bit further.   Let
P, a define a fan (Section 2.9) and let r, s be its asymptotes (see Fig. 2.28).

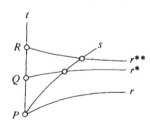

Figure 2.28   **Figure 2.29**

From P we drop the perpendicular b onto a.   The line b intersects a in a
point A.   Among the two rays into which a is split by A, one is on the same
side of b as r is; we denote it by r^*.   We say that r is **parallel** to r^* and to any
ray obtained from r^* by adding or taking away a line segment.   It is readily
seen that r^* is an asymptote of the fan determined by A and the carrier of
r so that r^* is parallel to r.   If r^* is parallel to r and r^{**} is parallel to r^*, we
assert that r^{**} also is parallel to r.   For a proof, we can assume that the
three rays emanate from collinear points P, Q, R (Fig. 2.29).   We distinguish
two cases: $[PQR]$ and $[PRQ]$.

If $[PQR]$, let s emanate from P and lie in the interior of $\angle(r, t)$, where Q is on t. Since r is parallel to r^*, the ray s intersects r^*, and, since r^* is parallel r^{**}, it meets r^{**}, too. Since s is arbitrary in the interior of $\angle(r, t)$, r is parallel to r^{**}. We say that any ray r is also **parallel** to itself and to any ray obtained from r by adding or subtracting a line segment. So we have proved that *the relation "parallel" is an equivalence relation on the set of all rays*. The equivalence classes under this equivalence relation are called **ends**.

Any ray obviously belongs to one and only one end. So if A is the end point of a ray r, we may consider the end in which r lies the second "end point" of r. So rays behave somewhat like line segments if ends are considered as points. An arbitrary point and an arbitrary end are joined by a unique ray, by definition of an end and the above statement on the relation "parallel."

Since every line joins two ends, we also may consider lines as generalized line segments. Applying these ideas to triangles, we obtain three kinds of new triangles: **simply**, **doubly**, and **trebly asymptotic triangles** (Fig. 2.30).

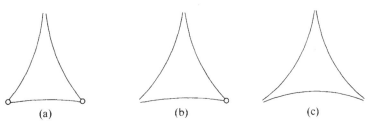

(a)   (b)   (c)

Figure 2.30

The above discussion on parallel rays and ends applies also to Euclidean planes, though it is not so interesting in this case as in the hyperbolic one. However, doubly and trebly asymptotic triangles do not exist in the Euclidean case.

Another specific feature of hyperbolic planes will now be discussed. We shall consider pencils of the first, second, and third kind as introduced in Section 1.16. Pencils of the third kind are now seen to be the carriers of all rays of an end. If a, b are two lines of a pencil, we generally call

$$S = R_a R_b$$

a **quasi-rotation**. For pencils of the first kind, S is a rotation; for pencils of the second kind, it is a translation. We consider the set of all images of some point P under quasi-rotations of the form $S = R_a R_b$, where a, b are arbitrary lines of a pencil. We call this set

a *circle*	if the pencil is of the first kind	(Fig. 2.31(a)),
a *hypercycle*	if the pencil is of the second kind	(Fig. 2.31(b)),
a *horocycle*	if the pencil is of the third kind	(Fig. 2.31(c)).

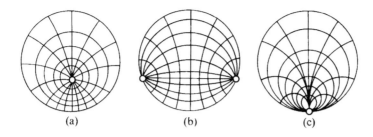

(a)   (b)   (c)

Figure 2.31

Exercises

1. Define betweenness for the lines that carry the rays of an end and show that Axioms O.1–O.3 are satisfied (for rays instead of points).

2. Show that for any trebly asymptotic triangle the vertices can be arbitrarily permuted by motions. (*Hint*: Find, by using Dedekind cuts for any such triangle, a point P with the property that for the feet A, B, C of the perpendiculars dropped from P onto the sides of the triangle,

$$(PA) \equiv (PB) \equiv (PC).)$$

3. Show that any two trebly asymptotic triangles are congruent (*Hint*: Make use of the points P constructed, as in Exercise 2).

4. Suppose that in a hyperbolic plane an area measure is introduced in such a way that two areas have the same measure if they can be divided into finitely or infinitely many mutually congruent pieces. Triangles are assumed to have finite measure. Prove Liebmann's theorem, which states: *Every asymptotic triangle possesses finite measure.* (This is clearly not true for Euclidean planes.) (*Hint*: Coxeter, 1961, p. 295): Let P_0C be the finite vertices of a simply asymptotic triangle. Let P_0, a be a fan as indicated in Figure 2.32. Split the two asymptotic rays of the fan into

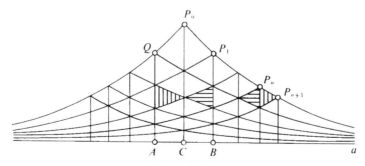

Figure 2.32

congruent segments $(P_n P_{n+1})$, $n = 0, 1, 2, \ldots$, and shift by using motions the (hatched) triangles $\triangle P_n P_{n+1} Q_{n+1}$, $\triangle P_n Q_{n+1} Q_n$ onto those in the pentagon $P_0 Q A B P_1$ in the figure.

5. Why do doubly asymptotic triangles not exist in a Euclidean plane?

6. Discuss the analogues of circles, hypercycles, and horocycles in ordinary Euclidean planes.

2.12   Euclidean Motions

We shall now investigate further properties of motions in ordered Euclidean planes. By a **Saccheri quadrangle** we mean a nondegenerate quadrangle $ABCD$ possessing right angles at A, B, and C. (See Fig. 2.33.)

Figure 2.33

Theorem 2.12.1   *In an arbitrary Euclidean plane, every Saccheri quadrangle is a rectangle.*

Proof CD does not intersect AB, since, otherwise, the plane would be elliptic (see Theorem 2.2.4). Therefore, CD is parallel to AB. For the same reason, the perpendicular p of AD through D is parallel to AB. So, by the Euclidean axiom of parallels, $p = CD$. ◆

Theorem 2.12.1 immediately implies the next one.

Theorem 2.12.2   *If two lines are parallel, they possess only common perpendiculars, which are also parallel to each other.*

The theorem permits the following analysis of translations: If T is a translation along a, then T is also a translation along any parallel of a. Hence there is a pencil of parallel lines each line of which is a fixed line under T. So T does not possess a fixed point unless it is the identity map I. Conversely, by Section 1.18, if a proper motion possesses no fixed point, it is a translation. Hence, if a proper motion possesses all lines of one parallel pencil as fixed lines, it is a translation. We call these fixed lines **traces** of the translation. If there exists a fixed line not in this pencil of traces, the translation is the identity map.

Suppose T is a translation $\neq$ I. If b is not a trace of T, we assert that $b' = (b)$T is parallel to b. In fact, suppose b and b' intersect in a point P.

Since the trace of T through P is mapped onto itself, P is a fixed point of T; hence $T = I$, contrary to our assumption. So a translation maps every line either onto itself or onto a parallel of itself. This implies that if $P' = (P)T \neq P$, then PP' is a trace of T.

Theorem 2.12.3 *The set of all translations is a subgroup of the group of all proper motions.*

Proof Let T, T' be translations. TT' is a proper motion. Suppose $(P)TT' = P$. Setting $Q = (P)T$, we have $(Q)T' = P$. If $Q \neq P$ then PQ is a trace of T and also a trace of T'. PQ and all parallels of PQ are fixed lines under T and T', and hence under TT'. Therefore, TT' is a translation. If $P = Q$, then $T = I = T'$, hence $TT' = I$. If T is a translation, T^{-1} possesses the traces of T as fixed lines; hence it is also a translation. ◆

The product of two rotations is, by Theorem 1.18.5, a rotation or a translation. Translations can, in fact, occur as products of rotations (see below). Therefore, the set of rotations is not a subgroup of the group of motions. The following theorem, however, is readily proved by using Theorem 1.18.5.

Theorem 2.12.4 *The set of all rotations and translations is a subgroup of the group of all motions; it is identical with the group of all proper motions.*

Exercises

1. Show that the product of an even number of half-turns is a translation.
2. Show that the group of all translations is simply transitive on the set of all points.
3. Prove that in a metric plane the opposite sides (lines) of a rectangle do not intersect.
4. Prove that a metric plane is Euclidean if every Saccheri quadrangle is a rectangle.
5. Show that a metric plane is Euclidean if and only if the vertices of any Saccheri quadrangle $ABCD$ satisfy

$$H_A H_B H_C H_D = I.$$

6. If a, c and b, d are pairs of opposite sides of a rectangle, investigate $R_a R_c R_b R_d$.

2.13 An Analysis of Motions

Now we pose a question that is very interesting from a theoretical point of view as well as from a practical one.

Given a motion M *in an ordered Euclidean plane; we know the images P',
Q', R' of three noncollinear points P, Q, R, respectively. Can we decide
what kind of motion M is and can we find a method of constructing the image
X' of any point X?*

The answer is positive; we present it in several steps: First consider the
flags P, r, α and P', r', α', where Q, Q' are on r, r' and R, R' on α, α', respectively (Fig. 2.34). If these flags are in the same orientation class (see

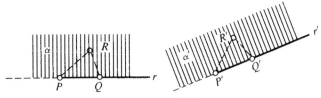

Figure 2.34

Theorem 2.5.6), M is a proper motion; otherwise it is an improper motion.
Let M be a proper motion, that is, a rotation or a translation. If two of
the points P, Q, R remain fixed, M is the identity map (see Theorem 2.4.5).
So let $P' \neq P$ and $Q' \neq Q$. If $PP'Q'Q$ is a **parallelogram**, as in Figure 2.35,

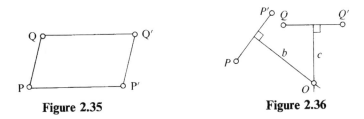

Figure 2.35   **Figure 2.36**

that is, if PP' is parallel to QQ' and PQ parallel to $P'Q'$, then let T be a
translation mapping P onto P'. From the properties of a translation we
find: $(Q)T = Q'$. So, by Theorem 2.4.5, T = M.

If $PP'Q'Q$ is not a parallelogram, M is not a translation; hence it is a
rotation. The center of rotation, O, is a point of intersection of the perpendicular bisectors b, c of (PP'), (QQ'), respectively, provided b, c are different
(Fig. 2.36). If $b = c$, then O is the intersection of PQ and $P'Q'$.

So we have analyzed the case of a proper motion completely. Let M be
improper; that is, by Theorem 1.18.1,

$$M = TR_a,$$

where T is a translation. P, Q, R cannot all remain fixed, since otherwise
M = I would be proper. So let $P \neq P'$. If $Q = Q'$ and $R = R'$, we can
set $QR = a$ and obtain M = R_a. If $Q \neq Q'$, let M, N be the midpoints of
(PP'), (QQ'), respectively. If $M \neq N$, let $a = MN$. If $M = N$, let a pass
through M and be parallel to PQ'. In both cases, let $\tilde{P} = (P')R_a$, and let
T be the translation mapping P onto $\tilde{P}$. Now it can easily be verified that,
in each case, M = TR_a.

To answer the second part of our initial question, let X be an arbitrary point. If M is a translation $\neq I$, suppose X is not on PP' (otherwise take

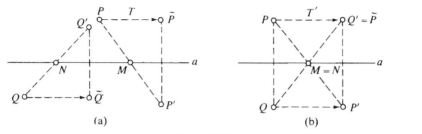

Figure 2.37

QQ' or RR' instead of PP'). There is an X' such that $PP'X'X$ is a parallelogram. Clearly $X' = (X)M$. If M is a rotation, we construct X' in the well-known manner using (OX) and one of the angles between b, c (compare Fig. 2.37). If, finally, M is a glide reflection TR_a, we can also find X' by the usual constructions.

Exercises

1. Let a motion M permute the vertices of an equilateral triangle. Which possibilities for M occur?

2. Describe all motions that interchange two given points.

3. Show that if M is a motion and if there exist points P, Q such that $PQP'Q'$ is a rectangle ($P' = (P)M$; $Q' = (Q)M$), then M is a half-turn or a glide reflection, depending on whether or not it has a fixed point.

4. Prove by using congruence properties that the perpendicular bisectors in a triangle meet in a point (compare Exercise 2, of Section 1.13.)

5. Show that the line joining the midpoints of two sides of a triangle is parallel to the third side (apply a half-turn).

6. *Orthocenter.* Drop from each vertex of a triangle the perpendicular onto the opposite side of the vertex. Show that these lines meet in a point, called the orthocenter of the triangle. (*Hint:* Draw a parallel to any side through its opposite vertex; use Exercises 4 and 5.) (It should be noted that the theorem of the orthocenter is true in any metric plane; the proof is somewhat involved; see Bachmann (1959).)

2.14   Euclidean Circles

Let C, X_0 be given points, $C \neq X_0$. The set of all points derived from X_0 by rotations about C is called a *circle* $\mathscr{C}$. If X is such a point, clearly $(CX_0) \equiv (CX)$. Conversely, if $(CX_0) \equiv (CX)$, then reflecting X_0, first in the angular bisector of $\measuredangle X_0CX$ and then in CX, yields a rotation about C that moves X_0 onto X. So a circle is also characterized as the set of all points X such that $(CX) \equiv (CX_0)$. We call C the **center** of $\mathscr{C}$ and (CX), for any X on $\mathscr{C}$, a **radius** of $\mathscr{C}$.

Theorem 2.14.1 *Three noncollinear points are on one and only one circle.*

Proof If P, Q, R are the given points, then the perpendicular bisectors of (PQ) and (QR) intersect in a point C, which is the center of a circle $\mathscr{C}$ passing through P, Q, R. Since, conversely, C must be on these perpendicular bisectors, the circle $\mathscr{C}$ is uniquely determined. ◆

By a **tangent** of $\mathscr{C}$ at P we mean a line that meets $\mathscr{C}$ at P and only at P.

Theorem 2.14.2 *A circle $\mathscr{C}$ with center C possesses a unique tangent at each point P on $\mathscr{C}$. This tangent is perpendicular to CP.*

Proof Let a be the perpendicular of CP at P (Fig. 2.38). By Theorem 2.8.2, $\mathscr{C}$ does not intersect a in a second point. So a is a tangent of $\mathscr{C}$. Let b be any line $\neq a$ passing through P. Drop a perpendicular c from C onto b. Clearly, P is not on c; hence $(P)R_c \neq P$ is a second point in which $\mathscr{C}$ and b intersect. ◆

Figure 2.38   **Figure 2.39**

Theorem 2.14.3 *If P is on the circle $\mathscr{C}$ and if Q is not on $\mathscr{C}$, then Q is either on the tangent of $\mathscr{C}$ at P or there is a unique circle $\mathscr{C}'$ passing through P, Q that intersects $\mathscr{C}$ only in P (Fig. 2.39).*

We call $\mathscr{C}'$ a **tangent circle** of $\mathscr{C}$ at P.

Proof If Q is not on the tangent line of $\mathscr{C}$ (center C) at P, then CP and the bisector of (PQ) meet in a point C', which provides the center of $\mathscr{C}'$. The uniqueness of C' can be seen here by a procedure similar to that used in the proof of Theorem 2.14.2. ◆

The **interior** of a circle $\mathscr{C}$ is defined as the set of all those points that are on a radius (CX) of $\mathscr{C}$ but not on $\mathscr{C}$ itself. (See Fig. 2.40.) If a line a passes

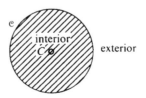

Figure 2.40

through a point of the interior of $\mathscr{C}$, we expect that it will meet $\mathscr{C}$ in two points. To prove this, however, one needs Dedekind's axiom, as we see next.

Theorem 2.14.4 *Any line passing through the interior of a circle $\mathscr{C}$ meets $\mathscr{C}$ in two different points.*

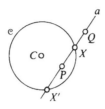

Figure 2.41

Proof Let P be interior to $\mathscr{C}$ and let a pass through P (Fig. 2.41). One easily finds a point Q on a that is exterior to $\mathscr{C}$. On (PQ) the set of points interior to $\mathscr{C}$ and the set of those exterior to $\mathscr{C}$ form a Dedekind cut. The cut point X is neither interior nor exterior to $\mathscr{C}$; hence it is on $\mathscr{C}$. By Theorem 2.14.2, a intersects $\mathscr{C}$ in a further point $X' \neq X$. ◆

Theorem 2.14.5 *If a circle $\mathscr{C}'$ passes through the interior and through the exterior of a circle $\mathscr{C}$, it intersects $\mathscr{C}$ in two different points.*

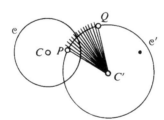

Figure 2.42

Proof Let P, Q lie on $\mathscr{C}'$ such that P is interior to $\mathscr{C}$ and Q is exterior to $\mathscr{C}$. (See Fig. 2.42.) If C' is the center of $\mathscr{C}'$, we call the intersection of $\mathscr{C}'$ and the angular region of $\sphericalangle QC'P$ a circular segment $\widehat{PQ}$. Every line that intersects (PQ) also intersects $\widehat{PQ}$, and conversely. Now a Dedekind-cut property on (PQ) carries over to one on $\widehat{PQ}$. One proceeds as follows: The lines $C'X$, where X is on $\widehat{PQ}$ and interior to $\mathscr{C}$, form a set $\mathscr{G}$; those with X on $\widehat{PQ}$ and exterior to $\mathscr{C}$ form a set $\mathscr{T}$. By intersecting $\mathscr{G}$ and $\mathscr{T}$ with (PQ), we can apply Dedekind's axiom to find a point Z on PQ such that $C'Z$ intersects $\widehat{PQ}$ in a point R neither interior nor exterior to $\mathscr{C}$. So R is on $\mathscr{C}$. Reflection at CC' provides a second point in which $\mathscr{C}$ and $\mathscr{C}'$ intersect. ◆

Theorems 2.14.4 and 2.14.5 are sometimes used as axioms, so that Axiom C is not needed.

Exercises

1. Given three points of a circle, find the center.

2. Show without using Dedekind's axiom that the interior of any circle is convex (for definition, see Exercise 1, Section 2.5).

3. State conditions for the centers and the radii of two circles so that the circles (a) intersect in two points, and (b) are tangent to each other.

4. How many tangents are common to two given circles (distinguish all possible cases).

5. Two circles are called perpendicular at P if they pass through P and have perpendicular tangents there. Show that if two circles are perpendicular at P, then they are perpendicular at a point $P' \neq P$.

6. Show that if three circles are given, each two of which are tangent to each other at points P, Q, R, the circle through P, Q, R is perpendicular to all three circles.

2.15 Arcs and Inscribed Angles of a Circle

Let α be the angular region of an angle $\sphericalangle(r, s)$ with vertex C. If $\mathscr{C}$ is a circle with center C, then the intersection of $\mathscr{C}$ and α is called an **arc** of $\mathscr{C}$. We assume α to be a proper angle. Then the arc is determined by the points P, Q in which r, s intersect $\mathscr{C}$, respectively. (See Fig. 2.43.) Two arcs $\overset{\frown}{PQ}$ and $\overset{\frown}{P'Q'}$ on $\mathscr{C}$ are said to be **directly congruent** if there is a rotation

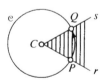

Figure 2.43

about C mapping P onto P' and Q onto Q'. They are called **oppositely congruent** if there is a reflection at a line through C mapping P onto P' and Q onto Q' (see Fig. 2.44).

Theorem 2.15.1 *Two arcs $\overset{\frown}{PQ}$ and $\overset{\frown}{P'Q'}$ on $\mathscr{C}$ are oppositely congruent if and only if PQ and $P'Q'$ are parallel.*

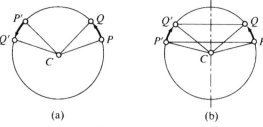

(a)   (b)

Figure 2.44

The proof is clear from the properties of a reflection. By a reflection in the perpendicular bisector of P, Q the arc $\overset{\frown}{PQ}$ is mapped onto $\overset{\frown}{PQ}$. Theorem 2.15.1, therefore, implies the next one.

Theorem 2.15.2 *Two arcs $\overset{\frown}{PQ}$ and $\overset{\frown}{P'Q'}$ on $\mathscr{C}$ are directly congruent if and only if PQ' and $P'Q$ are parallel.* (See Fig. 2.45.)

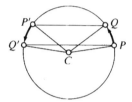

Figure 2.45

The following theorem is easily verified.

Theorem 2.15.3 *If a rotation about C maps P onto P' and Q onto Q', then $\overset{\frown}{PP'}$ is directly congruent to $\overset{\frown}{QQ'}$.*

If P, X, Q are three points on a circle $\mathscr{C}$, we say that the angle $\sphericalangle PXQ$ is **inscribed** in the arc $\overset{\frown}{PQ}$ provided X is not on $\overset{\frown}{PQ}$. ·

Theorem 2.15.4 Theorem of inscribed angles *All angles inscribed in the same arc are congruent.*

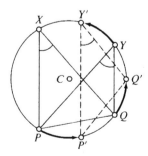

Figure 2.46

Proof Let X and Y be points on $\mathscr{C}$ but not on $\overset{\frown}{PQ}$. We rotate $\mathscr{C}$ such that the perpendicular of PY through C is mapped onto the perpendicular of PX through C. Let P', Q', Y' be the images of P, Q, Y, respectively. (See Fig. 2.46.) From Theorem 2.15.1, we deduce that $\overset{\frown}{PP'}$ is directly congruent to $\overset{\frown}{Y'X}$, and from Theorem 2.15.3 we know that $\overset{\frown}{PP'}$ and $\overset{\frown}{QQ'}$ are directly congruent. So $\overset{\frown}{QQ'}$ and $\overset{\frown}{Y'X}$ are directly congruent. Therefore, by Theorem 2.15.1, QX and $Q'Y'$ are parallel. This implies

$$\sphericalangle PXQ \equiv \sphericalangle P'Y'Q' \equiv \sphericalangle PYQ. \qquad \blacklozenge$$

Theorem 2.15.5 Thales' theorem *If (PQ) is a diameter of $\mathscr{C}$, any angle inscribed to $\overset{\frown}{PQ}$ is a right angle.*

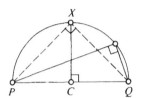

Figure 2.47

Proof By Theorem 2.15.4, it suffices to prove Theorem 2.15.5 for the angle at X on the perpendicular of PQ through C (see Fig. 2.47). A rotation about a right angle maps Q onto X and X onto P, hence CQ onto CX and CX onto $CP(=CQ)$. Since under a motion the angle between a line and its image line is the same for all lines (the angle being less than or equal to a right angle), we see that QX and the image XP of QX meet in a right angle. $\blacklozenge$

Corollary 2.15.6 *The center of the circle circumscribed about a right triangle lies on the hypotenuse.*

Exercises

1. Let $\sphericalangle(r, s)$ and $\sphericalangle(r', s')$ be congruent angles with apexes A and A', respectively, and let r intersect r' in R; also let s intersect s' in S. Show that A, A', R, S lie on a circle.

2. Show that if a circle can be circumscribed about a parallelogram, the parallelogram is a rectangle.

3. Given two circles $\mathscr{C}$, $\mathscr{C}'$, find circles tangent to both.

4. Given a point P not interior to a circle $\mathscr{C}$ and not on $\mathscr{C}$, find the tangents of $\mathscr{C}$ through P.

5. Let P, Q, X be different points on a circle $\mathscr{C}$ with center C such that X is not on $\overset{\frown}{PQ}$. Show that $\angle PCO$ is (in an obvious sense) twice as large as $\angle PXQ$.

6. Show that if P is a point on a circle $\mathscr{C}$, there exists a square inscribed in $\mathscr{C}$ with P as a vertex.

7. State and prove the well-known theorem relating to the chord and tangent of a circle.

8. *Nine-point circle.* Show that the following nine points of a triangle are on a circle (Fig. 2.48): The midpoints M_1, M_2, M_3 of the sides, the feet A_1, A_2, A_3 of the three altitudes and the midpoints O_1, O_2, O_3 of the segments joining H to the vertices of the triangle, where H is the orthocenter (that is, the point of intersection of the altitudes, compare Exercise 6 of Section 2.13). (*Hint*: Show that O_1O_2 (parallel to A_3M_3) is perpendicular to O_2M_1 (parallel to A_3O_3); apply Theorem 2.15.6.)

[2.16]

Order and Congruence

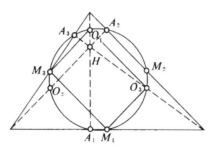

Figure 2.48

2.16 *Constructions by Ruler and Compass*

In our developing geometry we have laid much emphasis on the idea of mappings. This means that the point of view of constructions, on which traditional Euclidean geometry is largely based, has been moved somewhat into the background. Nevertheless, geometrical constructions form a vital element of ancient and modern geometry. In fact, there are close relationships between the ideas of mappings and the idea of constructions.

We restrict our attention to constructions by ruler and compass. These are not only historically the most important ones, but also modern draftsmanship consists mainly of constructions by ruler and compass. To be precise, we list the operations that are allowed.

(a) The ruler can be used to draw a line joining two given points and to intersect lines that are not parallel.

(b) The compass can be used to draw a circle about a given point C through a given point P.

So there are no "marks" on the ruler. We cannot, for example, use the ruler to check (by marks) whether two segments are congruent. The compass is considered to be "collapsible" in the following sense: We cannot place the spikes onto two given points, then lift the compass into the air, place one spike onto another point and draw a circle about it with the distance of the original points as radius: The compass "collapses" in the air. This technical restriction, however, does not mean a restriction of possible constructions, since we can place one end point of a segment onto a given point by a translation, the translations being carried out by ruler and compass (see below). In order to point out the allowed operations with ruler and compass, one speaks about constructions by the **unmarked ruler** and *collapsible compass*.

We assume an ordered Euclidean plane to be given in which the Dedekind axiom or, more generally, the line-circle theorem and the two-circle theorem are satisfied (see Section 2.14).

The following constructions are basic for our further discussion:

Construction 1 Given a point P on a line a. Draw a line through P perpendicular to a.

Solution Find Q, Q' on a such that P is the midpoint of (QQ'). Draw circles about Q and Q' with (QQ') as radius. These circles intersect in points X, X' that are different from P (by application of the two-circle theorem). (See Fig. 2.49.)

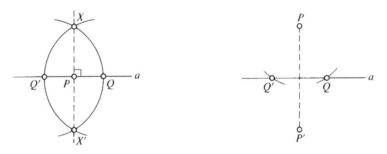

Figure 2.49   **Figure 2.50**

Construction 2 Given a point P not on a line a. Draw a line through P perpendicular to a.

Solution If Q is on a, draw a circle about P with (PQ) as radius. This circle intersects a in a second point, Q'. If $Q = Q'$, the line PQ is perpendicular to a. If $Q \neq Q'$, draw circles with radius (PQ) about Q and Q'. These intersect in a point $P' \neq P$ (two-circle theorem). PP' is perpendicular to a. (See Fig. 2.50.)

Construction 3 Given a line a and a point P not on a. Draw a line through P parallel to a.

Figure 2.51

Solution   This is done according to Figure 2.51, by an application of both Construction 2 and Construction 1.

Construction 2 also provides the solution of the following problem.

Problem 1   *Construct the image of P under reflection at a line a.*

As an application we present:

Problem 2   Given two points P, Q on one side of a line a. Find X on a such that the angle between PX and a is congruent to the angle between QX and a.

Solution   Find the image P' of P under reflection at a and intersect $P'Q$ with a.

A translation can be considered to be given by a point A and its image A'.

Problem 3   *Find the image P' of any point P under this translation.*

Solution   If P is not on AA', join P to A; draw a parallel of AA' through P and a parallel of AP through A'. The intersection of the latter two lines provides P'. If P is on AA', construct first the image of a point not on AA' and then the image of P (compare Section 2.13).

Problem 4   Given two nonparallel lines a, b and a line segment (PQ). Move (PQ) by a translation so that the image of P is on a and the image of Q is on b.

Solution   Translate (PQ) into $(P'Q')$ so that P' lies on a. Draw the parallel of a through Q' and intersect it by b, obtaining a point Q''. Translate $(P'Q')$ so that Q' is moved onto Q''.

A *rotation* can be considered given by the center C and an arc $\overset{\frown}{RS}$ (see Section 2.15), unless the rotation is a half-turn (carrying out the latter is trivial). Find the image P' of any point P under this rotation. We draw a circle about C with radius (CP). It intersects the rays through R, S with end point C in the points R^*, S^*, respectively. Draw a line a through C perpendicular to PS^* (if $P \neq S^*$; otherwise draw it perpendicular to the tangent at P). Construct the image of R^* under a reflection at a; call it P'. Then P' is the image of P under the given rotation about C. (See Fig. 2.52.)

Problem 5   Given two points P, Q interior to a circle $\mathscr{C}$, draw congruent and perpendicular chords through P, Q.

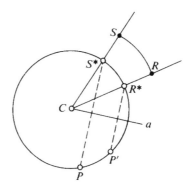

Figure 2.52

Solution   If C is the center of $\mathscr{C}$, rotate P about C through a right angle onto P'.   If $P' \neq Q$, then P', Q determine one of the chords; the perpendicular of $P'Q$ through P determines the other.   If $P' = Q$, draw chords through C and P, Q, respectively.

We list two more problems, one on half-turns and one on glide reflections.

Problem 6   Given a convex quadrangle $ABCD$ and a point P interior to it, find points Q, R, S, T on the (possibly extended) sides of the quadrangle such that $QRST$ is a parallelogram with center P.

Solution   Apply a half-turn about P and intersect the image of each side with the opposite side of the original quadrangle.

Problem 7 P is to be moved onto P' by a glide reflection TR_a, where $(P(P)T)$ is congruent to a given segment (PQ).   Find a.

Solution   We set $\tilde{P} = (P)T$.   Construct the midpoint M of (PP').   Draw a circle about M with radius (MP).   Draw a circle about P with radius (PQ).   Consider one of the points of intersection of these two circles as $\tilde{P}$. Draw a line a through M perpendicular to $P'\tilde{P}$ (or let $a = MP$ if $P' = \tilde{P}$).

As a curiosity it should be mentioned that all constructions that can be achieved by ruler and compass can be achieved by compass alone.   This is seen by constructing "inversions" in circles (see Chapter 6).   An analytic treatment of construction problems will follow at the end of the next chapter.

Exercises

1. Find the angular bisector of an angle.

2. Let P, Q be on one side of a line a.   Find X on a such that for a point $Y \neq X$ on a, $\sphericalangle XPY \equiv \sphericalangle AQY$.

3. Construct a trapezoid from four given segments, assuming the construction to be possible.

4. Let $\triangle ABC$ be an arbitrary triangle and let Z be on (AB). Find points Z', Z'' on (AC) and (BC), respectively, such that $\triangle ZZ'Z''$ is an equilateral triangle. (*Hint*: Rotate (AC) about Z through the angle of an equilateral triangle.)

5. Given a line a and points P, Q on opposite sides of a. Find X, Y on a such that (XY) is congruent to a given segment and such that

$$\sphericalangle PXY \equiv \sphericalangle YXQ.$$

6. Find the midpoint of a line segment by using only a compass.

Chapter Three

Affine Planes and Numbers

In the first two chapters we analyzed the concept of a planar motion (or "isometry"). For this, we needed metric properties of the plane, which were given by axioms on perpendicularity and on motions. Now we turn to a much wider class of geometrical mappings—that of affinities. For these planes, we no longer need metric properties. In fact, we can restrict ourselves to pure incidence properties. Consequently, the underlying axioms of a plane will concern only incidence of points and lines. We call these planes *affine planes*. However, we shall postulate the Euclidean axiom of parallels and thus eliminate elliptic and hyperbolic planes from our discussion. Insofar as mappings are concerned, this will be no loss, since it can be shown that in non-Euclidean geometry there is no generalization from motions to affinities as there is in Euclidean geometry.

Groups of affinities will show us a way to treat affine geometry with the tools of analytic geometry. In an ordinary Euclidean plane, analytic methods are usually introduced by choosing two nonparallel lines as "coordinate lines." The set of points on each of these lines is considered as representing the set of all real numbers. To every point of the plane there is assigned a pair of real numbers as "coordinates." Algebraic operations with these coordinates are used to solve geometrical problems. In this procedure, the operations of addition, multiplication, subtraction, and division are the main ones carried out, as, for example, in the simultaneous solution of linear equations. Other properties of real numbers, such as order and continuity properties, are rarely exhibited.

We shall see in this chapter that for a wide class of affine planes a coordinatization may also be carried out. One defines for the set of all points on a "coordinate line" two binary operations, called "addition" and "multiplication," carried out through the successive application of translations and dilatations (Section 3.6), respectively. Taking advantage of the group structure of the respective sets of translations and dilatations, one may operate algebraically with the points on coordinate lines in many ways; in the case of Euclidean lines they represent real numbers.

Moon photo courtesy of National Aeronautics and Space Administration.

We shall thus introduce a generalized system of numbers (skew field), in which the term "number" is related to ordinary numbers as are the elements of an abstract group.   So it will be seen that analytic geometry is not a privilege of ordinary Euclidean planes (and spaces).   It is a matter of all affine planes having enough properties of affine transformations.   Furthermore, our introduction of coordinates will not rest on the vague assumption that points on a line may be identified with real numbers.   Our "homemade" coordinates will flow out of clear-cut geometrical considerations.   After introducing coordinates by group-theoretic means, we shall study the role that Desargues' theorem plays in this introduction (Sections 3.9–3.11).   This will give us further insight into the geometrical basis on which algebraic methods in affine geometry rest.

3.1   *Affine Planes and Affinities*

We are given a set of elements called **points**, P, Q, ..., and a set of elements called **lines**, a, b, ... Furthermore, we assume a relation is given between points and lines ("P is on a" or "P is incident with a," and so on, compare Section 1.1), satisfying the following axioms.

Axiom A.1   *Two different points are on one and only one line.*

Axiom A.2   *On every line, there are at least two points.   There exist three noncollinear points.*

Axiom A.3   Euclidean axiom of parallels   *Given a line a and a point P not on a, there exists one and only one line passing through P and not intersecting a.*

Axioms A.1, A.2, and A.3 correspond to Axioms I.1, I.2, and I.3, respectively of Section 1.1.   We call the totality of points and lines satisfying Axioms A.1–A.3 an **affine plane**.   A collineation of an affine plane—that is, an incidence-preserving one-to-one mapping of the set of all points onto itself—is called an **affinity**.   Clearly, every Euclidean metric plane is a special affine plane, and, therefore, every motion in it is a special affinity.

Two lines that are either identical or nonintersecting are called **parallel**. They are **properly parallel** if they do not intersect.   Axiom A.3 readily implies that *"parallel" is an equivalence relation*, the equivalence classes being called **parallel pencils**.   So (as in the special case of Euclidean planes— compare Section 1.16) we may distinguish between pencils consisting of all lines through a given point (pencils of the first kind) and parallel pencils (pencils of the second kind).

Exercises

1. Show that (a) an affine plane possesses at least four points, no three of which are collinear, (b) there exists an affine plane containing precisely four points. (*Hint*: Call any pair out of four points a line.)

2. Show that if in an affine plane one line possesses precisely n points, then every line possesses precisely n points.

3. Show that if in an affine plane there exists a line possessing precisely two points, then it is the four point plane of Exercise 1(b).

4. Show explicitly that an affinity maps parallel lines onto parallel lines.

5. Show that the set of all affinities is a group.

6. Find all affinities of the four-point plane of Exercise 1(b).

7. Show that a mapping of the set of points onto itself is an affinity if it satisfies the following two conditions: Lines are mapped onto lines, and among the image points there exist three which are not collinear.

8. Prove that Axiom A.2 can be replaced by the following axiom: A.2′—On every line, there exists at least one point. There exist three noncollinear points.

9. Which of the metric planes in Chapter 1 are affine planes?

3.2 Dilatations and Translations

A *dilatation* is defined as an affinity that possesses a fixed point and maps every line onto a parallel of itself. A *translation* (or *parallel displacement*) is defined as an affinity which leaves either all points fixed or no point fixed and which maps every line onto a parallel of itself. The identity map I is both a dilatation and a translation. No other affinity has this property.

So we see that dilatations and translations, though of rather different intuitive meanings, differ only in the existence or nonexistence of a fixed point. This is a striking fact that becomes visible when we concentrate on incidence properties.

Let D be a dilatation and let F be a fixed point of D. Any line through F being mapped onto a parallel of itself must then be a fixed line, that is, it must be mapped onto itself.

Suppose, conversely, that a line a remains fixed under D and that a does not pass through F. Then every point P on a is the intersection of two fixed lines of D, namely, a and FP, and hence P is a fixed point. Joining an arbitrary point X to F and to a properly chosen point on a provides two fixed lines through X, so that all points remain fixed under D, hence D = I (identity map). Therefore, if D $\neq$ I, then D possesses the lines through one point, and only these, as fixed lines. This, in turn, implies the following theorem.

Theorem 3.2.1 *A dilatation that is not the identity map has one and only one fixed point, called the **center** of the dilatation. Every fixed line passes through the center.*

A fixed line of an affinity is also called a **trace**. So a dilatation possesses a pencil of lines (pencil of the first kind) as traces (Fig. 3.1). A translation possesses a parallel pencil of traces (Fig. 3.2).

Figure 3.1   **Figure 3.2**

Theorem 3.2.2 *The set $\mathscr{D}_C$ of all dilatations with a given center C is a group under "successive application" as group operation.*

Proof If D_1, D_2 are in $\mathscr{D}_C$, then $D_1 D_2$ leaves C fixed and it maps every line onto itself or a parallel of itself. Hence $D_1 D_2$ is in $\mathscr{D}_C$. The associative law is trivially satisfied (compare Section 1.8). The identity map I serves as unit element. The inverse map D^{-1} of a dilatation with center C is clearly again a dilatation with center C. Therefore, $\mathscr{D}_C$ is a group. ◆

Theorem 3.2.3 (a) *The set $\mathscr{T}$ of all translations is a group.* (b) *$\mathscr{T}$ is commutative if there exist translations not equal to I that have different pencils of traces.* (c) *The set $\mathscr{T}_a$ of all translations with a prescribed trace a is a subgroup of $\mathscr{T}$.*

We say that two translations are in **different** or **equal directions** if their pencils of traces are different or equal, respectively.

Proof $\mathscr{T}$ and $\mathscr{T}_a$ are shown to be groups in a way similar to that used in the proof of Theorem 3.2.2. This proves parts (a) and (c).

Let T, T′ be translations not equal to I in different directions and set $P_1 = (P)\text{T}$, $P_2 = (P)\text{T}'$, $P_3 = (P)\text{TT}'$, $P_3^* = (P)\text{T}'\text{T}$. Then, for any P, the lines PP_1 and $P_2 P_3^*$ are parallel and also PP_2 and $P_1 P_3$ are parallel, P_3 and P_3^* are both on $P_1 P_3$ and $P_2 P_3^*$; hence $P_3 = P_3^*$, and so $(P)\text{TT}' = (P)\text{T}'\text{T}$. This implies that TT′ = T′T. (See Fig. 3.3.)

If T, T′ have equal directions, there exists a translation T″ with direction different from that of T, T′. Then T and T′T″ are also in different

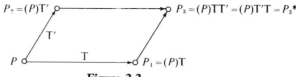

Figure 3.3

directions, and, by what we just have shown,

$$(TT')T'' = T(T'T'') = (T'T'')T$$
$$= T'(T''T) = T'(TT'') = (T'T)T'';$$

hence

$$TT' = T'T.$$

(See Fig. 3.4.) ◆

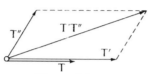

Figure 3.4

Theorem 3.2.4 *A point P is mapped onto a point P' by at most one dilatation with a prescribed center not equal to P, P' and by at most one translation.*

Proof Let D, D' have center C and map P onto P', where $P \neq C$. If X is not on CP, then $(X)D$ and $(X)D'$ are both on CX and both on the parallel of PX through P'. This implies $(X)D = (X)D'$ or $(X)D'D^{-1} = X$. Since also $(C)D'D^{-1} = C$, we conclude from Theorem 3.2.1 that $D'D^{-1} = I$ and thus $D' = D$. An analogous proof applies for translations. ◆

Theorem 3.2.5 *If T is a translation and A is an arbitrary affinity, then $A^{-1}TA$ is a translation T'. Any trace a of T is mapped by A onto a trace of T'.*

Proof Let x be an arbitrary line and denote $\hat{x} = (x)A^{-1}$. Then $\hat{x}$ is parallel to $(\hat{x})T = (x)A^{-1}T$. Since A maps parallel lines onto parallel lines, we conclude that $(\hat{x})A = x$ and $(\hat{x})TA = (x)A^{-1}TA$ are parallel. So $A^{-1}TA$ is either a dilatation or a translation.

Let a be a trace of T (Fig. 3.5). From

$$((a)A)A^{-1}TA = (a)TA = (a)A,$$

we see that $(a)A$ is a trace of $A^{-1}TA$. Any two such lines are parallel. So $A^{-1}TA$ is a translation T'. ◆

Figure 3.5

Exercises

1. Show that if an affinity maps every line onto a parallel of itself and if it has two fixed points, it is the identity map.

2. Show that any half-turn of a Euclidean metric plane is a dilatation.

3. Which of the affinities in Exercise 6, Section 3.1, are translations? Which are dilatations?

4. Show that the system of all dilatations and translations of an affine plane is a group.

5. Carry out in detail the proof of Theorem 3.2.3, parts (a), (c).

6. If D is a dilatation and A is an arbitrary affinity, show that $A^{-1}DA$ is also a dilatation.

3.3   *Stretches and Shears*

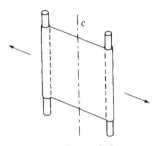

Figure 3.6

If a flat piece of rubber is stretched, as indicated in Figure 3.6, there is a middle line, c, that remains pointwise fixed. In an affine plane, we define the corresponding mapping as follows: A *stretch* Š is an affinity that leaves a line c pointwise fixed and maps each line of one parallel pencil not containing c onto itself. The pencil of fixed lines (traces) is called the *direction* of Š, and any element of this pencil is a *direction representative*; c is called the *axis* of Š.

A stretch is the counterpart of a dilatation in which "points" and "lines" switch roles. Whereas in a dilatation all lines through a point remain fixed, in a stretch, all points on a line remain fixed. Also the construction of images shows the following correspondence (we denote by xy the point of intersection of nonparallel lines x, y):

Dilatation	Stretch
Given a point P not equal to C and its image, P'.	Given a line p not equal to c and its image p'.
Let X not be on PC, where C is the center of the dilatation.	Let x not pass through pc and not be parallel to p or c, c being the axis of the stretch.
To obtain the image X' of X, find the image of the line XP and intersect it by XC (Fig. 3.7).	To obtain the image x' of x, find the image of the point xp and join it to xc (Fig. 3.8).

[3.3]

*Affine Planes
and Numbers*

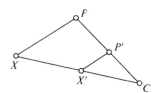

Figure 3.7

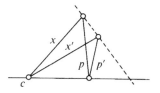

Figure 3.8

This correspondence (principle of duality) will be further illuminated in Chapter 5.

It should be noted that, in a Euclidean plane, a reflection is a special stretch ("factor -1"), corresponding to the fact that a half-turn in a Euclidean plane is a special dilatation (also "factor -1"). More generally, an involutoric stretch of an affine plane (that is, a stretch $\check{S}$ satisfying $\check{S}\check{S} = I$ and $\check{S} \neq I$) is called a *skew reflection* (Fig. 3.9; compare Exercise 2 of this section). Suppose

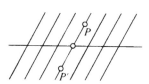

Figure 3.9

that an affinity leaves all points of a line c fixed and the image of any point X lies on the parallel of c through X. Then the affinity is called a *shear*, denoted by $\bar{S}$ (Fig. 3.10). Shears and stretches differ only in that the pencil

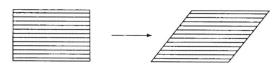

Figure 3.10

of fixed lines does or does not contain the axis c (the "$\smile$" or "$-$" in $\check{S}$ or $\bar{S}$ is to remind us of this difference).

Theorem 3.3.1 *A stretch or a shear with axis c has no other fixed points than those on c, unless it is the identity map.*

Proof Let S be either a stretch or a shear with axis *c*. If *F* is a fixed point not on *c*, let *f* be the direction representative of S through *F* (see Fig. 3.11). Join any point *P*, not on *c* or *f*, to *F*. The line *PF* either intersects *c* in a point *Q* or is properly parallel to *c*. In both cases, we conclude that *PF*

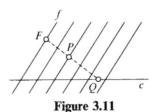

Figure 3.11

is a fixed line and thus, since *PF* is different from the trace of S through *P*, *P* is fixed. Now it is readily seen that S = I. ◆

Theorem 3.3.2 *If an affinity has two fixed points, F_1, F_2 and if it possesses a pencil of traces to which F_1F_2 belongs, then it is a shear with axis F_1F_2.*

Proof Let A be the given affinity and let *Q* be any point on $c = F_1F_2$. Suppose $Q' = (Q)A \neq Q$. Then for a line $q \neq c$ through *Q* either the image $q' = (q)A$ of *q* intersects *q* in a point *F* not on *c* or it is properly parallel to *q*. In the first case, consider the lines $q_1 = F_1F$; $q_2 = F_2F$ (Fig. 3.12(a)); in the second case, consider the parallels q_1 and q_2 of *q* through F_1 and F_2, respectively (Fig. 3.12(b)). Clearly q_1 and q_2 are

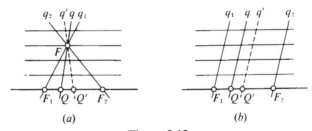

Figure 3.12

fixed lines. Since they "cut across" the pencil of traces, they even remain pointwise fixed. From this, it readily follows that A = I, which contradicts the assumption $Q' \neq Q$. Therefore, all points of *c* remain fixed, and A is a shear. ◆

Theorem 3.3.3 *There is at most one stretch or shear with a prescribed axis c and a prescribed trace t that maps a line p onto a line $p' \neq p$ (compare Theorem 3.2.4).*

Proof Suppose S_1, S_2 are two such stretches or shears. *p* is not a trace of either of them, since otherwise $p' = p$ would follow. $S_1S_2^{-1}$ is also a stretch or shear with axis *c* and *t* as a trace; it leaves *p* fixed. Since *p* "cuts across" the traces of S_1, S_2 (otherwise $p' = p$), every point of p is a fixed point under $S_1S_2^{-1}$. Therefore, by Theorem 3.3.1, $S_1S_2^{-1} = I$, and thus $S_2 = S_1$. ◆

The following theorem provides an interesting link between stretches and dilatations.

Theorem 3.3.4 *If* D *is a dilatation with center* C, *and if* Š *is a stretch with axis* c *such that* C *is on* c, *then*

$$D\check{S} = \check{S}D.$$

Proof If Figure 3.13 is used as a guide, the proof is accomplished easily. ◆

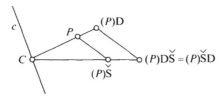

Figure 3.13

Theorem 3.3.5 *If* Š *is a stretch with* c *as its axis and* t *as one of its traces, then, for any affinity* A, *the mapping* $A^{-1}\check{S}A$ *is a stretch with axis* $(c)A$ *and* $(t)A$ *as one of its traces.*

Proof Let $P' = (P)A$ be an arbitrary point on $(c)A$. We obtain

$$(P')A^{-1}\check{S}A = (P)\check{S}A = (P)A = P';$$

so P' is a fixed point of $A^{-1}\check{S}A$. If $t' = (t)A$, then, correspondingly,

$$(t')A^{-1}\check{S}A = (t)\check{S}A = (t)A = t'.$$

Clearly, all parallel lines of t' are fixed lines of $A^{-1}\check{S}A$, too. Therefore, $A^{-1}\check{S}A$ is a stretch with axis $c' = (c)A$ and $t' = (t)A$ as a trace. ◆

Exercises

1. Prove that if an affinity has two fixed points, F_1, F_2, and if it posseses a pencil of traces to which F_1F_2 does not belong, then it is a stretch with axis F_1F_2.

2. Show that in an ordinary Euclidean plane any skew reflection is the product of a reflection and a shear.

3. Show that the set of all shears and translations with a prescribed pencil of traces is a group.

4. Show that the set of all stretches and shears with a prescribed pencil of traces is a group (interchange "points" and "lines" in Theorem 3.2.3(a)).

5. Find subgroups of stretches and shears analogous to $\mathscr{T}_a$ and $\mathscr{D}_c$ (see Section 3.2).

6. Is the set of all stretches, shears, and translations a group?

7. Carry out in detail the proof of Theorem 3.3.4.

3.4   *The Erlanger Programm of F. Klein*

Throughout this section, suppose that we are given an ordinary Euclidean plane. It is a special affine plane. The group $\mathscr{M}$ of all motions of this plane is a subgroup of the group $\mathscr{A}$ of all affinities of the plane. We now define two more such groups.

By a *similarity* we mean an affinity that maps perpendicular lines onto perpendicular lines. Clearly motions and dilatations are similarities. The product of two similarities and the inverse of a similarity are also similarities. Therefore, the set of all similarities is a group $\mathscr{S}$.

An *equiaffinity* is defined as a product of finitely many shears. Clearly, the equiaffine mappings again form a group $\mathscr{E}$.

The four groups mentioned above are linked according to the following diagram:

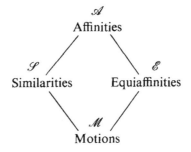

A geometrical property of a plane that remains unchanged under a mapping is called an *invariant* of this mapping. So perpendicularity of lines is an invariant of any similarity, and length is an invariant under motions. In both cases, the invariant applies to all elements of a group; we say it is an invariant of the respective group. It was the idea of F. Klein to subdivide geometry according to invariants of some standard groups of mappings. Geometry can thus be defined as theory of invariants under a standard group of mappings. This is the main content of Klein's *Erlanger Programm*. We list some examples in the following table:

Group	Invariants	Geometry
affinities	betweenness	affine geometry
similarities	perpendicularity; angles	geometry of similarities
equiaffinities	area	equiaffine geometry
motions	length	geometry of motions

Other geometries are defined in later chapters (projective, inverse geometry). Klein's definition of geometry provides a useful classification of geometrical theorems. It is, however, not to be made a dogma, since it overlooks many things in geometry that cannot be subsumed under the *Erlanger Programm*. Actually, Klein's program was issued for high school teaching. But it had a stronger influence on geometrical research than it had on the teaching of geometry. W. Blaschke, for example, remarks: "I spent a life applying Klein's program to differential geometry." It seems that, more recently, the *Erlanger Programm* is slowly finding the place in teaching that it deserves. A valuable application of Klein's ideas will be found in coordinate geometry: We can shift from a coordinate system to a more convenient one by applying a mapping of the group that is the defining group of the geometry under consideration.

Exercises

1. Find more invariants for the geometries listed above.
2. Show that if a similarity is an equiaffinity, it is a motion.
3. Find the geometries in which the two members of each family displayed are related to each other: (a) triangles and equilateral triangles, (b) circles and circles of radius 1, (c) rectangles and parallelograms.
4. Describe, according to the Erlanger Programm, the geometry whose group is that of all affinities with a given fixed point 0.
5. Find, in Exercise 4, a diagram of subgroups analogous to that in the text.
6. Express the geometry of a metric plane (Chapter 1) in terms of Klein's Erlanger Programm.

3.5 Transitivity Axioms

The axioms that we used for defining an affine plane are still rather weak. Concerning dilatations and translations, the situation is similar to that of rotations and translations in a metric plane. There need not be such a mapping transforming a prescribed line a onto a prescribed line a' (see Section 3.12). We now postulate an axiom analogous to the axiom of "free mobility" (Section 2.1).

Axiom A.4 *Transitivity axiom I. If P, P', C are on a line, where $P, P' \neq C$, then there exists a dilatation with center C that maps P onto P'.*

Axiom A.4′ Transitivity axiom II. If p, p′, c, and d are concurrent lines all different from each other, then there exists a stretch with axis c and trace d that maps p onto p′.

We call an affine plane satisfying Axiom A.4 and Axiom A.4′ a *transitive affine plane*. (We shall see in Section 3.11 that Axioms A.4 and A.4′ imply each other, so only one of them needs to be postulated.) The motivation for this name is seen if we rephrase Axiom A.4 as follows: For any C, the group $\mathscr{D}_c$ is transitive on the set of all points not equal to C on an arbitrary line through C. By Theorem 3.2.4, it is even simply transitive on any such set. For groups $\mathscr{T}_a$ we obtain, correspondingly, the following theorem.

Theorem 3.5.1 If P, $P′$ are any points, there exists one and only one translation mapping P onto $P′$.

Proof If $P = P′$, the identity map I translates P onto $P′$. So let $P \neq P′$ and let C_1, C_2 be different points such that C_1C_2 is properly parallel to $PP′$ and PC_1, $P′C_2$ intersect in a point Q (assume Q to exist; otherwise the

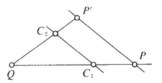

Figure 3.14

plane has only four points). Let the dilatation D_1 have center C_1 and map P onto Q. Furthermore, let the dilatation D_2 map Q onto $P′$ and have center C_2. (D_1, D_2 exist by Axiom A.4.) Now D_1D_2 maps P onto $P′$. D_1D_2 is either a dilatation or a translation (see Section 3.2). C_1C_2 and $PP′$ are fixed lines under D_1D_2 and are properly parallel. So, by Theorem 3.2.1, D_1D_2 cannot be a dilatation and is, therefore, a translation. The uniqueness of this translation follows from Theorem 3.2.4. ◆

Theorem 3.5.2 If P, $P′$ are two points not on c such that $P \neq P′$ and $PP′$ is not parallel to c, then there exists one and only one stretch that maps P onto $P′$ and possesses c as an axis.

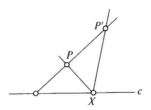

Figure 3.15

Let X lie on c, but not on PP' (Fig. 3.15). By Axiom A.4', there exists a stretch with direction PP' that maps XP onto XP' and hence P onto P'. By Theorem 3.3.3, this stretch is uniquely determined. ◆

Exercises

1. State and prove an analogue of Theorem 3.5.1 for shears.

2. Express Axiom A.4' and Theorem 3.5.2 in terms of transitivity properties of the groups introduced in Exercise 4, Section 3.3.

3. Given, in a transitive affine plane, three noncollinear points P, Q, R and three noncollinear points P', Q', R', show that there exists an affinity that maps P, Q, R, onto P', Q', R', respectively.

4. Define parallelograms and show that in a transitive affine plane any two parallelograms can be mapped onto each other by an affinity.

5. Prove that if in an affine plane there exist two points C, C', where $C \neq C'$, such that the groups $\mathscr{D}_C$, $\mathscr{D}_{C'}$ are transitive, then $\mathscr{D}_X$ is transitive for an arbitrary X on the line CC'.

6. Prove that Axiom A.4 need only be postulated for three noncollinear centers. (*Hint*: Use the result of Exercise 5.)

7. Show that the nine-point plane of Section 1.7 is a transitive affine plane.

3.6   Coordinates on a Line

For a deeper geometrical study of affine planes, it is useful to employ the tools of linear algebra. These enter geometry if we introduce coordinates, that is, pairs of "numbers" by which the points of the plane are represented. This must be done in such a way that lines are represented by linear equations.

What, however, is a "number" in our axiomatic geometry? We cannot expect it to be an ordinary number. Rather, we need a generalized number system such that the ordinary system of real numbers is a special case of a generalized number system. We introduce such a system by an appropriate set of axioms. So our procedure will be analogous to that of axiomatic geometry, which generalizes ordinary geometry.

A foundation for a generalized number system has already been laid by our introduction of the concept of a group. We remind ourselves that a group generalizes the system of integers under addition, as well as the system of positive rationals or reals under multiplication. Also, systems of mappings with "successive application" as "multiplication" appear as groups. The common nucleus of all these examples is an operation between each pair of elements providing again an element of the system under consideration and satisfying some basic axioms—the associative law, the existence of a unit element, and the existence of an inverse element of each element.

Now we need a generalized number system which simultaneously generalizes addition and multiplication—not just one of these operations as a group does. Such a system can be introduced as follows: We assume a set $\mathscr{F}$ to be given with two binary operations defined on it. We call one of these operations "*addition*" (+) the other "*multiplication*" ($\cdot$). We say $\mathscr{F}$ is a *skew field* under "+" and "$\cdot$" if the following three axioms are satisfied:

1. *$\mathscr{F}$ is a commutative group under "+" (unit element 0).*
2. *The set $\mathscr{F}^{\times}$ obtained from $\mathscr{F}$ by leaving out 0 is a group under "$\cdot$".*
3. *The left distributive law and the right distributive law are true:*

$$(a + b)c = ac + bc; \qquad c(a + b) = ca + cb.$$

So we have on $\mathscr{F}$ two group structures which are linked by the distributive laws. If $\mathscr{F}$ also yields the commutative law of multiplication, that is, if $\mathscr{F}^{\times}$ is a commutative group, we call $\mathscr{F}$ a *field*. (An example of a skew field that is not a field will be discussed in Exercise 6 of Section 3.13).

A skew field (or field) can also be viewed as a set in which the four fundamental operations of algebra—addition, subtraction, multiplication, division—can be carried out. Clearly, the ordinary number system has many additional properties. These are not needed, however, if only the fundamental operations are being used. These operations, in turn, are all we need in our present study of points and lines.

We have studied many geometrical examples of groups consisting of geometrical mappings as elements. Can we also build up a skew field by using geometrical mappings? The answer is positive, although the solution is more difficult than in the case of a group. If we considered a set of mappings as elements of the field, how would we obtain two operations on it? There is no intuitive approach to this problem as there is to the problem of defining a group. So we proceed somewhat differently. We consider the group $\mathscr{T}_a$ of all translations along a line a. Furthermore we choose a point 0 on a and consider the group $\mathscr{D}_0$ of all dilatations with center 0. Let E be a point not equal to 0 on a. Now a point $A \neq 0$ on a can be assigned uniquely to each element of $\mathscr{T}_a$ and to each element of $\mathscr{D}_0$:

$A \leftrightarrow$ translation T_A that maps 0 onto A;

$A \leftrightarrow$ dilatation D_A that maps E onto A;

0 is assigned as the identity element of $\mathscr{T}_a$ but not as a dilatation of D_0.

If we now choose the points of a as elements of the set that is to be a skew field, we can use the group operations of $\mathscr{T}_a$ and $\mathscr{D}_0$ as "addition" and "multiplication," respectively:

$$A + B = (0)\mathrm{T}_A\mathrm{T}_B = (A)\mathrm{T}_B,$$
$$A \cdot B = (E)\mathrm{D}_A\mathrm{D}_B = (A)\mathrm{D}_B \quad \text{if } A, B \neq 0,$$
$$A \cdot 0 = 0 \cdot B = 0 \quad \text{for any } A, B.$$

Notice that $A \cdot B = (A)D_B$ also for $A = 0$, $B \neq 0$. So, as in the case of a group, we take advantage of "successive application" as a binary operation here in two different ways. The remainder of this section shows that the operations thus obtained define a skew field on the set of points of a.

Theorem 3.6.1 *The points on a form a commutative group under addition.*

Theorem 3.6.2 *The points on a not equal to 0 form a group under multiplication.*

The connection between addition, multiplication, and the respective group operations is also reflected in the following formulas (which are direct conclusions from our definitions):

$$T_{A+B} = T_A T_B,$$
$$D_{A \cdot B} = D_A D_B \quad \text{for } A, B \neq 0.$$

The crucial point is to prove the distributive laws. We proceed as follows. In Theorem 3.2.5, we saw that $A^{-1}TA$ is a translation if T is a translation and A an arbitrary affinity. In particular, $D_C^{-1}T_B D_C$ is a translation for $C \neq 0$. Since

$$(0)D_C^{-1}T_B D_C = (0)T_B D_C = (B)D_C = B \cdot C,$$

we see that

$$D_C^{-1}T_B D_C = T_{B \cdot C}$$

or

$$T_B D_C = D_C T_{B \cdot C}. \tag{a}$$

Intuitively, it is clear that, in general, $T_B D_C \neq D_C T_B$. (a) gives more precise information on how a translation and a dilatation can be interchanged.

Theorem 3.6.3 *Addition and multiplication satisfy the left distributive law:*

$$(A + B) \cdot C = A \cdot C + B \cdot C.$$

Proof By definition and by (a) we find

$$(A + B) \cdot C = (A + B)D_C = (A)T_B D_C = (A)D_C T_{B \cdot C}$$
$$= A \cdot C + B \cdot C. \qquad \blacklozenge$$

If $\mathscr{D}_0$ were commutative (as it is in ordinary Euclidean geometry) the right distributive law would follow from Theorem 3.6.3. However, we can also prove it without $\mathscr{D}_0$ being commutative: Denote by $\check{S}_C$ the stretch with an axis $c \neq a$, c passing through 0, that maps E onto some point $C \neq 0$ on a (existence by Theorem 3.5.2). By Theorem 3.3.4, we find

$$D_A \check{S}_C = \check{S}_C D_A,$$

and hence

$$(A)\check{S}_C = (E)D_A \check{S}_C = (E)\check{S}_C D_A = (C)D_A = C \cdot A.$$

This proves the formula

$$(A)\check{S}_C = C \cdot A \quad \text{if } A, C \neq 0. \tag{b}$$

For $A = 0$, (b) is also true. From Theorem 3.2.5, we conclude that $\check{S}_C^{-1}T_B\check{S}_C$ is a translation. Since

$$(0)\check{S}_C^{-1}T_B\check{S}_C = (0)T_B\check{S}_C = (B)\check{S}_C,$$

we find, with the aid of (b),

$$\check{S}_C^{-1}T_B\check{S}_C = T_{(B)}\check{S}_C = T_{C \cdot B}.$$

This proves the following formula, which attests to the interchangeability of translations and stretches in the same way that (a) attested to the interchangeability of translations and dilatations:

$$T_B \cdot \check{S}_C = \check{S}_C \cdot T_{C \cdot B}. \tag{c}$$

Theorem 3.6.4 *Addition and multiplication satisfy the right distributive law, thus,*

$$C \cdot (A + B) = C \cdot A + C \cdot B.$$

Proof By definition and by (b), (c),

$$C \cdot (A + B) = (A + B)\check{S}_C = (A)T_B\check{S}_C = (A)\check{S}_C T_{C \cdot B} = (C \cdot A)T_{C \cdot B}$$
$$= C \cdot A + C \cdot B. \qquad \blacklozenge$$

Theorem 3.6.5 *The points on a, with addition and multiplication as defined above, form a skew field.*

In other words, Theorem 3.6 5 states that on every line a, coordinates with respect to any two points 0, E of a can be introduced, with the coordinates being taken from a skew field.

Exercises

1. Prove that the set of rationals (fractions) is a field under ordinary addition and multiplication.
2. Prove that the set of all numbers $a + b\sqrt{2}$, where a, b are rational, is a field.
3. Prove that the set of all complex numbers $a + bi$, where a, b are real and $i = \sqrt{-1}$, is a field under a formal extension of the operations "$+$" and "$\cdot$" to complex numbers.
4. Let p be a prime number. Show that under addition and multiplication, mod p, the numbers $0, 1, 2, \ldots, p - 1$ form a field, the groups involved being cyclic groups (compare Section 1.10).
5. Prove, from the axioms of a skew field, that

$$a \cdot 0 = 0.$$

6. Addition and multiplication can be illustrated as in Figures 3.16 and 3.17. Formula (a) is illustrated in Figure 3.18. Find a corresponding illustration for (c).

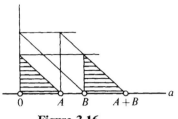

Figure 3.16

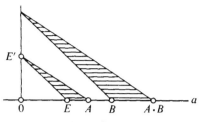

Figure 3.17

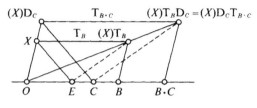

Figure 3.18

3.7   Coordinates in the Plane

On a line it has been possible to identify points and skew-field elements. In the plane, however, two coordinates will be needed. For this purpose we choose a second line, b, as carrier of skew-field elements. We assume b to intersect a at O. In order to distinguish points in the plane and skew-field elements on a or b, we shall denote the latter ones from now on by small letters, $x, y, \ldots$. Since E represents the "1" of the skew field, we denote it simply by 1. Also, O becomes zero.

The introduction of "numbers" on b can be simply done in the following way: We select a point $E' \neq 0$ on b. If a parallel line of EE' intersects a and b at points X, X', respectively, we assign to X and X' the same number x. Since a translation or a dilatation maps every line onto a parallel of itself, all skew-field operations on a are carried over automatically onto those on b. (See Fig. 3.19).

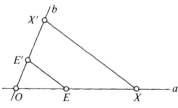

Figure 3.19

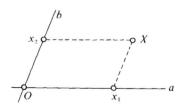

Figure 3.20

Let X be an arbitrary point of the plane. The parallels of b and a through X intersect a and b in numbers x_1 and x_2, respectively (Fig. 3.20). We call x_1 the **abscissa** of X and x_2 the **ordinate** of X. Together we call x_1 and x_2 the **coordinates** of X. In this way, we assign to every point X a unique pair of numbers (x_1, x_2). Conversely, to every such pair (x_1, x_2) there corresponds a unique point X. We obtain a one-to-one correspondence,

$$X \longleftrightarrow (x_1, x_2),$$

between points and pairs of skew-field elements. Identifying both, we write

$$X = (x_1, x_2).$$

Theorem 3.7.1 *Every line not parallel to a is represented by an equation*

$$ux_2 + x_1 = v, \tag{1}$$

and every line parallel to a is represented by an equation

$$x_2 = w \tag{2}$$

in the sense that a point is on the line if and only if its coordinates satisfy (1) *or* (2), *respectively.*

Proof

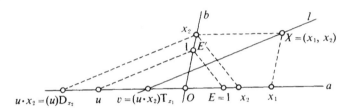

Figure 3.21

Let l be the given line. If it intersects a in a point ($=$ number) v, let u be the point in which the parallel of l through E' intersects a. As is seen with the aid of Figure 3.21,

$$v = (u \cdot x_2)\mathrm{T}_{x_1} = u \cdot x_2 + x_1;$$

so (1) is satisfied.

If l is parallel to a, then (2) holds, by definition of coordinates. If, conversely, x_1, x_2 satisfy (1) or (2), then (x_1, x_2) is readily seen to lie on l. ◆

We say that the affine plane has been **coordinatized** by a skew field $\mathscr{F}$. Multiplying (1) from the left by an arbitrary factor not equal to 0 does not alter the solutions of (1).

Theorem 3.7.2 *Every line is represented by an equation*

$$ux_1 + vx_2 = w,$$

where u and v are not both 0. Conversely, every such linear equation represents a line.

It should be noted, however, that a linear equation, as in Theorem 3.7.2, is not uniquely determined by a line; it can be multiplied by any factor not equal to 0. We turn now to an analytic representation of translations and dilatations.

Theorem 3.7.3 *Translations are represented by mappings*

$$(x_1, x_2) \longrightarrow (x_1 + c_1, x_2 + c_2), \tag{3}$$

c_1, c_2 *being constants.*

Dilatations with center $(0, 0)$ *are represented by*

$$(x_1, x_2) \longrightarrow (x_1 d, x_2 d) \tag{4}$$

where $d \neq 0$ *is a constant*

Proof Let X be mapped onto X' by a translation T. Since the parallel of a through X intersects the parallel of b through X' in a point Y, we split T into

$$T = T_1 T_2,$$

where $(X)T_1 = Y$ and $(Y)T_2 = X'$. Clearly T_1 leaves all ordinates fixed, and T_2 leaves all abscissas fixed. Let

$$(0)T_1 = c_1 \quad \text{and} \quad (0)T_2 = c_2.$$

Then T_1 is represented by

$$(x_1, x_2) \xrightarrow{T_1} (x_1 + c_1, x_2).$$

Since by parallels of EE' addition and multiplication are carried over from a onto b and, since T maps each line onto a parallel of itself, T_2 can be represented as follows:

$$(x_1 + c_1, x_2) \xrightarrow{T_2} (x_1 + c_1, x_2 + c_2).$$

Hence for $T = T_1 T_2$, we obtain:

$$(x_1, x_2) \xrightarrow{T} (x_1 + c_1, x_2 + c_2).$$

Conversely, any such mapping is readily seen to represent a translation. Now let X be mapped onto X' by a dilatation D with center $(0, 0)$, and set $(1)D = d$. Thus $D = D_d$. Since $(a)D = a$ and $(b)D = b$, and since every parallel of EE' is mapped onto a parallel of EE', we have:

$$(x_1, x_2) \xrightarrow{D} ((x_1)D_d, (x_2)D_d) = (x_1 d, x_2 d).$$

Conversely, any such mapping is readily seen to represent a dilatation. ◆

Exercises

1. Find explicitly the images of the lines (1) and (2): (a) under a translation (3), (b) under a dilatation (4). (Note that the skew field $\mathscr{F}$ need not be commutative.)

2. Show that if all dilatations with center $(0, 0)$ can be written in the form

$$(x_1, x_2) \longrightarrow (dx_1, dx_2),$$

the coordinate skew field $\mathscr{F}$ is commutative—that is, it is a field.

3. Find the coordinate skew field by which the "four-point plane" (see Exercise 1, page 99) is coordinatized.

4. Find the coordinate skew field by which the nine-point plane of Section 1.7, is coordinatized.

5. Find all groups of dilatations in the nine-point plane (see Exercise 4).

6. Describe geometrically the mapping defined by $(x_1, x_2) \to (x_1c, x_2d)$; $c \neq 0, d \neq 0, c \neq d$.

3.8   Skew Fields and Planes

In the preceding two sections we introduced a skew field in any transitive affine plane in order to describe algebraically such things as points, lines, translations, dilatations. The question of the converse arises—whether, for a given skew field, there exists an affine plane having this skew field as a coordinate field. The answer, yes, reveals a wonderful relationship between the geometrical concept of a transitive affine plane and the algebraic concept of a skew field.

Let $\mathscr{F}$ be a skew field. We call any pair (x_1, x_2) with x_1, x_2 in $\mathscr{F}$, a **point**. The set of points satisfying the linear equation

$$ax_1 + bx_2 = c, \qquad a, b \text{ not both} = 0, \tag{1}$$

is said to be a **line**. We wish to show that points and lines form an affine plane. To do so, we first carry over a theorem of elementary algebra to the algebra of skew fields.

Theorem 3.8.1   *A system of linear equations*

$$\begin{aligned} ax_1 + bx_2 &= c \\ dx_1 + ex_2 &= f \end{aligned} \tag{2}$$

possesses a unique solution (x_1, x_2) if and only if (at least) one of the following conditions is satisfied:

$$a \neq 0 \quad and \quad e - da^{-1}b \neq 0, \tag{3}$$

$$b \neq 0 \quad and \quad d - eb^{-1}a \neq 0. \tag{4}$$

Proof

I. Let (3) be satisfied. By obvious calculations, we find

$$\begin{aligned} x_2 &= (e - da^{-1}b)^{-1}(f - da^{-1}c), \\ x_1 &= a^{-1}[c - b(e - da^{-1}b)^{-1}(f - da^{-1}c)]. \end{aligned} \tag{5}$$

If (4) is satisfied, interchange x_1 and x_2, respectively. So we can assume (3) to be true. Then any solution of (2) is of the form (5) and hence is unique. Substituting (5) into (2) shows that (5) does provide a solution.

II. Let (2) possess a unique solution (x_1, x_2). If a and b were both zero, then c would also equal zero, and (2) would reduce to one equation and thus have more than one solution. So either $a \neq 0$ or $b \neq 0$; say $a \neq 0$. Suppose $e - da^{-1}b = 0$. Then $f - da^{-1}c = 0$, so that (2) can be written as

$$x_1 + a^{-1}bx_2 = a^{-1}c$$
$$dx_1 + da^{-1}bx_2 = da^{-1}c.$$

For $d = 0$ as well as for $d \neq 0$, this system reduces to one equation, implying that (2) has more than one solution, contrary to assumption. Therefore (3) is satisfied. ◆

Interchanging multiplication "from the right" and multiplication "from the left" in the proof of Theorem 3.8.1 gives the following theorem.

Theorem 3.8.2 *A system of linear equations*

$$y_1a + y_2b = c$$
$$y_1d + y_2e = f \tag{6}$$

has a unique solution if and only if one of the following conditions is satisfied:

$$a \neq 0 \quad and \quad ba^{-1}d - e \neq 0 \tag{7}$$
$$b \neq 0 \quad and \quad ab^{-1}e - d \neq 0. \tag{8}$$

Now we are ready to prove Theorem 3.8.3.

Theorem 3.8.3 *Points and lines form a transitive affine plane.*

Proof We must prove Axioms A.1–A.4, A.4' (see Sections 3.1 and 3.5). Let $P = (p_1, p_2)$ and $Q = (q_1, q_2)$. As can be verified easily, there exists at least one triplet (a, b, c) such that

$$ap_1 + bp_2 = c$$
$$aq_1 + bq_2 = c.$$

The line

$$ax_1 + bx_2 = c$$

joins P and Q.

We assert that two different lines possess at most one common point. Suppose, in fact, these lines are given by the system of equations (2), where a, b are not both $= 0$, and d, e are not both $= 0$. We may assume $a \neq 0$. If $e - da^{-1}b \neq 0$, then, by Theorem 3.8.1, the two lines possess a unique point of intersection. If $e - da^{-1}b = 0$, then $d \neq 0$, since otherwise it would follow that $e = 0$. Substituting $e = da^{-1}b$ in the second equation of (2) and multiplying from the left by ad^{-1} yields

$$ax_1 + bx_2 = ad^{-1}f$$

for the second line. With the aid of Theorem 3.7.1 and Figure 3.21, one can readily see that the two lines now are parallel. So there is at most one line joining two different points and A.1 is proved. Axiom A.3 also follows from the latter remarks.

Since, in $\mathscr{F}$, $1 \neq 0$, the points $(0, 0)$, $(0, 1)$, $(1, 0)$, $(1, 1)$ are all different. A simple calculation shows that no three of them are on a line. Any line will intersect two of the lines joining these four points, such that the points of intersection are different. So Axiom A.2 is true.

We define a **_dilatation_** D with center $(0, 0)$ by

$$(x_1, x_2) \xrightarrow{\ D\ } (x_1 d, x_2 d), \quad \text{where } d \neq 0.$$

To show that D is a dilatation in the sense of the definition given in Section 3.2, we consider the image of a line (l):

$$a x_1 d^{-1} + b x_2 d^{-1} = c.$$

Multiplying from the right by d yields the equation of a line parallel to the original line. $(0, 0)$ remains fixed. Clearly D is one-to-one, and thus is a dilatation as defined in Section 3.2.

To prove Axiom A.4 we can restrict ourselves to a discussion of transitivity on the x_1-axis (by parallels, it carries over to other lines through $(0, 0)$, by a translation to other centers). Let $(p, 0)$ and $(q, 0)$ be unequal to $(0, 0)$. The dilatation

$$(x_1, x_2) \longrightarrow (x_1 p^{-1} q, x_2 p^{-1} q)$$

maps the first point onto the second. One can prove A.4′ similarly (see Exercise 2). ◆

Affinities have been defined as one-to-one mappings of the set of all points onto itself such that lines are mapped onto lines. We represent a large class of affinities algebraically.

Theorem 3.8.4 *A linear mapping*

$$(x_1, x_2) \longrightarrow (u_1 x_1 + u_2 x_2, v_1 x_1 + v_2 x_2) \tag{9}$$

represents an affinity if and only if

$$u_1 \neq 0 \quad \text{and} \quad v_2 - v_1 u_1^{-1} u_2 \neq 0$$

or

$$u_2 \neq 0 \quad \text{and} \quad v_1 - v_2 u_2^{-1} u_1 \neq 0.$$

Proof By Theorem 3.8.1, (9) is a one-to-one mapping if and only if one of the conditions of Theorem 3.8.4 is satisfied. Lines are clearly mapped onto lines. ◆

In addition to the transitivity properties A4 and A4′ the following theorem provides another transitivity property.

Theorem 3.8.5 *Given two pairs P, Q and P′, Q′ of different points such that 0 is neither on PQ nor on P′Q′, there exists an affinity leaving 0 fixed and mapping*

$$P \longrightarrow P',$$
$$Q \longrightarrow Q'.$$

Proof Let $P = (p_1, p_2)$, $Q = (q_1, q_2)$. We assume first that $P' = (1, 0)$, $Q' = (0, 1)$.

By Theorem 3.8.4, it suffices to solve the systems

$$u_1 p_1 + u_2 p_2 = 1 \qquad v_1 p_1 + v_2 p_2 = 0$$
$$u_1 q_1 + u_2 q_2 = 0 \qquad v_1 q_1 + v_2 q_2 = 1 \tag{10}$$

simultaneously in u_1, u_2, v_1, v_2 such that either

$$u_1 \neq 0 \quad \text{and} \quad v_2 - v_1 u_1^{-1} u_2 \neq 0$$

or

$$u_2 \neq 0 \quad \text{and} \quad v_1 - v_2 u_2^{-1} u_1 \neq 0.$$

If $p_1 = 0$ or $p_2 = 0$, the existence of such a solution can be verified directly. So let $p_1 \neq 0$ and $p_2 \neq 0$. We have

$$p_1^{-1} q_1 \neq p_2^{-1} q_2,$$

since otherwise the dilatation

$$(x_1, x_2) \longrightarrow (x_1 p_1^{-1} q_1, x_2 p_1^{-1} q_1)$$

would map P onto Q, so that 0 would lie on PQ, contrary to assumption. Therefore, $p_2 p_1^{-1} q_1 - q_2 \neq 0$, and, by Theorem 3.8.2, the systems (10) possess unique solutions u_1, u_2 and v_1, v_2, where u_1, u_2 and v_1, v_2 are not both zero.

We assert that if $u_1 \neq 0$, then $v_2 - v_1 u_1^{-1} u_2 \neq 0$. Suppose this is not so. Then the upper right equation of (10) reads

$$v_1 p_1 + v_1 u_1^{-1} u_2 p_2 = 0.$$

Since $v_1 = 0$ would also imply $v_2 = 0$, we have $v_1 \neq 0$. Now

$$p_1 + u_1^{-1} u_2 p_2 = 0 \quad \text{or} \quad u_1 p_1 + u_2 p_2 = 0,$$

which contradicts the upper left equation of (10). Therefore, $v_2 - v_1 u_1^{-1} u_2 \neq 0$.

One can prove similarly that if $u_2 \neq 0$ then $v_1 - v_2 u_2^{-1} u_1 \neq 0$. Let $P' = (p_1', p_2')$, $Q' = (q_1', q_2')$ be any two different points. By what we just have proved, there exists an affinity mapping P', Q' onto $(1, 0)$, $(0, 1)$, respectively. The definition of an affinity implies that the inverse of an affinity is also an affinity and also that the product of two affinities is another affinity. Therefore, P, Q can be mapped via $(1, 0)$, $(0, 1)$ onto P', Q' by an affinity. ◆

Exercises

1. Show that any stretch with the coordinate axis b as axis and the coordinate axis a as a trace is represented by

$$(x_1, x_2) \longrightarrow (c x_1, x_2), \quad \text{where } c \neq 0.$$

(*Hint*: Use Equation (b) of Section 3.6.)

2. Prove Axiom A.4' explicitly.

3. Find an algebraic representation for a shear with $x_2 = 0$ as axis.

4. Find an algebraic representation for an arbitrary dilatation.

5. Find an algebraic representation for an arbitrary stretch.

6. Describe how the successive application of two of the above affinities (dilatations, translations, stretches, shears) is expressed algebraically.

3.9   Dilatations and Desargues' Theorem

We are now going to replace the transitivity Axiom A.4 by an incidence property of points and lines:

(D) Desargues' theorem (affine version) *Suppose each vertex of a triangle is joined by a line to a vertex of a second triangle such that these lines meet in a point. If two pairs of sides of the triangles are properly parallel, the same is true for the third pair* (Fig. 3.22).

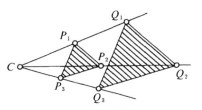

Figure 3.22

Theorem 3.9.1 *An affine plane satisfies Axiom* A.4 *if and only if* (D) *is true.*

Proof I. Let Axiom A.4 be satisfied and let $\triangle P_1P_2P_3$, $\triangle Q_1Q_2Q_3$ be the triangles under consideration such that P_1Q_1, P_2Q_2 and P_3Q_3 meet in a point C. Suppose P_1P_2 is properly parallel to Q_1Q_2 and P_1P_3 is properly parallel to Q_1Q_3. Clearly, C is different from all P_i and Q_i ($i = 1, 2, 3$), since, otherwise, P_1P_2 and Q_1Q_2 or P_1P_3 and Q_1Q_3 would coincide, contradicting the assumption that these lines are properly parallel. By Axiom A.4, there exists a dilatation with center C that maps P_1 onto Q_1. By definition of a dilatation, P_1P_2 is mapped onto Q_1Q_2 and P_1P_3 is mapped onto Q_1Q_3. Hence P_2, P_3 are mapped onto Q_2, Q_3, respectively, and thus P_2P_3 is properly parallel to Q_2Q_3.

II. Let (D) be true. Given three collinear points C, P, P', where $C \neq P, P'$, we are going to construct a dilatation with center C that maps P onto P'. Let X be a point not on CP. Then we define X' as the intersection of CX and the parallel of PX through P' (compare Section 3.3, Fig. 3.7). If Y is neither on CP nor on CX, then we define Y' also as the intersection of CY and the parallel of PY through P'. If we replace P, P' by X, X', when constructing the image of Y, we again obtain, by (D), the point Y'. If Z lies on CP and is not equal to C, we define Z' correspondingly with the aid of X, X' or Y, Y'. By (D) the result is again the same (see Fig. 3.23). Setting $C' = C$ we have found a mapping D,

$$X \longrightarrow X',$$

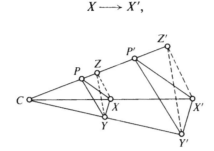

Figure 3.23

that is clearly a one-to-one mapping of the set of all points onto itself. By construction, D maps every line a onto a parallel of a. Since D has a fixed point, namely C, it is a dilatation. It satisfies $(P)D = P'$. ◆

It can be shown that Desargues' little theorem, which follows, is related to translations in the same way that (D) is related to dilatations (Exercise 1).

(d) Desargues' little theorem (affine version) *Suppose each vertex of a triangle is joined by a line to a vertex of a second triangle, such that these three lines are properly parallel. If two pairs of sides of the triangle are properly parallel, the same is true for the third pair* (Fig. 3.24).

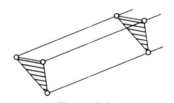

Figure 3.24

Exercises

1. Prove that in an affine plane each group $\mathscr{T}_a$ of translations (see Theorem 3.2.3) is transitive on a if and only if (d) is true (compare Theorem 3.5.1).

2. Is (D) true in the four-point plane?

3. Prove from (D) the converse of Desargues' theorem: Suppose $\triangle P_1P_2P_3$ and $\triangle Q_1Q_2Q_3$ are triangles with different vertices. If P_iP_j is properly parallel to Q_iQ_j for all pairs $i \neq j$, then P_1Q_1, P_2Q_2, P_3Q_3 meet in a point or P_1Q_1, P_2Q_2, P_3Q_3 are parallel.

4. Prove from (D) the "butterfly theorem": Let ABCD and A'B'C'D' be quadrangles whose vertices alternate between two lines a, b, as is indicated in Figure 3.25. If three pairs of respective sides of the quadrangles are parallel, so is the fourth.

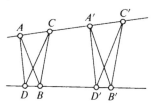

Figure 3.25

5. Show that (D) implies (d).

6. State and prove a converse of the "butterfly theorem" (See Exercise 4).

3.10   *Stretches and the Dual Theorem of Desargues*

In Section 3.3 we explained a relationship that exists between dilatations and stretches. Essentially, one has to interchange "points" and "lines" to proceed from one to the other. This relationship also becomes visible in a transition from Desargues' theorem to a "dual" theorem: If one replaces, in Figure 3.22, the points C, P_1, P_2, P_3, Q_1, Q_2, Q_3, by the lines c, p_1, p_2, p_3, q_1, q_2, q_3, respectively, and replaces the relation "to be on a line" by "pass through a point," one obtains Figure 3.26. The new theorem is stated below (in its converse form; compare Exercise 3 of Section 9).

(D*) *Dual converse theorem of Desargues (affine version) Suppose the vertices of two triangles with different vertices are joined by three properly parallel lines that provide a correspondence between these triangles. If two corresponding pairs of side lines intersect on a line c, the remaining two sides either intersect on c also, or are both parallel to c.*

Theorem 3.10.1 *In an affine plane, Axiom A.4' is true if and only if (D*) holds.*

Proof I. Let Axiom A.4' be true, and let p_1, p_2, p_3 be the sides of one triangle and let q_1, q_2, q_3 be the sides of the other, such that (in the terminology of Section 3.3) the line u joining p_1p_2 and q_1q_2, the line v joining p_1p_3 and q_1q_3, and the line w joining p_2p_3 and q_2q_3 are all properly parallel. Furthermore, assume that q_1p_1 and q_2p_2 are points on a line c. If c were

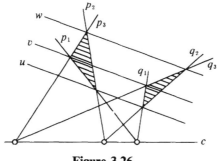

Figure 3.26

$= p_1$ or q_1, the lines p_1, p_2, p_3 or q_1, q_2, q_3 would coincide and, hence, would not form a triangle. So, by Axiom A.4′, there exists a stretch or shear Š with axis c and traces u, v, w mapping p_1 onto q_1. Since $p_1 p_2 = p_1 u$, the image of $p_1 p_2$ is $q_1 q_2$. Similarly,

$$(p_1 p_3)\check{S} = q_1 q_3.$$

Since $(p_2 c)\check{S} = (q_2 c)\check{S}$, it follows that $(p_2)\check{S} = q_2$. This implies $(p_2 p_3)\check{S} = q_2 q_3$, since $p_2 p_3 = p_2 w$ and $q_2 q_3 = q_2 w$. Hence $(p_3)\check{S} = q_3$. If p_3 and q_3 are not parallel, they intersect in a point $p_3 q_3$, which is fixed under Š and thus is on c. If p_3 and q_3 are parallel, and if p_3 intersects c, then cp_3 is a fixed point of Š, and q_3 passes through cp_3 so $p_3 = q_3$ follows, which is a contradiction. Therefore, p_2, q_3 and c are parallel.

II. Assume now that (D*) is true. Let c, p, p' pass through a point. We construct a stretch Š with axis c and a prescribed trace t not parallel to c, such that $(p)\check{S} = p'$. (If t does not exist, A.4′ is trivially true.) Let

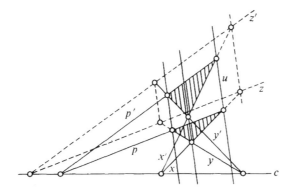

Figure 3.27

x be a line not passing through cp and not parallel to p. We draw a parallel u of t through px. The image x' of x is now defined as the line joining cx and up' (see Fig. 3.27). If y is a line not parallel to p or x, we can apply an analogous construction in order to obtain y'. (D*) guarantees that, in this construction, p, p' can be replaced by x, x'; the result remains the same. If z is parallel to p (Fig. 3.27), we find z' by using x, x' or y, y', obtaining again the same result. Finally, we set $c' = c$. Now

$$x \longrightarrow x'$$

is a one-to-one mapping of the set of all lines onto itself. Points of intersection are mapped onto points of intersection. Therefore, we have defined an affinity $\check{S}$ which leaves c pointwise fixed and possesses all parallels of t as traces. So $\check{S}$ is a stretch satisfying

$$(p)\check{S} = p'. \qquad \blacklozenge$$

Exercises

1. Illustrate (D*) by the following experiment: Fix three strings of rubber on two sticks, such that one stick serves as axis c in Figure 3.26 and the strings as lines p_1, p_2, p_3. Draw Figure 3.26 so that the rubber strings are placed onto p_1, p_2, p_3. Pull the second stick in such a way that p_1, p_2, p_3 are shifted onto q_1, q_2, q_3, respectively.

2. State the analogue (d*) of (D*) with c parallel to t and prove a transitivity theorem for shears corresponding to that for translations in Exercise 1, Section 3.9, (Fig. 3.28).

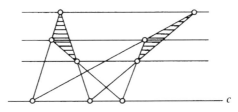

Figure 3.28

3. Relax the condition in (D*) that any two pairs of corresponding sides of the triangles intersect. Find the new case of (D*) and relate it to (d) in Section 3.9.

4. State and prove the dual theorem of Desargues, assuming (D*) to be true.

5. Let two properly parallel lines be given and a point P lying on neither of them. Find a parallel of the lines through P by using only the following two operations: (a) join two points by a line, (b) intersect nonparallel lines.

6. Prove (D*) algebraically in an affine plane over a skew field. (*Hint:* Introduce suitable coordinates.)

3.11 Equivalence of Desargues' Theorem and Its Dual

Theorem 3.11.1 *In an affine plane,* (D) *implies* (D*).

[3.11]

Proof

*Affine Planes
and Numbers*

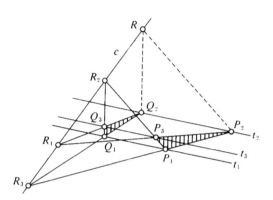

Figure 3.29

Let $\triangle P_1 P_2 P_3$ and $\triangle Q_1 Q_2 Q_3$ be the triangles occurring in (D*) (Fig. 3.29). We can assume $P_i \neq Q_i$ ($i = 1, 2, 3$), since otherwise (D*) becomes trivial. We denote by R_3 the intersection of $P_1 P_2$ and $Q_1 Q_2$, by R_2 the intersection of $P_1 P_3$ and $Q_1 Q_3$. Furthermore, we set $c = R_2 R_3$. If $P_2 P_3$ and $Q_2 Q_3$ are both parallel to c, then (D*) is true. So we can assume that $P_2 P_3$ or $Q_2 Q_3$, say $P_2 P_3$, intersects c in a point R_1. Finally, we draw a parallel of $P_1 P_3$ through P_2 and intersect it with c in a point R. We apply (D) to $\triangle Q_1 R_2 P_1$ and $\triangle Q_2 R P_2$. Then $Q_1 R_2$ is seen to be parallel to $Q_2 R$. Let Q_3' be the intersection of $R_1 Q_2$ and $P_3 Q_3$. If we apply (D) to $\triangle R P_2 Q_2$ and $\triangle R_2 P_3 Q_3'$, then it follows that $R_2 Q_3'$ is parallel to $R Q_2$ and hence, by the Euclidean axiom of parallels, $R_2 Q_3 = R_2 Q_3'$. This, in turn, implies $Q_3' = Q_3$. Therefore, c, $P_2 P_3$, and $Q_2 Q_3$ meet in a point R_1. ◆

Theorem 3.11.2 *In an affine plane,* (D*) *implies* (D).

Proof In (D) let $\triangle P_1 P_2 P_3$ and $\triangle Q_1 Q_2 Q_3$ be the triangles. $P_i \neq Q_i$ ($i = 1, 2, 3$) follows from the assumption that $P_1 P_2$ is properly parallel to $Q_1 Q_2$ and that $P_1 P_3$ is properly parallel to $Q_1 Q_3$. Suppose $P_2 P_3$ were not parallel to $Q_2 Q_3$. Then $P_2 P_3$ and $Q_2 Q_3$ would intersect in a point S, which, evidently, is not on $P_1 P_3$ or $Q_1 Q_3$ (draw a "false" figure!). Let l be the parallel of $P_1 P_3$ through S, and let l and $P_1 P_2$ intersect in a point T. Applying (D*) to the $\triangle S P_3 Q_3$ and $\triangle T P_1 Q_1$ shows that $S Q_3$ and $T Q_1$ meet in a point of $P_2 Q_2$, hence in Q_2. This implies $T Q_1 = Q_1 Q_2$, which contradicts our assumption that $P_1 P_2$ and $Q_1 Q_2$ are parallel. Therefore, $P_2 P_3$ is parallel to $Q_2 Q_3$. ◆

Theorem 3.11.3 In an affine plane, the transitivity axioms A.4 and A.4′ imply each other.

The proof follows from Theorems 3.9.1, 3.10.1, 3.11.1, and 3.11.2. The chain of conclusions can be illustrated by the following diagram ($\Leftrightarrow$ means "implies and is implied by"):

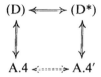

An affine plane in which (D) (and therefore also (D*)) holds is called a *Desarguesian affine plane*, which is identical to a transitive affine plane:

Desarguesian affine plane = transitive affine plane.

The following theorem is a fundamental result of our investigation.

Theorem 3.11.4 An affine plane can be coordinatized by a skew field if and only if it satisfies Desargues' theorem.

Since any skew field can be used to construct a transitive affine plane (see Section 3.8), we have the following one-to-one correspondence:

Desarguesian planes $\longleftrightarrow$ Skew fields.

Exercises

1. Using the characterization of multiplication given in Exercise 6, Section 3.6, deduce directly from the "butterfly theorem" of Exercise 4, Section 3.9 that $(A \cdot B) \cdot C = A \cdot (B \cdot C)$.

2. Show that in a Desarguesian affine plane the following theorem is true: Let P, Q, R, S be four points no three of which are collinear, and let P', Q', R', S' be four points no three of which are collinear. If the lines PQ, PR, PS, QR, QS are parallel to $P'Q'$, $P'R'$, $P'S'$, $Q'R'$, $Q'S'$, respectively, then RS is parallel to $R'S'$.

3. Carry out explicitly the proof of Theorem 3.11.4.

4. In a Desarguesian plane, find two shears the product of which is a translation.

5. In a Desarguesian plane, find three shears $\bar{S}_1$, $\bar{S}_2$, $\bar{S}_3$ the product, $\bar{S}_1\bar{S}_2\bar{S}_3$, of which is a reflection in a point (dilatation with "factor" -1). (*Hint:* Introduce coordinates and find shears with axes through $(0, 0)$ such that

$$(0, 1) \xrightarrow{\bar{S}_1} (-1, 1) \xrightarrow{\bar{S}_2} (-1, -3) \xrightarrow{\bar{S}_3} (0, -1).)$$

6. Prove that the set of all shears with axes parallel to a given line a and the translations along a form a commutative group.

3.12   A Non-Desarguesian Plane

One may ask whether or not Desargues' theorem can be derived from the axioms of an affine plane—that is, whether or not *all* affine planes are Desarguesian. We shall show that this is not so by presenting an example of a "non-Desarguesian" plane.

As the set of points, we choose an ordinary Euclidean plane coordinatized by the field of real numbers. So we may consider all pairs (x_1, x_2) of real numbers to be points. Lines, however, are defined in a new way—namely, as a set of solutions of one of the following equations or system of equations:
If $a \geq 0$,

$$ax_2 + x_1 = b \tag{1}$$

and

$$x_2 = c. \tag{2}$$

If $a < 0$,

$$\begin{aligned} 2ax_2 + x_1 &= b \quad \text{for } x_2 \geq 0 \\ ax_2 + x_1 &= b \quad \text{for } x_2 < 0. \end{aligned} \tag{3}$$

(1) and (2) are lines in the original sense. However, (3) represents "lines" that are bent when they cross the x_1 axis. One easily verifies that all axioms of an affine plane are satisfied (see Exercise 1 of this section). Figure 3.30 illustrates that Desargues' theorem (in its converse form) is not true.

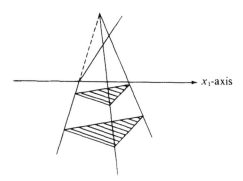

x_1-axis

Figure 3.30

It should be noted that from the axioms of a three-dimensional affine space (see Section 4.1) Desargues' theorem can very well be deduced. In the last two decades, non-Desarguesian planes have become a favorite subject of research and have unveiled nice relationships between algebraic and geometric facts.

Exercises

1. Show explicitly that the system of points and lines discussed in this section is an affine plane.

2. In the above non-Desarguesian plane, find all translations and dilatations.

3. Replace, in the list of axioms defining a skew field, the left distributive law by the following axiom: If $a \neq b$ and c is arbitrary, then $-ax + bx = c$ has a solution in x. A more general algebraic structure is obtained, called an **almost field** $\mathscr{F}$. An example is discussed in the next exercise. Show that one obtains an affine plane (non-Desarguesian!) if pairs (x_1, x_2) of elements of $\mathscr{F}$ are taken as points, and if lines are defined by

$$ax_2 + x_1 = b$$

and

$$x_2 = c.$$

4. Show that the following system is an almost field (See Exercise 3): Consider all linear polynomials $a + bj$ with variable j and $a, b \in \mathscr{F}$. Define addition and multiplication as follows:

$$(a + bj) + (c + dj) = (a + c) + (b + d)j$$
$$(a + bj)(c + dj) = (ac + bd) + (bc - ad)j$$

So, in particular, $j^2 = -1$. Show that all laws of a skew field except the left distributive law $(u + v)w = uw + vw$ are satisfied.

5. In Exercise 4, if multiplication is defined as

$$(a + bj)(c + dj) = (ac - bd) + (ad + bc)j$$

show that all laws of a field are satisfied. (Complex numbers over $\mathscr{F}$.)

6. Are any two affine planes with the same finite number of points isomorphic in the sense that their points can be related by an incidence-preserving one-to-one mapping?

3.13 Commutativity and Pappus' Theorem

In addition to Desargues' theorem, another closure theorem plays an important role in affine (and projective) geometry. Two of its features interest us most at present: It implies Desargues' theorem and it turns the coordinate skew field into a field—that is, it makes the multiplication in the skew field commutative. Since there exist skew fields that are not fields (see

Exercise 6 upcoming), we conclude that Desargues' theorem does *not* imply the new theorem, called Pappus' theorem.

(P) *Pappus' theorem (affine version)*   Let P, Q, R lie on a, and let $\bar{P}$, $\bar{Q}$, $\bar{R}$ lie on b, such that a, b do not intersect in one of these six points.   If $P\bar{Q}$ is properly parallel to $\bar{P}Q$ and if $Q\bar{R}$ is properly parallel to $\bar{Q}R$, then $P\bar{R}$ is parallel to $\bar{P}R$ (Fig. 3.31).

[3.13]

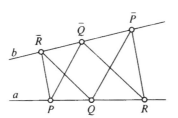

Figure 3.31

Theorem 3.13.1   *Pappus' theorem* (P) *implies that all groups* $\mathcal{D}_C$ *of dilatations are commutative.*

Proof   Let D map P onto Q and let D' map Q onto R, where D, D' $\in \mathcal{D}_C$ and both different from the identity map.   If $\bar{P}$, $\bar{Q}$, $\bar{R}$ are chosen as in Figure 3.31 on a line through C, then clearly D' maps $\bar{R}$ onto $\bar{Q}$ and D maps $\bar{Q}$ onto $\bar{P}$.   Since, by (P), $P\bar{R}$ and $\bar{P}R$ are parallel, it follows that DD' and D'D both map $\bar{R}$ onto $\bar{P}$.   By Theorem 3.2.4, this implies that

$$DD' = D'D. \qquad \blacklozenge$$

Theorem 3.13.2   (*Hessenberg*) *Pappus' theorem,* (P), *implies Desargues' theorem,* (D).

Proof   Let $\triangle P_1P_2P_3$ and $\triangle Q_1Q_2Q_3$ be the triangles that occur in (D).   It can be assumed that $P_i \neq Q_i$ $(i = 1, 2, 3)$.   Denote P_1Q_1, P_2Q_2, P_3Q_3 by l_1, l_2, l_3, respectively, and let C be their intersection.   (See Fig. 3.32.) Suppose P_1P_2 is parallel to Q_1Q_2 and P_1P_3 is parallel to Q_1Q_3.   If P_2P_3 and Q_2Q_3 are both parallel to P_1Q_1, then they are parallel to each other and there is nothing to prove.   So we can assume that P_2P_3 or Q_2Q_3, say P_2P_3, is not parallel to P_1Q_1.   We draw a parallel of P_2P_3 through C; it intersects P_1P_3 in a point S, which is not on P_1Q_1.   Furthermore, a parallel of P_1Q_1 through P_3 intersects P_2Q_2 in a point R and SQ_1 in a point T.

The proof of (D) is now achieved in the following steps:

(a) In $SCP_1P_2P_3R$, it follows by (P) that SR is parallel to P_1P_2.
(b) In $SCQ_1Q_3TP_3$, it follows by (P) that SC is parallel to Q_3T.
(c) In SCQ_1Q_2TR, it follows by (P) that SC is parallel to Q_2T.

Then from $Q_3T = Q_2T = Q_2Q_3$ and the fact that SC is parallel to P_2P_3, we conclude that P_2P_3 is parallel to Q_2Q_3.   The main difficulty lies in checking all assumptions that are needed for applying (P) in (a), (b), (c):

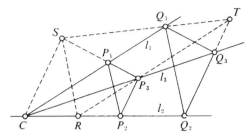

Figure 3.32

Ad (a): We must show that S, P_1, P_3 is a triplet of different points not on P_2Q_2 and C, R, P_2 is a triplet of different points not on P_1P_3. This is all clear from our construction. S in particular, cannot lie on P_2Q_2, since otherwise SC would pass through P_2 and not be parallel to P_2P_3.

Ad (b): If S were on P_3Q_3, then CS would pass through P_3 and hence would not be parallel to P_2P_3. Everything else is clear from the construction.

Ad (c): Suppose T is on P_2Q_2. Then from the fact that RP_3 is parallel to P_1Q_1, we conclude $T = R$. Hence $SQ_1 = SR$ and, since SR is parallel to P_1P_2 which is parallel to Q_1Q_2, it follows that $T = R = Q_2$. Now $Q_2Q_3 = TQ_3$ is, by (b), parallel to SC and hence, by construction, it is parallel to P_2P_3. So P_2P_3 parallel to Q_2Q_3 follows directly. If R or Q_2 is on SQ_1, then again $T = R$ follows and we conclude as above. All other cases are clear from the construction. ◆

Theorem 3.13.3 *If an affine plane can be coordinatized by a field, then Pappus' theorem is satisfied.*

Proof From the transitivity property A.4 of the groups $\mathcal{D}_C$, we can deduce Pappus' theorem by reversing the main steps in the proof of Theorem 3.13.1.
◆

An affine plane in which (P) holds is called a **Pappian plane**. We have the fundamental correspondence (analogous to that in Section 3.11):

$$\text{Pappian planes} \longleftrightarrow \text{fields.}$$

Exercises

1. In a Desarguesian plane if (P) is true for points P, Q, R on a prescribed line, prove that (P) is true in general.

2. Show that (P) follows from Desargues' theorem in the special case that the line joining $(P\bar{Q})(Q\bar{R})$ and $(\bar{P}Q)(\bar{Q}R)$ intersects a and b in the same point.

3. Let a Desarguesian plane over a skew field with $1 + 1 \neq 0$ be given. Show that (P) can be proved for the special case that a and b are parallel.

4. Carry out explicitly the proof of Theorem 3.13.3.

5. Find an alternative proof of Theorem 3.13.3 by using Figure 3.17.

6. Let a, b, c, d be arbitrary real numbers. Consider all linear polynomials $a + bi + cj + dk$ with variables i, j, k; they are called quaternions. Define addition and multiplication as follows.

$$(a + bi + cj + dk) + (a' + b'i + c'j + d'k)$$
$$= (a \overset{+}{} a') + (b + b')i + (c + c')j + (d + d')k$$
$$(a + bi + cj + dk) \cdot (a' + b'i + c'j + d'k)$$
$$= (aa' - bb' - cc' - dd')$$
$$+ (ab' + ba' + cd' - c'd)i + (ac' + a'c + db' - d'b)j$$
$$+ (ad' + a'd + bc' - b'c)k.$$

(So, in particular, $i^2 = j^2 = k^2 = -1$, $ij = k$, $jk = i$, $ki = j$, $ji = -k$, $kj = -i$, $ik = -j$.) Prove that the quaternions form a skew field which is not commutative.

3.14   *Isometries and Motions*

We assume the ordinary field $\mathscr{R}$ of real numbers to be given (for a definition of $\mathscr{R}$, see Appendix 3.III). In this section we shall show how, by the fundamental notion of distance, the affine plane over $\mathscr{R}$ can be turned into a Euclidean metric plane. In Appendices 3.I–3.III, additional properties of this plane are stated. Furthermore, it is shown there that, conversely, every Euclidean metric plane with these additional properties can be co-ordinatized by the field of real numbers. Here we wish to stress the following point. In the affine plane over $\mathscr{R}$, motions can be characterized as *isometries*, which are defined as distance-preserving affinities. We remark that this characterization is also possible for more general fields, though not for all fields defining Euclidean metric planes.

We define the *distance* between two points $X = (x_1, x_2)$ and $Y = (y_1, y_2)$ of the affine plane over $\mathscr{R}$ by

$$d(X, Y) = \sqrt{(x_1 - y_1)^2 + (x_2 - y_2)^2},$$

the root being non-negative.

Theorem 3.14.1   *If X, Y is mapped onto X', Y', respectively, by a translation, then*

$$d(X, Y) = d(X', Y').$$

Proof Let

$$X' = (x_1', x_2') = (x_1 + c, x_2 + d)$$

and

$$Y' = (y_1', y_2') = (y_1 + c, y_2 + d).$$

Then

$$d(X', Y') = \sqrt{(x_1 + c - y_1 - c)^2 + (x_2 + d - y_2 - d)^2}$$
$$= \sqrt{(x_1 - y_1)^2 + (x_2 - y_2)^2}$$
$$= d(X, Y). \qquad \blacklozenge$$

Theorem 3.14.2 *The distance function d has the following properties:*

(1) $d(X, Y) = d(Y, X)$,

(2) $d(X, Y) = 0$ if and only if $X = Y$,

(3) $d(X, Y) + d(Y, Z) \geq d(X, Z)$ for any X, Y, Z.

Proof (1) and (2) are clear by definition. (3) is seen as follows: Because of Theorem 3.14.1, we may assume $Y = (0, 0)$. So we must show

$$\sqrt{x_1^2 + x_2^2} + \sqrt{z_1^2 + z_2^2} \geq \sqrt{(x_1 - z_1)^2 + (x_2 - z_2)^2}.$$

We begin with the trivial inequality

$$(x_1 z_2 - x_2 z_1)^2 \geq 0.$$

Squaring out and adding

$$x_1^2 z_1^2 + 2\, x_1 x_2 \cdot z_1 z_2 + x_2^2 z_2^2 = (x_1 z_1 + x_2 z_2)^2$$

on both sides, we obtain

$$x_1^2 z_1^2 + x_1^2 z_2^2 + x_2^2 z_1^2 + x_2^2 z_2^2 = (x_1^2 + x_2^2)(z_1^2 + z_2^2) \geq (x_1 z_1 + x_2 z_2)^2,$$

from which it follows that

$$\sqrt{x_1^2 + x_2^2}\, \sqrt{z_1^2 + z_2^2} \geq -x_1 z_1 - x_2 z_2;$$

and thus

$$(\sqrt{x_1^2 + x_2^2} + \sqrt{z_1^2 + z_2^2})^2 \geq (x_1 - z_1)^2 + (x_2 - z_2)^2.$$

From this inequality, the assertion (3) is easily concluded. $\blacklozenge$

Theorem 3.14.3 *Three points are collinear if and only if they can be denoted by X, Y, Z such that*

$$d(X, Y) + d(Y, Z) = d(X, Z).$$

Proof We may assume $Y = (0, 0)$ and also that $z_1 \neq 0$. As is seen from the proof of Theorem 3.14.2, the equality sign in (3) holds if and only if

$$x_1 z_2 - x_2 z_1 = 0$$

or

$$x_2 = \frac{z_2}{z_1} x_1.$$

If x_1, x_2 are considered as variables, this equation represents a line through $(0, 0)$, (z_1, z_2), (x_1, x_2). $\blacklozenge$

We call two lines a, b **perpendicular** if they intersect in a point P and if there exist points $A \neq P$ and $B \neq P$ on a and b, respectively, such that

$$d^2(A, P) + d^2(P, B) = d^2(A, B)$$

(we use $d^2(X, Y)$ instead of $(d(X, Y))^2$).

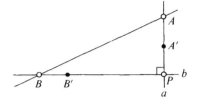

Figure 3.33

Theorem 3.14.4 *If a and b intersect in P and are perpendicular, then for any two points A', B', on a, b, respectively,*

$$d^2(A', P) + d^2(P, B') = d^2(A', B').$$

Proof It is sufficient to prove the assertion for $P = (0, 0)$. We may set

$$A' = (a'_1, a'_2) = (a_1 c, a_2 c)$$

and

$$B' = (b'_1, b'_2) = (b_1 e, b_2 e).$$

From

$$a_1^2 + a_2^2 + b_1^2 + b_2^2 = (a_1 - b_1)^2 + (a_2 - b_2)^2,$$

we obtain

$$a_1 b_1 + a_2 b_2 = 0$$

as a necessary and sufficient condition for the lines AP and BP to be perpendicular. So also

$$a'_1 b'_2 + a'_2 b'_2 = ce(a_1 b_1 + a_2 b_2) = 0,$$

from which the assertion is readily deduced. ◆

Theorem 3.14.5 *Any two perpendiculars of a given line are parallel.*

Proof Suppose two perpendiculars, p, p' of a intersect each other in P and intersect a in points A, A'. Then

$$d^2(P, A) = d^2(P, A') + d^2(A, A') = d^2(P, A) + 2d^2(A, A')$$

from which we conclude $d(A, A') = 0$; hence $A = A'$.

If $A \neq P$ we obtain $p = p' = AP$. If $A = P$, suppose $p \neq p'$. Then a line can be found which intersects p, p', a in different points, Q, Q', B, respectively. Q' can be assumed to lie between Q and B. By Theorem 3.14.4 and Theorem 3.14.3 we have

$$d^2(Q', A) + d^2(A, B) = d^2(Q', B)$$

and

$$d^2(Q, A) + d^2(A, B) = (d(B, Q') + d(Q', Q))^2;$$

hence

$$d^2(Q, A) = d^2(Q', A) + 2d(Q', B)d(Q, Q') + d^2(Q, Q')$$
$$\geq d^2(Q', A) + 2d(Q', A)d(Q, Q') + d^2(Q, Q')$$
$$= (d(Q', A) + d(Q, Q'))^2$$

Therefore, $d(Q, A) \geq d(Q', A) + d(Q, Q')$ where, by Theorem 3.14.2, the equality sign holds. So, by Theorem 3.14.3, the points Q, Q', A are collinear; hence $A = B$, which is a contradiction. ◆

Theorem 3.14.6 *Given a line a and a point P, there exists one and only one line b passing through P and perpendicular to a.*

Proof Again we may assume $P = (0, 0)$. First, let a pass through P, and let $A = (a_1, a_2) \neq (0, 0)$ lie on a. We easily find a point $B = (b_1, b_2) \neq (0, 0)$ such that $a_1 b_1 + a_2 b_2 = 0$, so that PA is perpendicular to PB (see proof of Theorem 3.14.4).

If a does not pass through P, consider the parallel a' of a through P and the perpendicular b of a' through P. By Theorem 3.14.5, a coincides with the perpendicular of b at the point where a and b intersect. The uniqueness of the perpendicular b follows immediately from Theorem 3.14.5.

◆

Now we define a **reflection** R_a at a line a as an involutoric stretch with axis a and traces perpendicular to a.

Theorem 3.14.7 *Reflections are distance preserving.*

Proof Let P be an arbitrary point and let A be the foot of P on a (Fig. 3.34.) There exists a translation T mapping A onto $O = (0, 0)$. $T^{-1}R_a T$ is

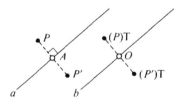

Figure 3.34

clearly again a reflection R_b with $b = (a)T$ passing through O. The image of $(P)T$ under R_b is obtained by replacing the coordinates of $(P)T$ by their negatives. Hence

$$d(P, A) = d((P)T, O) = d((P)TR_b, O) = d((P)TR_b T^{-1}, A) = d((P)R_a, A).$$

Let Q be a point $\neq P$ such that PQ intersects a in a point R. Denote the foot of Q on a by B and set $(P)R_a = P'$, $(Q)R_a = Q'$. We may assume

$$d(P, Q) + d(Q, R) = d(P, R).$$

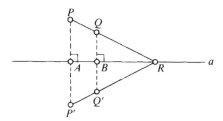

Figure 3.35

Then

$$d(P, R) = \sqrt{d^2(P, A) + d^2(A, R)} = \sqrt{d^2(P', A) + d^2(A, R)} = d(P', R).$$

Similarly, $d(Q, R) = d(Q', R)$; hence

$$d(P, Q) = d(P, R) - d(Q, R) = d(P', R) - d(Q', R) = d(P', Q').$$

The proof is readily extended to the case that PQ is parallel to a. ◆

The next theorem is a consequence of Theorem 3.14.7.

Theorem 3.14.8 *Reflections map perpendicular lines onto perpendicular lines.*

So the name "reflection" is justified in the sense of Section 1.3. It is readily seen that any two lines can be mapped onto each other by a reflection and that the following theorem is true.

Theorem 3.14.9 *The affine plane over $\mathcal{R}$ is a Euclidean metric plane.*

Every product of finitely many reflections is called a ***motion***. A distance-preserving affinity is called an ***isometry***.

Theorem 3.14.10 *Every motion is an isometry, and every isometry is a motion.*

Proof The first part of this theorem is clear by Theorem 3.14.7 and the definition of motions. Let, therefore, an isometry $\mathcal{I}$ be given, and let P', Q', R' be the respective images of three noncollinear points P, Q, R. We map P onto P' by a translation T (which obviously can be represented as the product of two reflections). We apply a reflection R_a at the perpendicular $a(Q)TQ'$ through P', so that $(Q)R_a = Q'$. If also $(R)R_a = R'$, we set $TR_a = M$. Otherwise, we apply a further reflection at $b = PQ'$ and set $TR_aR_b = M$. In both cases M is a motion mapping P, Q, R onto P', Q', R', respectively. We assert that M is the isometry $\mathcal{I}$, or, equivalently, $\mathcal{I}M^{-1}$ is the identity map. Otherwise $S \neq S^*$ for points S and $S^* = (S)\mathcal{I}M^{-1}$, both S and S^* having the same distance from each of the lines P, Q, and R. This readily implies $S = S^*$, contradicting the assumption $S \neq S^*$. ◆

Exercises

1. List again all axioms of a Euclidean metric plane and check all details in the proof of Theorem 3.14.9.

2. Define half-turns by using the distance concept. (Prove the mapping to be a half-turn!)

3. Prove directly by using the distance concept that if an isometry has two fixed points, it is either the identity map or a reflection.

4. Can the results of this section also be proved for the affine plane over the field of rational numbers?

5. A function d defined on all pairs of a set M is called a **distance function** if all its values are non-negative real numbers and if it satisfies properties (1)–(3) of Theorem 3.14.2. M is then called a **metric space**. Which of the following "trivial" functions turn any set M into a metric space?

   (a) $d_1(X, Y) = 0$ for any two elements X, Y of M,

   (b) $d_2(X, Y) = \begin{cases} 0 & \text{if } X = Y \\ 1 & \text{if } X \neq Y. \end{cases}$

6. The affine plane over $\mathscr{R}$ can also be made a metric space (see Exercise 5) by introducing the following distance function:

   $$d(X, Y) = |x_1 - y_1| + |x_2 - y_2|.$$

   Prove properties (1)–(3) of Theorem 3.14.2 and find the "unit circle" under this distance function—that is, the set of all points X for which $d(X, O) = 1$.

7. Is Theorem 3.14.3 true for the distance function of Exercise 6?

3.15 Fields Constructed by Ruler and Compass

Let the affine plane over the field $\mathscr{R}$ of real numbers be given. We assume the coordinate axes c_1, c_2 to be perpendicular to each other. For the purposes of this section, we need the concept of a **subfield** of a field $\mathscr{F}$, which is defined as a subset $\mathscr{F}_0$ of $\mathscr{F}$ that is a field under the operations of $\mathscr{F}$. Subfields of $\mathscr{R}$ include the field $\mathscr{R}_0$ of all rationals (fractions). $\mathscr{R}$, in turn, is a subfield of the field of all complex numbers.

Let $\mathscr{R}_0$ be given. If $r_1, r_2 \in \mathscr{R}_0$, we call the point

$$P = (r_1, r_2)$$

a **rational point**. If $Q = (s_1, s_2)$ is another rational point, the line

$$-\frac{r_1 - s_1}{r_2 - s_2} x_2 + x_1 = -\frac{r_1 - s_1}{r_2 - s_2} r_2 + r_1, \quad \text{if } r_2 \neq s_2, \tag{1}$$

or the line

$$x_2 = r_2, \quad \text{if } r_2 = s_2, \tag{2}$$

joins P and Q. It is, in any case, a linear equation with rational coefficients. We call a line represented by such a linear equation a **rational line.** Clearly, the intersection of two nonparallel rational lines is a rational point. So we find, in terms of constructions, that the family of all rational points and lines is closed under constructions by the ruler only.

An interesting observation lies in the fact that, given five rational points in a sufficiently general position, we can construct any rational point from these five points by using the ruler only (Exercise 1 of this section).

Let now a number

$$\sqrt{d_1}$$

be given, with rational d_1 such that $d_1 > 0$ and $\sqrt{d_1}$ is not rational. By applying the elementary operations of algebra to

$$\mathscr{R}_0 \cup \{\sqrt{d_1}\},$$

we obtain numbers of the form

$$\frac{r + s\sqrt{d_1}}{t + u\sqrt{d_1}}, \quad r, s, t, u \text{ rational}; \ t, u \text{ not both } 0.$$

Checking all defining properties of a field shows immediately that these numbers form a subfield

$$\mathscr{R}_{d_1}$$

of $\mathscr{R}$. We obtain the inclusion

$$\mathscr{R}_0 \subset \mathscr{R}_{d_1} \subset \mathscr{R}.$$

We can consider $\sqrt{d_1}$ to be constructed by drawing a circle about O with a rational radius $1/2(d_1 + 1)$ and intersecting this circle with the line $x_2 = 1/2(d_1 - 1)$:

$$x_1^2 + x_2^2 = 1/4(d_1 + 1)^2$$
$$x_1 = 1/2(d_1 - 1)$$

Applying such a compass construction a second time, we can find a real number d_2 not contained in $\mathscr{R}_{d_1}$. Applying all elementary rules of algebra to

$$\mathscr{R}_{d_1} \cup \{d_2\},$$

we obtain a field $\mathscr{R}_{d_2}$ that contains $\mathscr{R}_{d_1}$ as a subfield. Continuing in this way, we obtain a chain of fields

$$\mathscr{R}_0 \subset \mathscr{R}_{d_1} \subset \mathscr{R}_{d_2} \subset \cdots \subset \mathscr{R}_{d_{n-1}} \subset \mathscr{R}_{d_n} \subset \cdots, \tag{1}$$

such that $\mathscr{R}_{d_{n-1}}$ is contained in $\mathscr{R}_{d_n}$, $n = 1, 2, 3 \cdots$, and such that every d_n is a square root of some element of $\mathscr{R}_{d_{n-1}}$. So $\mathscr{R}_{d_n}$ contains elements like

$$1 + \sqrt{5 - \sqrt{\tfrac{1}{2} + 3\sqrt{2}}} \quad (n \geq 3).$$

Since $\mathcal{R}_0$ has only a countably infinite number of elements, we can choose the sequence $d_1, d_2, \cdots$ in such a way that every number obtained from a rational number by applying finitely often rational operations and drawing square roots belongs to some $\mathcal{R}_{d_n}$.

Theorem 3.15.1 *If a number y is obtained from rational points through constructions by ruler and compass, then there exists a chain (1) such that y lies in $\mathcal{R}_{d_n}$ but not in $\mathcal{R}_{d_{n-1}}$.*

We denote by $\mathcal{R}_{\surd}$ the family of all numbers that belong to at least one $\mathcal{R}_{d_n}$. Clearly $\mathcal{R}_{\surd}$ does not depend on the special choice of (1). Checking again all defining rules of a field yields that $\mathcal{R}_{\surd}$ is a field.

Theorem 3.15.2 *The set of all numbers constructed from the rationals by using ruler and compass is a field $\mathcal{R}_{\surd}$.*

This theorem can also be verified directly by showing that if x and y can be constructed by ruler and compass, so can $x + y$, $x - y$, $x \cdot y$, and x/y ($y \neq 0$). In figure 3.36, we indicate how $1/y$ is constructed from y. (Remember from Section 2.16 that dropping perpendiculars and drawing parallels of a line through given points can be carried out by using ruler and compass.)

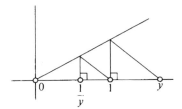

Figure 3.36

Exercises

1. Show that all rational points can be constructed by ruler alone if the points $(0, 0)$, $(0, 1)$, $(1, 0)$, $(1, 1)$, and $(2, 1)$ are given. Construct explicitly the point $(7, 3)$.

2. Construct $x + y$, $x - y$, $x \cdot y$, and x/y by using ruler and compass if x, y are given. Assume $x \neq y$; $y \neq 0$.

3. Show that all elements of the field $\mathcal{R}_{d_1}$ can be written in the form $a + b\sqrt{d_1}$, a, b rational.

4. Does the field of rational numbers have a proper subfield?

5. Find a field which has no proper subfield.

6. Show that the results of Section 3.14 can be obtained if $\mathcal{R}$ is replaced by $\mathcal{R}_{\surd}$.

3.16 Impossibility of the Trisection of the Angle and the Duplication of the Cube

For more than two thousand years, people have tried to find a general method for the trisection of the angle by ruler and compass and for the construction of a line segment of length $\sqrt[3]{2}$, that is, the edge of a cube whose volume is 2. Though it has been known for about one hundred years that neither problem has a solution, attempts are still made by amateur mathematicians to solve them. We shall present now an impossibility proof due to D. Laugwitz ([1962], p. 54), which avoids the more complicated machinery of modern algebra that is usually used in such proofs (compare also Moise, Chapter 19).

We assume you know the meaning of cos α and that the identity

$$\cos 3\alpha = 4 \cos^3 \alpha - 3 \cos \alpha$$

holds. Furthermore, we need some elementary algebraic tools: Given a cubic polynomial equation

$$a_3 x^3 + a_2 x^2 + a_1 x + a_0 = 0 \qquad (1)$$

with rational coefficients a_3, a_2, a_1, a_0, we prove the following theorem.

Theorem 3.16.1 *Equation* (1) *possesses a root that can be constructed from rationals by using ruler and compass if and only if it possesses a rational root.*

Proof If (1) possesses a rational root, this root can be constructed trivially even with a ruler alone. Let (1) have a root x_1 that can be constructed by ruler and compass. There is a field $\mathscr{R}_{d_n}$ as introduced in Section 3.15 such that x_1 is in $\mathscr{R}_{d_n}$ but not in $\mathscr{R}_{d_{n-1}}$. We can assume that among all roots of (1) that can be constructed by ruler and compass, x_1 possesses the smallest number n in $\mathscr{R}_{d_n}$. Then, in particular, no root of (1) is in $\mathscr{R}_{d_{n-1}}$. We assert that $n = 0$, which is to say that x_1 is rational. For a proof, suppose $n > 0$. Then

$$x_1 = a + b\sqrt{d}$$

for some a, b, d in $\mathscr{R}_{d_{n-1}}$, $\sqrt{d}$ not in $\mathscr{R}_{d_{n-1}}$, $b \neq 0$. Substituting in (1) shows that

$$x_2 = a - b\sqrt{d}$$

is also a root of (1), different from x_1.

By Viëta's theorem (which says that $x_1 + x_2 + x_3 = -a_2$; one easily proves it by expanding $(x - x_1)(x - x_2)(x - x_3)$), we have

$$x_3 = -x_1 - x_2 - a_2 = -2a - a_2$$

as the third root of (1), which is in $\mathscr{R}_{d_{n-1}}$, contrary to our assumption. Hence $n = 0$. ◆

We now multiply (1) by the common denominator of a_3, a_2, a_1, a_0 and obtain an equation

$$b_3x^3 + b_2x^2 + b_1x + b_0 = 0$$

with integer coefficients b_3, b_2, b_1, b_0. We multiply this equation by b_3^2 and substitute $y = b_3x$, obtaining the equation

$$y^3 + c_2y^2 + c_1y + c_0 = 0, \qquad (2)$$

c_2, c_1, c_0 also being integers.

Theorem 3.16.2 *All rational roots of* (2) *are integers that divide* c_0.

Proof Let m/n be a root of (2) in reduced form (m, n integers). Then

$$m^3/n = -c_0n^2 - c_2m^2 - c_1mn$$

is an integer; hence $n = \pm 1$, and thus m/n is an integer. Any integer root y of (2) divides c_0, since

$$c_0 = y(-y^2 - c_2y - c_1). \qquad \blacklozenge$$

We can always check in finitely many steps whether or not an equation (2) has an integer root. Therefore, Theorems 3.16.1 and 3.16.2 imply Theorem 3.16.3.

Theorem 3.16.3 *There exists a method that allows one to decide in finitely many steps whether* (1) *does or does not have a root that can be constructed by ruler and compass.*

The method consists of changing (1) into the form (2) and testing all divisors of c_0 to determine whether or not they are roots of (2).

Problem of the trisection of an angle:

If an angle θ is given, $\cos \theta$ can be considered to be given, too. So if

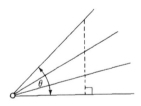

Figure 3.37

$\theta/3$ can be constructed by ruler and compass, the same is true for $\cos \theta/3$. Letting $x = 2 \cos \theta/3$, we conclude that (see the above identity)

$$x^3 - 3x - 2 \cos \theta = 0$$

must have an integer root for any θ. Choosing, in particular, $\theta = \pi/3$, we see that

$$x^3 - 3x - 1 = 0$$

must have an integer root. However, this is not so, since $+1$ and -1 are the only integer divisors of -1 and neither is a root. Therefore, $\pi/3$ cannot be trisected by ruler and compass.

Problem of the duplication of the cube:

The problem of the duplication of a cube leads directly to the question of whether or not

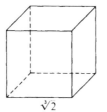

$\sqrt[3]{2}$

Figure 3.38

$$x^3 - 2 = 0$$

has an integer root. It has none, since ± 1 and ± 2 are not roots. So there is no possibility of constructing $\sqrt[3]{2}$ by ruler and compass.

Exercises

1. Prove explicitly Viëta's theorem (see proof of Theorem 3.16.1).

2. Show that $\sqrt[8]{9}$ cannot be constructed by ruler and compass.

3. Show that an isosceles triangle cannot always be constructed by ruler and compass if the angular bisector $b_\alpha = b_\beta$ and the altitude a_c are given as lengths (see Fig. 3.39). (*Hint:* Prove

$$c \sin \alpha = b_\alpha \sin \frac{3\alpha}{2} \quad \text{and} \quad c \sin \alpha = 2a_c \cos \alpha;$$

set $x = 2 \sin \alpha/2$. Choose $b_\alpha = 2$ and a_c a prime > 2.)

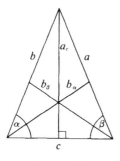

Figure 3.39

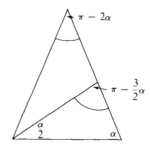

Figure 3.40

4. Show that an isosceles triangle cannot always be constructed by ruler and compass if $a = b$ and $b_\alpha = b_\beta$ are given (Fig. 3.40). (*Hint:* Prove

$$c \sin \alpha = b_\alpha \sin 3\alpha/2 \quad \text{and} \quad c \sin \alpha = a \sin 2\alpha.$$

Set $x = \sin^2 \alpha/2$; use Fig. 3.40.)

5. Deduce from Exercise 2 that there is no general method of constructing a (general) triangle by ruler and compass if the following triplet of pieces is given $\{b_\alpha, b_\beta, b_\gamma\}$ ($b_\alpha, b_\beta, b_\gamma$ are angular bisectors; see Fig. 3.41).

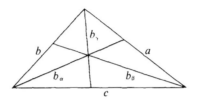

Figure 3.41

6. Repeat Exercise 5, replacing $\{b_\alpha, b_\beta, b_\gamma\}$ by $\{a, b, b_\alpha\}$.

Chapter Four

Affine Space and Vector Space

The first three chapters laid a foundation of planar geometry—both synthetic and analytic. We turn now to study a spatial geometry, beginning with a system of axioms that defines affine 3-space; we shall show how skew-field coordinates can be introduced in every such space. It is a surprising fact that Desargues' theorem can be deduced from very elementary axioms of 3-space. This is not so for Pappus' theorem. So the concept of a skew field, as introduced in Section 3.6 turns out to be an adequate algebraic concept for a general analytic treatment of affine (and also projective) geometry.

The main goal of this chapter, however, is to develop some of the very powerful tools of linear algebra by which a number of geometrical problems can be attacked elegantly. In particular, we shall develop some vector and matrix calculus. Linear algebra permits us to prove a large number of geometrical theorems simultaneously for planar and spatial geometry and also for geometry of higher dimensions. Nevertheless, we shall also deal with a number of specific three-dimensional facts.

4.1   Axioms of Affine Space and Desargues' Theorem

We consider here three nonempty sets of objects, the elements of which are called *points* P, Q, ...; *lines* a, b, ...; and *planes* α, β, ..., respectively. We assume there exists an incidence relation between points and lines, points and planes, and lines and planes that is symmetric and transitive in the following sense. If P is incident with a (or α), then a (or α) is incident with P; if a is incident with α, then α is incident with a; if P is incident with a and a is incident with α, then P is incident with α. The incidence relation we describe again by various expressions (compare Section 1.1). The system of points,

Photo of calcite crystals.

lines, and planes is called an **affine space** or, more precisely, an **affine 3-space,** if the following five axioms are satisfied.

Axiom S.1 *Two different points lie on one and only one line.*

Axiom S.2 *Three points that are not on a line are on a uniquely determined plane.*

Axiom S.3 *If a point P is not on a given line a, there exists a unique line b passing through P and spanning a plane with a but not intersecting a.*

Axiom S.4 *If two planes intersect in a point, they intersect in a line.*

Axiom S.5 *There are at least two points on each line. There exist three noncollinear points on each plane. Not all points lie in the same plane.*

By Axioms S.1, S.4, and S.5, a line is incident with a plane if two of its points lie on the plane. By Axioms S.1 and S.5, every line can be identified with the set of points lying on it. The same is true for every plane, by Axioms S.2 and S.5. The incidence relation can, therefore, be interpreted by set-theoretic inclusion. Axioms S.2 and S.5 imply, furthermore, that a line and a point not on this line determine a unique plane on which they both lie. Two different lines that intersect in a point determine a unique plane which passes through both lines.

Two lines in the same plane are called **parallel** if they are equal or do not intersect. Also, nonintersecting or identical planes are called **parallel.** If equality is excluded, we use the term **properly parallel** instead of parallel. If a plane intersects two parallel planes, the lines of intersection are parallel.

Theorem 4.1.1 *The relation "parallel" for lines is an equivalence relation.*

Proof The relation clearly is reflexive and symmetric. To show that it is transitive, let a be parallel to b and let b be parallel to c. We can assume a, b, and c to be different, since, otherwise, transitivity follows from reflexivity and symmetry. a, b determine a plane α; and b, c determine a plane β. We join a point C of c to a by a plane γ (Fig. 4.1). If $\beta = \gamma$, then a, b, and c are all in β and, by Axiom S.3, a is parallel to c. So let

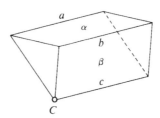

Figure 4.1

β and γ intersect in a line c' (see Axiom S.4). If c' were to intersect b, we would have $\alpha = \gamma = \beta$, contrary to assumption. Therefore, by

Axiom S.3, $c' = c$. If c' were to intersect a, the point of intersection would be common to α and β and so would be common to a and b. Therefore, a and $c' = c$ do not have a point in common, and both lie in γ, which means that a is parallel to c. ◆

By Axioms S.1, S.3, and S.5, we have (see Section 3.1) the following theorems.

Theorem 4.1.2 *The points and lines incident with a given plane α form an affine plane.*

Theorem 4.1.3 *Given a plane α and a point Q (on α or not), there exists a unique plane β that passes through Q and is parallel to α.*

Proof Let a, a' be two arbitrarily chosen lines of α that intersect in a point P of α. By Axiom S.3, there exist lines b, b' parallel to a, a', respectively, both passing through Q. They span a plane β. Suppose β intersects α in a point R. Then we draw lines c and c' through R parallel to a and a' and, hence, (by Theorem 4.1.1) parallel to b and b', respectively. From Theorem 4.1.2, we conclude that c intersects a' and b' and that c' intersects a and b. This readily implies, by Axiom S.2 and Theorem 4.1.2, that $\alpha = \beta$. Therefore, α and β either do not intersect at all, or they are identical—that is, α is parallel to β.

Suppose a second plane, β', also passes through Q and is parallel to α. It intersects a plane through Q and a in a line that is parallel to a and, hence, in b. Similarly, β' intersects a plane through Q and a' in b'. Since b and b' span β, we find $\beta' = \beta$. ◆

We call two triangles ***perspective*** if there exist three lines through a point C, or properly parallel, each of which joins a vertex of one of the triangles to a vertex of the other. If C exists, we say the triangles are ***C-perspective.***

We now prove the spatial analogue of Desargues' theorem (D) or (d) (see Section 3.9) (Figs. 4.2 and 4.3).

Theorem 4.1.4 (***Spatial theorem of Desargues***) *If two triangles not both contained in the same plane are perspective, such that two pairs of corresponding sides are properly parallel, the third pair of corresponding sides is also properly parallel.*

Figure 4.2   **Figure 4.3**

Proof From the assumption, we find immediately (compare the proof of Theorem 4.1.3) that the two given triangles lie in properly parallel planes α, β. The pair of lines not assumed parallel lie, by perspectivity, in a plane γ. Since γ intersects α, β in parallel lines, we conclude the theorem. ◆

Now we show (D) to hold in any plane.

Theorem 4.1.5 (**Desargues' theorem** (**D**)) *If two triangles of the same plane are C-perspective and if two pairs of corresponding sides are properly parallel, the sides of the third pair are also properly parallel.*

Proof Let $\triangle PQR$ and $\triangle P'Q'R'$ be the given triangles, where PQ is properly parallel to $P'Q'$ and PR is properly parallel to $P'R'$ (Fig. 4.4). By assumption, PP', QQ', and RR' meet in a point C. Let D be a point not in the plane α that contains the triangles. In the plane CPD, the line parallel to CD through P meets DP' in a point P'' not on α. Similarly, the parallels of CD through Q, R meet DQ', DR' in points Q'', R'', respectively. Clearly, $P''Q''$ is parallel to both PQ and $P'Q'$ (see

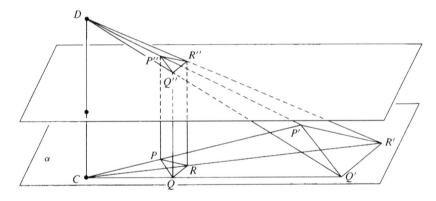

Figure 4.4

proof of Theorem 4.1.1), and $P''R''$ is parallel to PR and $P'R'$. Therefore, by Theorem 4.1.4, $Q''R''$ is parallel to QR and to $Q'R'$. By Theorem 4.1.1, QR is parallel to $Q'R'$ and also properly parallel, since, by assumption, $Q \neq Q'$ and $R \neq R'$. ◆

Exercises

1. Describe the "minimal model" of an affine 3-space (two points on every line). Count the number of points, lines, and planes.

2. Give a direct proof of (d) without using (D) (see Section 3.9).

3. Show that Axiom S.5 can be replaced by the following axiom.
   S.5′ There are at least two points on each line. There is a point on

each plane. There exist five points, no four of which are in the same plane.

4. Show that any two planes have the same number of points and also that any two lines have the same number of points, provided this number is finite.

5. Prove that if a, b, c are any three lines, no two of which are parallel, there exists a line d intersecting a, b, and c.

6. We say that three points P, Q, R are projected from C onto P', Q', R', respectively, if there exist lines through C passing through the corresponding pairs of points. Show that any three noncollinear points can be mapped onto any other three noncollinear points by the successive application of several projections.

4.2   *Translations and Vectors*

A ***translation*** is defined as a one-to-one incidence-preserving map of the set of all points onto itself, such that every line is mapped onto a parallel of itself and such that there exists a line all parallels of which are fixed lines. Since Desargues' little theorem (d) holds as a consequence of (D) (see Section 3.9; compare Exercise 2 of the preceding section) and since the spatial form of (d) is true, by Theorem 4.1.4, we are in a position to construct translations as follows.

Given two points P, P', we wish to find a translation that maps P onto P'. If $P = P'$, the identity does it. So let $P \neq P'$. If X is not on PP', then its image X' is determined by parallels of PP' and PX through X and P', respectively (Fig. 4.5). The image of a further point Y can be constructed

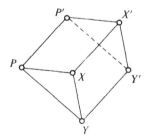

Figure 4.5

in the same way, starting either from PP' or from XX'. If Y is on PP', we start from XX'. If Y is neither on PP' nor on XX', we obtain the same Y' in both constructions, by Theorem 4.1.4 or (d).

Theorem 4.2.1 *The set of all translations under successive application is a commutative group.*

Proof From the definition of a translation, we obtain immediately the group property. To show commutativity, one proceeds as in the proof of Theorem 3.2.3. ◆

A translation is determined if only one point P and its image point P' are given. This shift from P to P' by a translation is sometimes indicated by an arrow pointing from P to P' (Fig. 4.6). The arrow is known as a *vector*

Figure 4.6

and is used in physics to express forces, velocities, and other entities possessing length and direction. To be precise, we define a *vector* as an ordered pair of points (P, P'). Two vectors (P, P') and (Q, Q') are said to belong to the same class if and only if there exists a translation mapping P onto Q

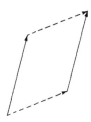

Figure 4.7

and P' onto Q'. A member of a class of vectors is called a *free vector,* which is often referred to simply as a *vector.* So there is a one-to-one correspondence:

$$\text{free vector} \longleftrightarrow \text{translation.}$$

The distinction between vectors and free vectors has to be kept in mind; we shall not bother to introduce different symbols for them. It becomes clear from the context whether a vector or a free vector is in force. We use bold letters,

$$\mathbf{p}, \mathbf{q}, \cdots,$$

for vectors.

The successive application of translations can be interpreted as addition of the corresponding free vectors. If $\mathbf{a}$ corresponds to the translation T and $\mathbf{a}'$ corresponds to the translation T', we set (see Fig. 4.8)

$$\text{T} \circ \text{T}' \longleftrightarrow \mathbf{a} + \mathbf{a}'. \tag{1}$$

The identity map corresponds to the *zero vector* $\mathbf{0}$, which represents the class of all pairs (P, P).

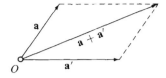

Figure 4.8

Let O be a point chosen once for all. Every free vector has a representative initiating from O. We add two such vectors by adding the corresponding free vectors and then taking again the representative which initiates from O. Theorem 4.2.2 is a result of Theorem 4.2.1.

Theorem 4.2.2 *The set of all vectors initiating from the same point O is a commutative group under vector addition.*

The correspondence (1) is a good example of the following general group-theoretic concept. Let $\mathscr{G}_1$, $\mathscr{G}_2$ be groups and let φ be a one-to-one mapping of $\mathscr{G}_1$ onto $\mathscr{G}_2$ such that the image of the product of two elements a, b equals the product of the image elements:

$$(a \circ b)\varphi = (a)\varphi \odot (b)\varphi. \tag{2}$$

Here the operations "$\circ$" and "$\odot$" may be different (compare Exercise 6 below). Then φ is called an *isomorphism* of $\mathscr{G}_1$ and $\mathscr{G}_2$. If such an isomorphism exists between $\mathscr{G}_1$ and $\mathscr{G}_2$, we say that $\mathscr{G}_1$ and $\mathscr{G}_2$ are ***isomorphic***. *So the group of translations and the vector group of all vectors initiating from O are isomorphic*, the structure relation (2) being given by (1).

Exercises

1. Let $P_0, P_1, \ldots, P_n$ be any points. Set $\mathbf{a}_1 = (P_0, P_1)$, $\mathbf{a}_2 = (P_1, P_2)$, $\ldots$, $\mathbf{a}_n = (P_{n-1}, P_n)$, $\mathbf{a}_{n+1} = (P_n, P_0)$ and show (for corresponding free vectors)

$$\mathbf{a}_1 + \mathbf{a}_2 + \cdots + \mathbf{a}_{n+1} = \mathbf{0}.$$

2. Describe the geometrical figure obtained from three vectors by adding each two and all three of them.

3. Describe all planes that are fixed planes under a given translation.

4. (a) Show that any translation defined for a plane α (as an affine plane) can be extended to a translation of the whole space. (b) Prove this extension to be uniquely determined.

5. Show that the group of all translations of a plane α is isomorphic to the group of all translations of the space possessing α as a fixed plane.

6. Show that the group of all real numbers under addition is isomorphic to the group of all positive real numbers under multiplication. (Use $(x)\varphi = \log x$.)

7. Find all translations of the minimal model in Exercise 1 of Section 4.1.

4.3   *Dilatations and Scalar Multiplication*

Dilatations can be constructed in a way that is strictly analogous to the construction of translations.   These are defined again (as in the planar case, see Section 3.2) as incidence-preserving, one-to-one maps of the set of all points onto itself, mapping every line onto a parallel of itself such that there exists a fixed point called the *center* of the dilatation.

Theorem 4.3.1 *The set of all dilatations with the same center O forms a group under successive application.*

It should be noted, however, that we cannot conclude the commutativity of this group without a further axiom being imposed on the affine 3-space.

We choose a point O and a plane α through O.   Every dilatation of the space with center O induces a dilatation with center O in α.   Conversely, every dilatation of α with center O can be extended to a dilatation of the whole space (by our construction process; compare Exercise 3 below).   So there is a one-to-one correspondence between the set $\mathscr{D}_0^{\text{space}}$ of all dilatations of the space with center O and the set $\mathscr{D}_0$ of all dilatations of α with center O. Furthermore, since $\mathscr{D}_0^{\text{space}}$ and $\mathscr{D}_0$ are groups, we see that $\mathscr{D}_0^{\text{space}}$ is isomorphic to $\mathscr{D}_0$.

In α there is again an isomorphism between the group $\mathscr{F}_0$ and the multiplicative group $\mathscr{F}^{\times}$ of the skew field $\mathscr{F}$ that can be introduced on a line c_1 through O, according to Section 3.6.   That is,

$$\mathscr{D}_0 \text{ is isomorphic to } \mathscr{F}^{\times}.$$

The multiplication in $\mathscr{F}$ can also be interpreted in the following way: Let $\mathbf{p} = (O, P)$ and let p be the number represented by P.   Let a be any other number of $\mathscr{F}$.   We denote the vector $(O, \tilde{P})$, where $\tilde{P}$ represents ap, by $a\mathbf{p}$. So the multiplication ap in $\mathscr{F}$ appears as multiplication of a vector $\mathbf{p}$ by a "scalar" a.   If $a \neq 0$, the assignment

$$\mathbf{p} \longrightarrow a\mathbf{p}$$

can be considered a dilitation of α and thus of the space.   We write $a\mathbf{q}$ for the image of any vector $\mathbf{q} = (O, Q)$ undert his dilatation.   Furthermore, we

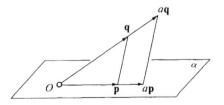

Figure 4.9

Theorem 4.3.2 *Between any vector* **p** *beginning at O and an arbitrary
element a of $\mathscr{F}$, there is a multiplication* **ap** *defined, providing another
vector beginning at O.   The following rules are satisfied:*

(1) $(a + b)\mathbf{p} = a\mathbf{p} + b\mathbf{p}$,   [4.3]
(2) $(ab)\mathbf{p} = a(b\mathbf{p})$,
(3) $1 \cdot \mathbf{p} = \mathbf{p}$.   *Affine Space
                                                                                    and
Our construction of dilatations gives us a final theorem.   Vector Space*

Theorem 4.3.3 *Scalars and vectors satisfy the rule*

(4) $a\mathbf{p} + a\mathbf{q} = a(\mathbf{p} + \mathbf{q})$ (Fig. 4.10).

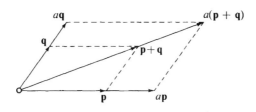

Figure 4.10

The rules (1)–(4) will prove fundamental for our analytic treatment of
affine geometry.

Exercises

1. Prove, with the aid of Theorem 4.3.2, that $(-1) \cdot \mathbf{p} = -\mathbf{p}$ for any **p**.

2. Find the vector $\mathbf{x} = (O, X)$ pointing to the "center" of a parallelogram
   spanned by two vectors $\mathbf{p} = (O, P)$ and $\mathbf{q} = (O, Q)$.   Does this vector
   exist for any $\mathscr{F}$?

3. (a) Show that any dilatation defined for a plane α (as an affine plane) can
   be extended to a dilatation of the space.   (b) Prove this extension to be
   uniquely determined.

4. Show that the set of all dilatations (arbitrary centers) and translations is
   a group.   Is this group commutative?

5. Find a necessary and sufficient condition for two dilatations to have a
   translation as their product.

6. Show that dilatations induce projections in the sense of Exercise 6,
   Section 4.1.   Conversely, is any projection induced by a dilatation?

7. Find all dilatations in the "minimal model" of Exercise 1, Section 4.1.

4.4   Vector Space and Coordinates

Let us summarize the results we have obtained on vector algebra so far. We have found a system of *vectors*, denoted by letters **p**, **q**, ..., and a system of *scalars*, denoted by letters *a*, *b*, ... Vectors can be *added*; scalars can be *added* and *multiplied*; and, finally, scalars and vectors can be *multiplied*. These algebraic operations obey the following laws:

(a) *Vectors form a commutative group under vector addition.*
(b) *Scalars form a skew field (see Section 3.6) under addition and multiplication.*
(c) *For each scalar a and each vector **p** the product a**p** is defined to be a vector. This scalar multiplication is associative and distributive in the sense of rules (1) and (2) of Theorem 4.3.2 and rule (4) of Theorem 4.3.3. Furthermore, 1 · a is identical to a, by rule (3) of Theorem 4.3.2.*

A system of vectors and scalars satisfying (a)–(c) is called a *left vector space*, or simply a *vector space*.   If, in particular, the skew field is a field $\mathscr{K}$, we can also write **p**a instead of a**p**.

So we have obtained a rich algebraic system that allows us to apply many standard algebraic operations to geometrical problems.   Note that all this has been deduced from a few simple geometrical axioms (S.1–S.5) of affine 3-space, and once again the power of geometrical thought and the beauty of its simplicity are revealed.   Though our axioms of a vector space have been chosen from properties of affine 3-spaces, it is not true that, conversely, every vector space corresponds to an affine 3-space.   We have arrived at a very general concept as will be seen now.

We call a set $\mathbf{a}_1, \ldots, \mathbf{a}_n$ of vectors *linearly independent* if any relation

$$x_1\mathbf{a}_1 + x_2\mathbf{a}_2 + \cdots + x_n\mathbf{a}_n = \mathbf{0}$$

implies $x_1 = x_2 = \cdots = x_1 = 0$.   Vectors   are   not   linearly   independent are said to be *linearly dependent.* If $\mathbf{a}_1, \ldots, \mathbf{a}_n$ are linearly independent and if every vector **x** can be expressed as a linear combination

$$\mathbf{x} = x_1\mathbf{a}_1 + x_2\mathbf{a}_2 + \cdots + x_n\mathbf{a}_n,$$

we say that $\mathbf{a}_1, \mathbf{a}_2, \ldots \mathbf{a}_n$ form a *basis* of the vector space.

Our next goal is to show that the number *n* does not depend on the special choice of a basis.   First we prove a lemma.

Lemma 4.4.1 *Let a vector space be given with a basis* $\mathbf{a}_1, \ldots, \mathbf{a}_n$. *If* $\mathbf{b}_1, \ldots,$ $\mathbf{b}_r$ *are linearly independent, there exist vectors* $\mathbf{b}_{r+1}, \ldots, \mathbf{b}_n$ *such that* $\mathbf{b}_1, \ldots,$ $\mathbf{b}_n$ *is a basis of the given vector space.*

Proof In case every basis vector $\mathbf{a}_i$ ($i = 1, \ldots, n$) depends linearly on $\mathbf{b}_1,$ $\ldots, \mathbf{b}_r$, we have $\mathbf{b}_1, \ldots, \mathbf{b}_r$ as a basis.   If not, let $\mathbf{a}_{i_1}$ be linearly independent of $\mathbf{b}_1, \ldots, \mathbf{b}_r$.   We set $\mathbf{b}_{r+1} = \mathbf{a}_{i_1}$.   Repeating the same

argument for $r + 1$ instead of r we find that either $\mathbf{b}_1, \ldots, \mathbf{b}_{r+1}$ form a
basis of the space or there exists a vector $\mathbf{a}_{i_2}$ that does not depend on $\mathbf{b}_1$,
$\ldots, \mathbf{b}_{r+1}$. In the latter case, we set $\mathbf{b}_{r+2} = \mathbf{a}_{i_2}$. After a finite number of
steps the a_i will be used up and the set $\mathbf{b}_1, \ldots, \mathbf{b}_r$ will be extended to a set
$\mathbf{b}_1, \ldots, \mathbf{b}_m$ such that every $\mathbf{a}_i$ depends on $\mathbf{b}_1, \ldots, \mathbf{b}_m$. Since every vector
depends on $\mathbf{a}_1, \ldots, \mathbf{a}_n$ it depends on $\mathbf{b}_1, \ldots, \mathbf{b}_m$, too. So $\mathbf{b}_1, \ldots, \mathbf{b}_m$ is a
basis of the space.

We must show that $m = n$. This is done by an "exchange procedure."
Up to a different numbering, we may assume that $\mathbf{b}_1$ does not depend on
$\mathbf{a}_2, \ldots, \mathbf{a}_n$. Then $\mathbf{b}_1, \mathbf{a}_2, \ldots, \mathbf{a}_n$ is easily seen to be a basis of the space.
There is now a second $\mathbf{b}_i$, say $\mathbf{b}_2$, that does not depend on $\mathbf{b}_1, \mathbf{a}_3, ., \mathbf{a}_n$.
Then $\mathbf{b}_1, \mathbf{b}_2, \mathbf{a}_3, \ldots, \mathbf{a}_n$ is a basis. Continuing this way, all $\mathbf{a}_i$'s are
exchanged for $\mathbf{b}_i$'s. Therefore, $n \leq m$. Interchanging the roles of $\mathbf{a}_i$
and $\mathbf{b}_i$, we obtain $m \leq n$; hence $m = n$. ◆

Lemma 4.4.1 implies that the number n of basis elements does not depend
on the basis itself. So it is specific for the vector space. We call it the
dimension of the space. If a basis $\mathbf{a}_1, \ldots, \mathbf{a}_n$ is given, we sometimes write,
instead of

$$\mathbf{x} = x_1\mathbf{a}_1 + \cdots + x_n\mathbf{a}_n,$$

simply

$$\mathbf{x} = (x_1, \ldots, x_n), \qquad \textit{row vector,}$$

or

$$\mathbf{x} = \begin{pmatrix} x_1 \\ \vdots \\ x_n \end{pmatrix}, \qquad \textit{column vector.}$$

$x_1, \ldots, x_n$ are called **coordinates** of $\mathbf{x}$. If $\mathbf{x} = (O, X)$ we set also $X =
(x_1, \ldots, x_n)$ and call $x_1, \ldots, x_n$ the coordinates of the point X.

The basis vectors are assigned to the following row vectors:

$$\begin{aligned}
\mathbf{a}_1 &= (1, 0, \ldots, 0) \\
\mathbf{a}_2 &= (0, 1, \ldots, 0) \\
&\vdots \\
\mathbf{a}_n &= (0, 0, \ldots, 1).
\end{aligned}$$

We want to show that affine 3-space can be considered a vector space in
which any basis consists of exactly 3 vectors. In fact, one sees that there
exist three linearly independent vectors $\mathbf{a}_1, \mathbf{a}_2, \mathbf{a}_3$. Consider any vector
$\mathbf{x} = (O, X)$. $\mathbf{a}_1$ and $\mathbf{a}_2$ span a plane which we call the (x_1, x_2)-**plane.**
Similarly the (x_1, x_3)-**plane** and the (x_2, x_3)-**plane** are defined. In Figure 4.11,
it is readily seen (using Axiom S.5) that we obtain numbers x_1, x_2, x_3 such
that

$$\mathbf{x} = x_1\mathbf{a}_1 + x_2\mathbf{a}_2 + x_3\mathbf{a}_3.$$

Therefore, $\mathbf{a}_1, \mathbf{a}_2, \mathbf{a}_3$ is a basis of the given space and is considered as the
vector space of all vectors beginning at O.

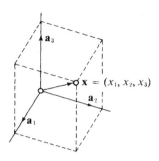

Figure 4.11

Theorem 4.4.2 *An affine 3-space can be considered (after selecting an origin O) as a vector space of dimension 3.*

If we consider only a Desarguesian affine plane, every basis consists of precisely two vectors, as the next theorem states.

Theorem 4.4.3 *Any Desarguesian affine plane can be considered (after selecting an origin O) to be a vector space of dimension 2.*

A more formal way of introducing a vector space is to consider combinations

$$(x_1, \ldots, x_n)$$

of n elements of $\mathcal{K}$, where n is fixed, and define

$$(x_1, \ldots, x_n) + (y_1, \ldots, y_n) = (x_1 + y_1, \ldots, x_n + y_n),$$
$$a(x_1, \ldots, x_n) = (ax_1, \ldots, ax_n).$$

The laws of a vector space are easily deduced from these definitions and the rules of a skew field.

Note that we have not yet introduced in the vector space anything like perpendicularity. So all "coordinate systems" are affine coordinate systems in the sense that the basis vectors need not be in some sense perpendicular.

The formal construction of vector spaces shows that there exist vector spaces of arbitrarily high dimension.

Exercises

1. Verify the rules of a vector space for the formal vectors $(x_1, \ldots, x_n)$ as defined above.
2. Let $\mathcal{K}$ be a field of characteristic 2 (that is, $1 + 1 = 0$). Show that in any parallelogram of an affine plane coordinatized by $\mathcal{K}$ the diagonals are parallel.

3. Show that vectors $\mathbf{p}_1, \ldots, \mathbf{p}_r$ are linearly dependent if and only if one of them can be expressed as a linear combination of the others.

4. Give an explicit proof of the fact that an affine 3-space is coordinatized by a 3-dimensional vector space.

5. Show that any field can be considered a one-dimensional vector space over itself.

6. Show that the set of infinite sequences $(x_1, x_2, \ldots)$ of field elements can be turned into a vector space in the same way as has been done for the finite sequences $(x_1, \ldots, x_n)$.

7. Consider the subspace of the vector space introduced in Exercise 6 consisting of all sequences $(x_1, x_2, \ldots)$ in which only finitely many of the x_i's are different from O. Find a set of vectors spanning this subspace such that any finitely many of them are linearly independent ("basis"!).

4.5 Subspaces

Let $\mathbf{p}, \mathbf{q}$ be two vectors that are linearly independent. If $\mathbf{p} = (O, P)$, $\mathbf{q} = (O, Q)$, then O, P, Q span a plane OPQ. Any multiple $a\mathbf{p}$ or $b\mathbf{q}$ lies in this plane. The same is true for $a\mathbf{p} + b\mathbf{q}$, as the parallelogram construction tells us (compare Section 4.2). Any point X of OPQ can be obtained as such a linear combination, as can be seen if one draws parallels of OP and OQ through X, respectively. So $\mathbf{p}, \mathbf{q}$ form a basis of the plane OPQ. The numbers a, b can be considered as coordinates of X with respect to the basis $\mathbf{p}, \mathbf{q}$. In particular, the ordinary coordinates x_1, x_2 of the x_1, x_2 plane refer to the basis vectors $\mathbf{a}_1, \mathbf{a}_2$ that we chose together with $\mathbf{a}_3, \ldots, \mathbf{a}_n$ as a basis of the whole space.

A line through O is given by a vector equation

$$\mathbf{x} = a\mathbf{p},$$

where a is an arbitrary element of $\mathscr{K}$ and $\mathbf{p}$ is a fixed vector $\neq \mathbf{0}$. Lines and planes through O are called **subspaces** of the vector space with origin O. Also, the whole space may be called a subspace of the vector space. Generally, we call the totality of vectors

$$\mathbf{x} = x_1\mathbf{p}_1 + x_2\mathbf{p}_2 + \cdots + x_r\mathbf{p}_r,$$

where $\mathbf{p}_1, \mathbf{p}_2, \ldots, \mathbf{p}_r$ are fixed vectors and $x_1, \ldots, x_r$ each attain all elements of $\mathscr{K}$, a **subspace** of a given vector space. If $r = 2$ and $\mathbf{p}_1, \mathbf{p}_2$ are linearly independent vectors, this subspace is called a **plane.**

We must distinguish carefully between an affine space and the vector space of all vectors (O, X), where O is a particularly chosen point. In a sense, selecting O means introducing a coordinate system into affine space, since any point X is represented by a vector (O, X). However, to say that we are introducing a "coordinate system," usually means not only that we are introducing such a vector space but that we are also introducing a basis $\mathbf{a}_1, \ldots, \mathbf{a}_n$ of this vector space, so that

$$(O, X) = x_1\mathbf{a}_1 + \cdots + x_n\mathbf{a}_n.$$

The distinction between affine space and vector space can be seen especially in the different systems of subspaces; all subspaces of a vector space pass through O. In affine space, however, there are lines and planes that do not pass through the origin.

However, we can represent all subspaces of the affine space—for example, all points, lines, planes, and the whole space—by using a vector space introduced in it. Any such subspace can be obtained from a subspace of the vector space by a translation. If, for example,

$$\mathbf{x} = t\mathbf{p}$$

is the equation of a line of the vector space ($\mathbf{p} \neq \mathbf{0}$ being a fixed vector), then

$$\mathbf{x} = t\mathbf{p} + \mathbf{q} \tag{1}$$

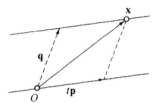

Figure 4.12

is the equation of a general line of the affine space. $\mathbf{q}$, in this case, represents a translation; similarly,

$$\mathbf{x} = t\mathbf{p} + s\mathbf{q} + \mathbf{r} \tag{2}$$

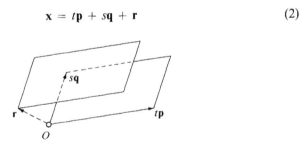

Figure 4.13

is the general vector equation of a plane in affine space ($\mathbf{p}$, $\mathbf{q}$, linearly independent).

With respect to a basis (coordinate system), (1) breaks down into n equations,

$$x_1 = tp_1 + q_1$$
$$\cdots \cdots \cdots \cdots$$
$$x_n = tp_n + q_n,$$

also known as a ***parametric representation*** of the line (t being the parameter).

Exercises

1. Show that a line of an affine space passing through P and Q $(P \neq Q)$ can be represented by

$$\mathbf{x} = (1 - t)\mathbf{p} + t\mathbf{q},$$

where $\mathbf{p} = (O, P)$ and $\mathbf{q} = (O, Q)$. Find in an affine 3-space the equation for the line through $P = (3, -1, 5)$, $Q = (0, 1, 2)$ (with respect to a coordinate system).

2. Show that for linearly independent $\mathbf{p}$, $\mathbf{q}$

$$\mathbf{x} = t\mathbf{p} + s\mathbf{q}$$

represents a line through P, Q where $\mathbf{p} = (O, P)$, $\mathbf{q} = (O, Q)$ if and only if t, s are restricted by the condition $t + s = 1$.

3. Find (as an analogy to Exercise 1) an equation for the plane determined by three noncollinear points P, Q, R.

4. Let $\mathbf{p}$, $\mathbf{q}$ be linearly independent and let

$$\mathbf{x} = (1 - t)2\mathbf{p} + t3\mathbf{q},$$
$$\mathbf{y} = (1 - s)\mathbf{p} - s\mathbf{q}$$

be two lines (in a plane through O). Find the point of intersection of the two lines. (*Hint*: Use the definition of linear independence to obtain equations in s and t.)

5. Let $\mathscr{K}$ be the field of real numbers. Show vectorially that in any triangle $\triangle PQR$ (defined in the plane PQR), the bisectors meet in a point. (*Hint*: $R = O$ can be assumed. Set $\mathbf{p} = (O, P)$, $\mathbf{q} = (O, Q)$ and use the idea of Exercise 4 to find the points of intersection of the bisectors.)

6. Solve the spatial analogue of the problem in Exercise 5, $\mathscr{K}$ again being the field of real numbers. A tetrahedron T is a set of four points not in a plane together with the line-segment edges and faces of the triangles spanned by these points (Fig. 4.14). Join the centroid of each face (common point of bisectors) to the vertex of T not contained in the face. Show that all four lines thus obtained meet in a point.

[4.5]

Affine Space
and
Vector Space

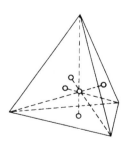

Figure 4.14

4.6   Inner Product

In this section we choose the field of real numbers, $\mathcal{R}$, as the co-ordinate skew field. The familiarity with trigonometry that we assume amounts only to some elementary laws of sines and cosines.

We begin with affine 3-space. The basis vectors $\mathbf{a}_1$, $\mathbf{a}_2$, $\mathbf{a}_3$ can be chosen such that they are unit vectors perpendicular in the ordinary sense. As

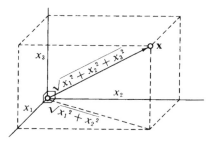

Figure 4.15

a natural extension of the distance of two points in the plane (see Section 3.14), we assign to any pair of points $X = (x_1, x_2, x_3)$, $Y = (y_1, y_2, y_3)$ the *distance*

$$d(X, Y) = \sqrt{(x_1 - y_1)^2 + (x_2 - y_2)^2 + (x_3 - y_3)^2}.$$

(See Fig. 4.15.)   When applying it to vectors $\mathbf{X} = (O, X)$, we call

$$|\mathbf{x}| = \sqrt{x_1^2 + x_2^2 + x_3^2}$$

the *length* of $\mathbf{x}$. The term under the root is also abbreviated by writing $\mathbf{x}^2$ or $\mathbf{x} \cdot \mathbf{x}$.   More generally, we write

$$\mathbf{x} \cdot \mathbf{y} = x_1 y_1 + x_2 y_2 + x_3 y_3$$

and call this real number the ***inner product*** of $\mathbf{x}$ and $\mathbf{y}$.   It has the following geometrical meaning: Let $\mathbf{x} \neq \mathbf{0}$, $\mathbf{y} \neq \mathbf{0}$ and let θ be the angle between $\mathbf{x}$

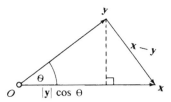

Figure 4.16

and **y**, $0 \leq \theta \leq \pi$ (Fig. 4.16). From the law of cosines, we have

$$|\mathbf{x} - \mathbf{y}|^2 = |\mathbf{x}|^2 + |\mathbf{y}|^2 - 2|\mathbf{x}|\,|\mathbf{y}|\cos\theta.$$

Since $|\mathbf{x} - \mathbf{y}|^2 = (\mathbf{x} - \mathbf{y})^2 = \mathbf{x}^2 - 2\mathbf{x}\cdot\mathbf{y} + \mathbf{y}^2 = |\mathbf{x}|^2 - 2\mathbf{x}\cdot\mathbf{y} + |\mathbf{y}|^2$, we find

$$\mathbf{x}\cdot\mathbf{y} = |\mathbf{x}|\,|\mathbf{y}|\,\cos\theta. \tag{1}$$

Theorem 4.6.1 *The absolute value of the inner product of two vectors* **x**, **y** *is equal to the length of the projection of* **y** *onto* **x** *times the length of* **x**.

A particularly important case of an inner product is that of $\theta = \pi/2$, in which case we call the vectors **x**, **y** *perpendicular*.

Theorem 4.6.2 *Two vectors* **x**, **y** *are perpendicular if and only if*

$$\mathbf{x}\cdot\mathbf{y} = 0.$$

We can also write (1) as

$$\cos\theta = \frac{\mathbf{x}\cdot\mathbf{y}}{|\mathbf{x}|\,|\mathbf{y}|},$$

thus obtaining a formula for calculating the angle between two vectors **x**, **y** if their coordinates are given.

Example $\mathbf{x} = (1, 0, 1),\qquad \mathbf{y} = (1, -1, 0).$

$$\cos\theta = \frac{1}{\sqrt{2}\cdot\sqrt{2}} = \frac{1}{2}\ \left(\text{that is, } \theta = \frac{\pi}{3}\right).$$

For the basis $\mathbf{a}_1$, $\mathbf{a}_2$, $\mathbf{a}_3$ we have $|\mathbf{a}_1| = |\mathbf{a}_2| = |\mathbf{a}_3| = 1$ and $\mathbf{a}_1\cdot\mathbf{a}_2 = \mathbf{a}_1\cdot\mathbf{a}_3 = \mathbf{a}_2\cdot\mathbf{a}_3 = 0$. With respect to this basis, an arbitrary vector **x** can be represented by

$$\mathbf{x} = (\mathbf{x}\cdot\mathbf{a}_1,\ \mathbf{x}\cdot\mathbf{a}_2,\ \mathbf{x}\cdot\mathbf{a}_3).$$

If we set

$$\mathbf{x}\cdot\mathbf{a}_1 = |\mathbf{x}|\cos\theta_1,\qquad \mathbf{x}\cdot\mathbf{a}_2 = |\mathbf{x}|\cos\theta_2,\qquad \mathbf{x}\cdot\mathbf{a}_3 = |\mathbf{x}|\cos\theta_3,$$

then $\cos\theta_1$, $\cos\theta_2$, $\cos\theta_3$ are called *direction cosines* of **x** (Fig. 4.17).

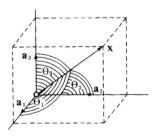

Figure 4.17

An interesting application of inner products is also provided by the following: We wish to show that *the altitudes of any triangle meet in a point.* We can assume that O is one vertex of the triangle. Let P, Q be the other vertices (Fig. 4.18). We set $\mathbf{p} = (O, P)$, $\mathbf{q} = (O, Q)$. Let S be the point in

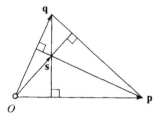

Figure 4.18

which the altitudes through P and Q meet and let $\mathbf{s} = (O, S)$. Then we have, by Theorem 4.6.2,

$$(\mathbf{p} - \mathbf{s})\mathbf{q} = 0,$$
$$(\mathbf{q} - \mathbf{s})\mathbf{p} = 0.$$

By a simple calculation we obtain

$$(\mathbf{p} - \mathbf{q})\mathbf{s} = 0,$$

which proves the line OS to be perpendicular to PQ.

Exercises

1. Which pairs of the following vectors are perpendicular? $(1, 2, 0)$, $(1, 1, 1)$, $(0, 0, 3)$, $(2, -1, 1)$.
2. Calculate the angles in the triangle with vertices $(1, 1, 1)$, $(0, 1, 1)$, $(1, 1, 0)$.
3. State and prove a spatial analogue of the above theorem on the altitudes of a triangle.
4. Show vectorially that the angular bisectors of a triangle meet in a point.
5. Is the set of vectors a group under scalar multiplication?
6. Show that the distance function d is invariant under translations.
7. Let c_1, c_2, c_3 be positive constants. Show that the distance function d defined by

$$d(X, Y) = \sqrt{c_1(x_1 - y_1)^2 + c_2(x_2 - y_2)^2 + c_3(x_3 - y_3)^2}$$

satisfies the conditions (1)–(3) of Section 3.14.
8. Does the result of Exercise 7 remain valid if c_1, c_2, c_3 are not all positive?

4.7   *Area and Spar Product*

Having introduced length, or "one-dimensional measure," in Section 4.6, we proceed now to discuss two-dimensional measure or "area" in an affine plane. We assume the plane to be Pappian, that is, coordinatized by a field (see Sections 3.6 and 3.13). We shall restrict our attention to the areas of parallelograms, which may be considered "bricks" for affine figures.

Let $\mathbf{x}, \mathbf{y}$ be two linearly independent vectors of the two-dimensional vector space which serves as a coordinate system for the given plane; basis $\mathbf{a}_1, \mathbf{a}_2$. The points represented by the vectors $\mathbf{o}, \mathbf{x}, \mathbf{y}, \mathbf{x} + \mathbf{y}$ are vertices of a parallelogram. We say: $\mathbf{x}$ and $\mathbf{y}$ *span* this parallelogram. For reasons that will become obvious in this section, it is useful to distinguish the parallelogram spanned by $\mathbf{x}$ and $\mathbf{y}$ from that spanned by $\mathbf{y}$ and $\mathbf{x}$. In a sense, this means "orientation" of the parallelogram (however, we do not make use of orientation here). An alternative name for parallelogram is spar (taken from mineralogy). We use it here for "oriented" parallelograms. To be precise, every ordered pair of linearly independent vectors $\mathbf{x}, \mathbf{y}$ is called a *spar* $\mathbf{x}, \mathbf{y}$.

Let $\mathscr{S}_0$ be the spar $\mathbf{a}_1, \mathbf{a}_2$. We assign to $\mathscr{S}_0$ (as an "affine unit square") the area 1:

$$\alpha(\mathscr{S}_0) = 1.$$

If $x_1 \neq 0$ and $x_2 \neq 0$, let $\mathscr{S}_1$ be the spar $x_1\mathbf{a}_1, x_2\mathbf{a}_2$. We set

$$\alpha(\mathscr{S}_1) = x_1 \cdot x_2$$

and call this number the *area* of $\mathscr{S}_1$. If, in case of an ordinary Euclidean plane, $\mathbf{a}_1, \mathbf{a}_2$ form an "orthonormal basis," this definition of area coincides with the usual one (Fig. 4.19). In the case of the field of real numbers

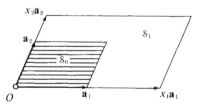

Figure 4.19

$\alpha(\mathscr{S}_1)$ may become negative—namely, if $x_1 < 0$ or $x_2 < 0$ (but not both). Then we denote the spar $x_2\mathbf{a}_2, x_1\mathbf{a}_1$ by $\mathscr{S}_1^*$ and set

$$\alpha(\mathscr{S}_1^*) = -x_1x_2.$$

A motivation of our further development of area may be given by the so-called *Cavalieri's principle*: "If two sets can be intersected by a pencil of parallel lines such that every line intersects both sets in subsets of equal

1-dimensional measure, then both sets have equal area." In particular, two sets have this property if one can be obtained from the other by applying a shear (compare Fig. 4.20).

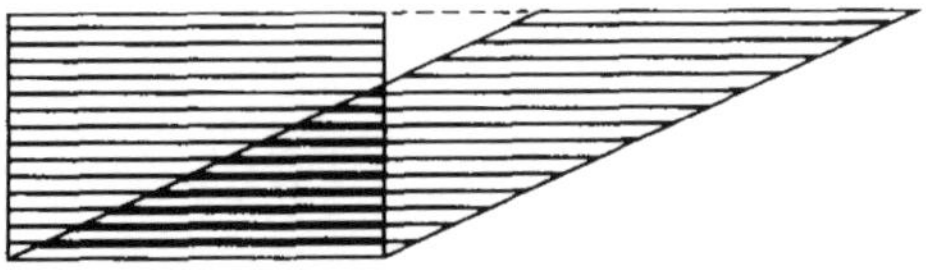

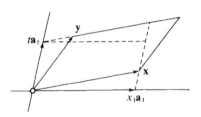

Figure 4.20

Let $\mathscr{S}$ be an arbitrary spar **x**, **y**. We again denote the spar **y**, **x** by $\mathscr{S}^*$. We may assume that neither is **x** a multiple of $\mathbf{a}_2$ nor is **y** a multiple of $\mathbf{a}_1$ (otherwise we interchange the names of **x** and **y**). We express **y** as a linear combination of $\mathbf{a}_2$ and **x** (Fig. 4.21):

$$\mathbf{y} = s\mathbf{x} + t\mathbf{a}_2.$$

Figure 4.21

By a shear, with the $\mathbf{a}_2$-axis as axis, we can map $x_1\mathbf{a}_1$ onto **x**. By a second shear which leaves **x** and all its multiples fixed, we can map $t\mathbf{a}_2$ onto **y**. Combining these shears we map the spar $x_1\mathbf{a}_1$, $t\mathbf{a}_2$ onto $\mathscr{S}$. So it makes sense to define

$$\alpha(\mathscr{S}) = x_1 t.$$

We wish to express this area by the coordinates of $\mathbf{x} = (x_1, x_2)$ and $\mathbf{y} = (y_1, y_2)$: From

$$y_1 = sx_1 + t0,$$
$$y_2 = sx_2 + t1,$$

we obtain

$$x_1 t = x_1(y_2 - sx_2) = x_1 y_2 - sx_1 x_2 = x_1 y_2 - y_1 x_2.$$

Instead of $\alpha(\mathscr{S})$, we write [**x**, **y**]. So

$$\alpha(\mathscr{S}) = [\mathbf{x}, \mathbf{y}] = x_1 y_2 - y_1 x_2.$$

We call this number the ***spar product*** of **x** and **y**. A traditional name for the term on the right is ***determinant*** (of two rows and two columns), symbolically written as

$$\begin{vmatrix} x_1 & y_1 \\ x_2 & y_2 \end{vmatrix}.$$

Given a basis $\mathbf{a}_1, \mathbf{a}_2$ we may, therefore, identify determinants with spar products. We set

$$[\mathbf{x}, \mathbf{y}] = \begin{vmatrix} x_1 & y_1 \\ x_2 & y_2 \end{vmatrix} = 0$$

in case $\mathbf{x}, \mathbf{y}$ are linearly dependent. By calculation,

$$\alpha(\mathscr{S}^*) = [\mathbf{y}, \mathbf{x}] = y_1 x_2 - x_1 y_2 = -[\mathbf{x}, \mathbf{y}] = -\alpha(\mathscr{S}).$$

So $\alpha(\mathscr{S}_1^*) = -\alpha(\mathscr{S}_1)$ extends in a natural way to $\alpha(\mathscr{S}^*) = -\alpha(\mathscr{S})$.

We can state five rules about spar products, all of which follow immediately from $[\mathbf{x}, \mathbf{y}] = x_1 y_2 - y_1 x_2$:

$$[\mathbf{x}, \mathbf{y}] = -[\mathbf{y}, \mathbf{x}], \tag{1}$$

$$[\mathbf{x}, \mathbf{x}] = 0, \tag{2}$$

$$[\mathbf{x}_1 + \mathbf{x}_2, \mathbf{y}] = [\mathbf{x}_1, \mathbf{y}] + [\mathbf{x}_2, \mathbf{y}], \tag{3}$$

$$[\mathbf{x}_1, \mathbf{y}_1 + \mathbf{y}_2] = [\mathbf{x}, \mathbf{y}_1] + [\mathbf{x}, \mathbf{y}_2], \tag{4}$$

$$[a\mathbf{x}, \mathbf{y}] = [\mathbf{x}, a\mathbf{y}] = a[\mathbf{x}, \mathbf{y}]. \tag{5}$$

Now let $PQRS$ be an arbitrary parallelogram, its vertices being represented by vectors $\mathbf{p}, \mathbf{q}, \mathbf{r}, \mathbf{s}$, respectively. Several spars can be assigned to this parallelogram (Fig. 4.22), namely $\mathbf{r} - \mathbf{s}, \mathbf{p} - \mathbf{s}$ or $\mathbf{q} - \mathbf{p}, \mathbf{s} - \mathbf{p}$ or $\mathbf{r} - \mathbf{q}$,

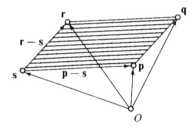

Figure 4.22

$\mathbf{p} - \mathbf{q}$, and so forth. Since $\mathbf{r} - \mathbf{s} = \mathbf{q} - \mathbf{p}$ (by definition of a parallelogram), we have

$$[\mathbf{r} - \mathbf{s}, \mathbf{p} - \mathbf{s}] = [\mathbf{q} - \mathbf{p}, \mathbf{p} - \mathbf{s}] = -[\mathbf{q} - \mathbf{p}, \mathbf{s} - \mathbf{p}]$$
$$= [\mathbf{q} - \mathbf{p}, \mathbf{q} - \mathbf{r}] = [\mathbf{p} - \mathbf{q}, \mathbf{r} - \mathbf{q}],$$

and so forth. To every parallelogram, two numbers that are negatives of each other are assigned in this way as the area of the parallelogram.

If, in particular, the coordinate field is the field of real numbers, we can choose the positive of these two numbers as the (positive) area of the parallelogram. This positive area, in turn, can be used to define an orientation of the plane (see Exercise 3).

By definition, the area of a parallelogram is invariant under translations. We wish it also to be invariant under all shears. This can, in fact, be verified. It suffices therefore to assume that the axis of the shear passes through O, for the following reason: If $\bar{S}$ is an arbitrary shear with axis c, let T be a translation mapping c onto a line c' through O. Now $T\bar{S}T^{-1}$ is a shear $\bar{S}'$ with axis c' (see Section 3.3), and hence

$$\bar{S} = T^{-1}\bar{S}'T.$$

Invariance of area under T, T^{-1}, $\bar{S}'$ now implies invariance of area under $\bar{S}$. Before showing that any $\bar{S}'$ leaves area unchanged, we present an analytic representation of $\bar{S}'$:

Theorem 4.7.1 *A shear having the line* $\mathbf{y} = t\mathbf{a}$ *as axis* ($\mathbf{a} \neq \mathbf{0}$ *fixed*) *can be represented by*

$$\mathbf{x}' = \mathbf{x} + d[\mathbf{a}, \mathbf{x}]\mathbf{a} \qquad (6)$$

with d a fixed element of the coordinate field (for $d = 0$ we obtain the identical map as a special shear).

Proof First we show that (6) is a shear. In fact, any line

$$\mathbf{x} = t\mathbf{p} + \mathbf{q} \quad (\mathbf{p} \neq \mathbf{0})$$

is mapped onto

$$t\mathbf{p} + \mathbf{q} + d[\mathbf{a}, t\mathbf{p} + \mathbf{q}]\mathbf{a} = t(\mathbf{p} + d[\mathbf{a}, \mathbf{p}]\mathbf{a}) + \mathbf{q} + d[\mathbf{a}, \mathbf{q}]\mathbf{a}.$$

We know that $\mathbf{p} + d[\mathbf{a}, \mathbf{p}]\mathbf{a}$ cannot be $=0$, since, otherwise,

$$-d[\mathbf{a}, \mathbf{p}][\mathbf{a}, \mathbf{p}] = [\mathbf{p}, \mathbf{p}] = 0;$$

and so either $d = 0$ or $[\mathbf{a}, \mathbf{p}] = 0$, both leading to

$$\mathbf{p} = -d[\mathbf{a}, \mathbf{p}]\mathbf{a} = \mathbf{0},$$

a contradiction. Therefore, lines are mapped onto lines. Clearly, $\mathbf{y} = t\mathbf{a}$ and all its parallel lines are fixed lines; $\mathbf{y} = t\mathbf{a}$ remains pointwise

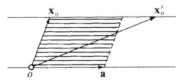

Figure 4.23

fixed. So (6) represents a shear. Now let a point $\mathbf{x}_0$ and its image $\mathbf{x}_0'$ be given. We determine a number d by

$$d = \frac{[\mathbf{x}_0' - \mathbf{x}_0, \mathbf{x}_0]}{[\mathbf{a}, \mathbf{x}_0]^2}.$$

Using this d we obtain a shear (6) mapping $\mathbf{x}_0$ onto $\mathbf{x}_0'$. Since, by Section 3.3, a shear is determined by the axis and a pair point-image not on the axis, this shear is identical with the given shear. This proves that the

group of all shears with the same axis is transitive on the set of all lines
different from the axis and passing through a definite point of the axis. ◆

167

Theorem 4.7.2 *The area of any spar, that is, the spar product, is invariant
under any shear with axis passing through O.*

Proof Substituting (6) in $[\mathbf{x}', \mathbf{y}']$, we obtain (using rules (1)–(5)):

$$\begin{aligned}
[\mathbf{x}', \mathbf{y}'] &= [\mathbf{x} + d[\mathbf{a}, \mathbf{x}]\mathbf{a}, \mathbf{y} + d[\mathbf{a}, \mathbf{y}]\mathbf{a}] \\
&= [\mathbf{x}, \mathbf{y}] + d[\mathbf{a}, \mathbf{y}][\mathbf{x}, \mathbf{a}] + d[\mathbf{a}, \mathbf{x}][\mathbf{a}, \mathbf{y}] + d^2[\mathbf{a}, \mathbf{x}][\mathbf{a}, \mathbf{y}][\mathbf{a}, \mathbf{a}] \\
&= [\mathbf{x}, \mathbf{y}].
\end{aligned}$$

◆

An important application of Theorem 4.7.2 follows. So far, area has been
defined only with respect to a given basis $\mathbf{a}_1$, $\mathbf{a}_2$. Now $\mathbf{a}_1$, $\mathbf{a}_2$ can be replaced
by any two vectors obtained from $\mathbf{a}_1$, $\mathbf{a}_2$ by applying a number of shears.
Hereby the area remains unchanged. Let the coordinate field not have
characteristic 2 (that is, $1 + 1 \neq 0$). If $\triangle$ is any triangle, the vertices of
which are given by $\mathbf{p}$, $\mathbf{q}$, $\mathbf{r}$, then either

$$\alpha(\triangle) = \tfrac{1}{2}[\mathbf{q} - \mathbf{p}, \mathbf{r} - \mathbf{p}]$$

or its negative may be considered as the **area** of $\triangle$ (Fig. 4.24). Again we
can replace $\mathbf{q} - \mathbf{p}, \mathbf{r} - \mathbf{p}$ by $\mathbf{p} - \mathbf{q}, \mathbf{r} - \mathbf{q}$, and so on, and obtain $\alpha(\triangle)$ or
$-\alpha(\triangle)$.

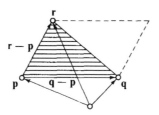

Figure 4.24

For a useful algebraic application of spar product, let $\mathbf{u} = (u_1, u_2)$,
$\mathbf{v} = (v_1, v_2)$, $\mathbf{c} = (c_1, c_2)$ be given vectors. We wish to find x, y such that

$$x\mathbf{u} + y\mathbf{v} = \mathbf{c}.$$

In other words, we wish to solve the system of linear equations

$$\begin{aligned}
xu_1 + yv_1 &= c_1 \\
xu_2 + yv_2 &= c_2
\end{aligned}$$

simultaneously in x, y. If $[\mathbf{u}, \mathbf{v}] \neq 0$, we find, by calculating the values of
$[\mathbf{c}, \mathbf{v}]$ and $[\mathbf{c}, \mathbf{u}]$, respectively,

$$x = \frac{[\mathbf{c}, \mathbf{v}]}{[\mathbf{u}, \mathbf{v}]}, \qquad y = \frac{[\mathbf{u}, \mathbf{c}]}{[\mathbf{u}, \mathbf{v}]}.$$

This is called **Cramer's rule** (for the 2-dimensional case).

Exercises

1. Calculate the area of a triangle with vertices $(0, 1)$, $(3, 1)$, $(1, -1)$.

2. In a Euclidean plane, show that the area of a spar changes its sign if a reflection is applied.

3. Let the coordinate field be the field $\mathscr{R}$ of real numbers. By using spar products, define an orientation of triangles such that the relation "equally oriented" is an equivalence relation with precisely two equivalence classes.

4. (a) In a Euclidean plane, show that a point P lies on a line a if and only if, for every shear $\bar{S}$ with axis a and the half-turn H_P, we have

$$\bar{S}H_P = H_P\bar{S}.$$

(b) Can the statement of (a) be extended to arbitrary Desarguesian planes?

5. Show that two lines a, b are parallel if and only if, for any two shears $\bar{S}_1$, $\bar{S}_2$ with axes a, b, respectively, we have

$$\bar{S}_1\bar{S}_2 = \bar{S}_2\bar{S}_1.$$

6. In a Euclidean plane, show that every rotation about a point R is the product of three shears with axis through R. (*Hint*: Choose R as origin of a Cartesian coordinate system, and the coordinate axes as axes of two of the shears. Applying the given rotation to the x_1-axis provides a third axis of a shear.)

4.8 Volume and Spar Product

The ideas developed in Section 4.7 for a planar area can be carried over to 3-space. Let the coordinate field again be commutative and let $\mathbf{a}_1$, $\mathbf{a}_2$, $\mathbf{a}_3$ be a basis of the coordinatizing vector space.

The spatial analogue of a parallelogram is that of a *parallelepiped*. We define it as a system of 8 points whose coordinate vectors can be expressed as follows, after an appropriate translation: $\mathbf{o}, \mathbf{x}, \mathbf{y}, \mathbf{z}, \mathbf{x} + \mathbf{y}, \mathbf{x} + \mathbf{z}, \mathbf{y} + \mathbf{z}, \mathbf{x} + \mathbf{y} + \mathbf{z}$, where $\mathbf{x}, \mathbf{y}, \mathbf{z}$ are supposed to be three linearly independent vectors. Any ordered triplet of three linearly independent vectors $\mathbf{x}, \mathbf{y}, \mathbf{z}$ we also call a *spar* $\mathbf{x}, \mathbf{y}, \mathbf{z}$ (Fig. 4.25). To the spar $\mathscr{S}_0$: $\mathbf{a}_1, \mathbf{a}_2, \mathbf{a}_3$, we assign a volume of 1:

$$v(\mathscr{S}_0) = 1;$$

for the spar $\mathscr{S}_1$: $x_1\mathbf{a}_1, x_2\mathbf{a}_2, x_3\mathbf{a}_3$, we set

$$v(\mathscr{S}_1) = x_1 x_2 x_3.$$

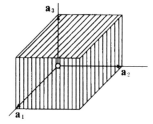

Figure 4.25

Let $\mathbf{x}$, $\mathbf{y}$, $\mathbf{z}$ be an arbitrary spar $\mathscr{S}$. We can assume $\mathbf{a}_3$ not to lie in the plane spanned by $\mathbf{x}$ and $\mathbf{y}$ (otherwise we interchange the roles of $\mathbf{x}, \mathbf{y}, \mathbf{z}$). We express $\mathbf{z}$ as

(∗)
$$\mathbf{z} = s\mathbf{x} + t\mathbf{y} + u\mathbf{a}_3$$

By a (*spatial*) *shear*, we mean a one-to-one mapping of the set of all points onto itself that maps planes onto planes, leaves one plane α pointwise fixed, and possesses all parallel planes of α as fixed planes. α is called the *axis* of the shear. If

$$\mathbf{x} = (x_1, x_2, x_3), \qquad \mathbf{y} = (y_1, y_2, y_3),$$

we set

$$\mathbf{x}' = (x_1, x_2, 0), \qquad \mathbf{y}' = (y_1, y_2, 0).$$

We map the spar $\mathbf{x}', \mathbf{y}', u\mathbf{a}_3$ onto $\mathbf{x}, \mathbf{y}, u\mathbf{a}_3$ by a shear and $\mathbf{x}, \mathbf{y}, u\mathbf{a}_3$ again onto $\mathbf{x}, \mathbf{y}, \mathbf{z}$ by a second shear (compare Fig. 4.26; the existence of these shears follows from Theorem 4.8.1 below; see Exercise 1).

So, motivated again by the spatial version of "Cavalieri's principle" (see Section 4.7), we define

$$[\mathbf{x}, \mathbf{y}, \mathbf{z}] = [\mathbf{x}', \mathbf{y}', u\mathbf{a}_3] = u(x_1 y_2 - x_2 y_1)$$

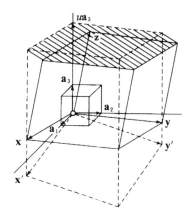

Figure 4.26

and call this number the spar *product* $[\mathbf{x}, \mathbf{y}, \mathbf{z}]$ of $\mathbf{x}, \mathbf{y}, \mathbf{z}$, or the *volume* of $\mathscr{S}$:

$$v(\mathscr{S}) = [\mathbf{x}, \mathbf{y}, \mathbf{z}].$$

We wish again to calculate $[\mathbf{x}, \mathbf{y}, \mathbf{z}]$ in terms of the coordinates of $\mathbf{x}, \mathbf{y}, \mathbf{z}$. To achieve this, we let

$$\hat{\mathbf{x}} = (x_1, x_2), \qquad \hat{\mathbf{y}} = (y_1, y_2), \qquad \hat{\mathbf{z}} = (z_1, z_2)$$

and solve the three coordinate equations obtained from $(*)$ in terms of spar products in the sense of Section 4.7.

$$z_1 = sx_1 + ty_1$$
$$z_2 = sx_2 + ty_2$$
$$z_3 = sx_3 + ty_3 + u.$$

We obtain

$$s = \frac{[\hat{\mathbf{z}}, \hat{\mathbf{y}}]}{[\hat{\mathbf{x}}, \hat{\mathbf{y}}]}, \qquad t = \frac{[\hat{\mathbf{x}}, \hat{\mathbf{z}}]}{[\hat{\mathbf{x}}, \hat{\mathbf{y}}]};$$

therefore,

$$u = z_3 - sx_3 - ty_3 = z_3 - \frac{[\hat{\mathbf{z}}, \hat{\mathbf{y}}]}{[\hat{\mathbf{x}}, \hat{\mathbf{y}}]} x_3 - \frac{[\hat{\mathbf{x}}, \hat{\mathbf{z}}]}{[\hat{\mathbf{x}}, \hat{\mathbf{y}}]} y_3,$$

from which we calculate

$$[\mathbf{x}, \mathbf{y}, \mathbf{z}] = [\hat{\mathbf{y}}, \hat{\mathbf{z}}]x_3 - [\hat{\mathbf{x}}, \hat{\mathbf{z}}]y_3 + [\hat{\mathbf{x}}, \hat{\mathbf{y}}]z_3$$
$$= (y_1 z_2 - y_2 z_1)x_3 - (x_1 z_2 - x_2 z_1)y_3 + (x_1 y_2 - x_2 y_1)z_3.$$

Symbolically, we write

$$[\mathbf{x}, \mathbf{y}, \mathbf{z}] = \begin{vmatrix} x_1 & y_1 & z_1 \\ x_2 & y_2 & z_2 \\ x_3 & y_3 & z_3 \end{vmatrix}$$

and call it the ***determinant*** of $\mathbf{x}, \mathbf{y}, \mathbf{z}$. In case $\mathbf{x}, \mathbf{y}, \mathbf{z}$ do not span a spar— that is, they are linearly dependent—we let

$$[\mathbf{x}, \mathbf{y}, \mathbf{z}] = 0.$$

As an example, we have

$$[\mathbf{a}_1, \mathbf{a}_2, \mathbf{a}_3] = \begin{vmatrix} 1 & 0 & 0 \\ 0 & 1 & 0 \\ 0 & 0 & 1 \end{vmatrix} = 1.$$

By direct computation, one easily verifies:

$$[\mathbf{x}, \mathbf{y}, \mathbf{z}] = -[\mathbf{y}, \mathbf{x}, \mathbf{z}] = [\mathbf{y}, \mathbf{z}, \mathbf{x}], \tag{1}$$

$$[\mathbf{x}, \mathbf{x}, \mathbf{z}] = [\mathbf{x}, \mathbf{y}, \mathbf{x}] = [\mathbf{x}, \mathbf{y}, \mathbf{y}] = 0, \tag{2}$$

$$[\mathbf{x} + \mathbf{x}', \mathbf{y}, \mathbf{z}] = [\mathbf{x}, \mathbf{y}, \mathbf{z}] + [\mathbf{x}', \mathbf{y}, \mathbf{z}], \tag{3}$$

$$[\mathbf{x}, \mathbf{y} + \mathbf{y}', \mathbf{z}] = [\mathbf{x}, \mathbf{y}, \mathbf{z}] + [\mathbf{x}, \mathbf{y}', \mathbf{z}], \tag{4}$$

$$[\mathbf{x}, \mathbf{y}, \mathbf{z} + \mathbf{z}'] = [\mathbf{x}, \mathbf{y}, \mathbf{z}] + [\mathbf{x}, \mathbf{y}, \mathbf{z}'], \tag{5}$$

$$[a\mathbf{x}, \mathbf{y}, \mathbf{z}] = [\mathbf{x}, a\mathbf{y}, \mathbf{z}] = [\mathbf{x}, \mathbf{y}, a\mathbf{z}] = a[\mathbf{x}, \mathbf{y}, \mathbf{z}]. \tag{6}$$

Only (1), (3), and (6) need to be shown explicitly; the others follow by applying (1). In order to show that spar products are invariant under shears with the axis passing through 0, we first characterize those shears.

Theorem 4.8.1 *Any shear with a plane*

$$\mathbf{y} = t\mathbf{a} + s\mathbf{b}$$

as axis (**a**, **b** *linearly independent*) *can be represented by*

$$\mathbf{x}' = \mathbf{x} + [\mathbf{a}, \mathbf{b}, \mathbf{x}](d\mathbf{a} + e\mathbf{b}), \tag{7}$$

with d, e being fixed numbers.

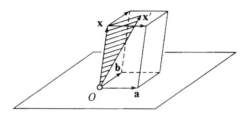

Figure 4.27

Proof The proof is achieved analogously to that of Theorem 4.7.1. First, one verifies that planes are mapped onto planes. Then, one shows that all parallel planes of the axis $\mathbf{y} = t\mathbf{a} + s\mathbf{b}$ are fixed planes. For points $\mathbf{y}$ on the axis, we find

$$\begin{aligned}
\mathbf{y}' &= \mathbf{y} + [\mathbf{a}, \mathbf{b}, t\mathbf{a} + s\mathbf{b}](d\mathbf{a} + e\mathbf{b}) \\
&= \mathbf{y} + (t[\mathbf{a}, \mathbf{b}, \mathbf{a}] + s[\mathbf{a}, \mathbf{b}, \mathbf{b}])(d\mathbf{a} + e\mathbf{b}) \\
&= \mathbf{y} \quad \text{(by (2))}. \qquad \blacklozenge
\end{aligned}$$

Theorem 4.8.2 *The volume of a spar remains invariant under all shears with axis through 0.*

Proof If we substitute $\mathbf{x}'$, $\mathbf{y}'$, $\mathbf{z}'$ as given by Equation (7) in $[\mathbf{x}', \mathbf{y}', \mathbf{z}']$, an easy calculation (using Equations (1)–(6)) shows

$$[\mathbf{x}', \mathbf{y}', \mathbf{z}'] = [\mathbf{x}, \mathbf{y}, \mathbf{z}]. \quad \blacklozenge$$

The volume of a parallelepiped can now be defined analogously to the area of a parallelogram by using spar products. Again, by proceeding in different ways, we see that the volume of the parallelepiped is given by a pair of numbers that are negatives of each other.

We mention two further volume formulas that are applications of spar products. We assume the coordinate field to have neither characteristic 2 nor characteristic 3 (that is, $1 + 1 \neq 0$, $1 + 1 + 1 \neq 0$): If $\mathbf{x}, \mathbf{y}, \mathbf{z}$ span a skew triangular prism $\mathscr{P}$, as shown in Figure 4.28, we assign to $\mathscr{P}$ the volume

$$v(\mathscr{P}) = \tfrac{1}{2}[\mathbf{x}, \mathbf{y}, \mathbf{z}].$$

Figure 4.28

Figure 4.29

Similarly, if **x**, **y**, **z** span a tetrahedron $\mathcal{T}$ as is indicated in Figure 4.29, we assign to $\mathcal{T}$ the volume

$$v(\mathcal{T}) = \tfrac{1}{6}[\mathbf{x}, \mathbf{y}, \mathbf{z}].$$

Exercises

1. Carry out the calculations in the proof of Theorem 4.8.1.

2. Find the volumes of the spar, the prism, and the tetrahedron spanned by the vectors which point from $(1, -1, 0)$ to $(1, 1, 1)$, $(2, -1, 1)$, $(0, 1, 3)$, respectively.

3. Show that the volume of a spar changes at most the sign if the vectors that span it are chosen in different ways as sides of a parallelepiped.

4. State explicitly the definition of the volume of a parallelepiped.

5. Determine the shear with axis spanned by $\mathbf{a} = (1, -1, 1)$, $\mathbf{b} = (-1, 0, 1)$ that maps the x_1-axis onto the x_2-axis.

6. State and prove the spatial analogue of Exercise 3, Section 4.7.

7. Let α be a plane passing through 0 and let P, P' be points on a line that does not intersect α.
   (a) Show by using Theorem 4.8.1 that there exists a shear with axis α that maps P onto P'.
   (b) Show by direct geometrical arguments (compare Section 3.3) that this shear is uniquely determined.

4.9   *Spar Products and Determinants*

The procedure applied in the preceding two sections may be extended to vector spaces of arbitrary dimension n.   Rather than carry this through,

we shall define immediately spar products and determinants and show that
the definitions make sense.

By a *spar* $\mathbf{x}_1, \ldots, \mathbf{x}_n$ spanned by linearly independent vectors $\mathbf{x}_1, \ldots, \mathbf{x}_n$,
we mean the ordered set of these vectors. We assign to each spar $\mathbf{x}_1, \ldots, \mathbf{x}_n$
an element $[\mathbf{x}_1, \ldots, \mathbf{x}_n]$ of the coordinate field $\mathcal{K}$ such that the following
rules are satisfied:

$$[\mathbf{x}_1, \ldots, \mathbf{x}_i + \mathbf{x}_i', \ldots, \mathbf{x}_n] = [\mathbf{x}_1, \ldots, \mathbf{x}_i, \ldots, \mathbf{x}_n]$$
$$+ [\mathbf{x}_1, \ldots, \mathbf{x}_i', \ldots, \mathbf{x}_n] \quad \text{for } i = 1, \ldots, n. \quad (1)$$

$$[\mathbf{x}_1, \ldots, \mathbf{x}_i, \ldots, \mathbf{x}_k, \ldots, \mathbf{x}_n] = -[\mathbf{x}_1, \ldots, \mathbf{x}_k, \ldots, \mathbf{x}_i, \ldots, \mathbf{x}_n]$$
$$\text{for } i, k = 1, \ldots, n; i \neq k. \quad (2)$$

$$[\mathbf{x}_1, \ldots, a\mathbf{x}_1, \ldots, \mathbf{x}_n] = a[\mathbf{x}_1, \ldots, \mathbf{x}_n] \quad \text{for } i = 1, \ldots, n \text{ and any } a \text{ of } \mathcal{K}. \quad (3)$$

We call the number $[\mathbf{x}_1, \ldots, \mathbf{x}_n]$ the *spar product* of $\mathbf{x}_1, \ldots, \mathbf{x}_n$.

Does there exist, for any given dimension, a function satisfying Equations
(1)–(3)—that is, does the spar product exist? In order to show that this is
so, let a basis $\mathbf{a}_1, \ldots, \mathbf{a}_n$ be introduced and let $x_{i1}, \ldots, x_{in}$ be the components
of $\mathbf{x}_i$ with respect to this basis. For $n = 2$, we use the definition

$$\begin{vmatrix} x_{11} & x_{12} \\ x_{21} & x_{22} \end{vmatrix} = x_{11}x_{22} - x_{21}x_{12}$$

and generally we define in a "recursive" way:

$$\begin{vmatrix} x_{11} & \cdots & x_{1k} \\ \vdots & & \vdots \\ x_{k1} & \cdots & x_{kk} \end{vmatrix} = x_{11} \begin{vmatrix} x_{22} & \cdots & x_{2k} \\ \vdots & & \vdots \\ x_{k2} & \cdots & x_{kk} \end{vmatrix} - x_{12} \begin{vmatrix} x_{21} & x_{23} & \cdots & x_{2k} \\ \vdots & & & \vdots \\ x_{k1} & x_{k3} & \cdots & x_{kk} \end{vmatrix}$$

$$+ \cdots + (-1)^{k+1} x_{1k} \begin{vmatrix} x_{2k} & \cdots & x_{2,k-1} \\ \vdots & & \\ x_{k1} & \cdots & x_{k,k-1} \end{vmatrix}.$$

The ith term on the right is obtained by deleting the first row and the ith
column in the square array on the left and multiplying by $(-1)^{i+1} x_{1i}$.
The term on the left is called a *determinant* of k rows. So after n steps we
arrive at the definition of

$$[\mathbf{x}_1, \ldots, \mathbf{x}_n] = \begin{vmatrix} x_{11} & \cdots & x_{1n} \\ \vdots & & \vdots \\ x_{n1} & \cdots & x_{nn} \end{vmatrix}.$$

Clearly the determinants satisfy rules (1)–(3).

This existence proof for spar products provides immediately a way of
calculating spar products explicitly. There are, however, a number of ways
to evaluate determinants. We do not enter these technical details.

$$\begin{vmatrix} 1 & 0 & 2 & -1 \\ 2 & 2 & 1 & 1 \\ 0 & 1 & 0 & 0 \\ -1 & 2 & 1 & 4 \end{vmatrix} = 1 \begin{vmatrix} 2 & 1 & 1 \\ 1 & 0 & 0 \\ 2 & 1 & 4 \end{vmatrix} - 0 + 2 \begin{vmatrix} 2 & 2 & 1 \\ 0 & 1 & 0 \\ -1 & 2 & 4 \end{vmatrix} - (-1) \begin{vmatrix} 2 & 2 & 1 \\ 0 & 1 & 0 \\ -1 & 2 & 1 \end{vmatrix}$$

$$= 1 \cdot (2 \cdot 0 - 1 \cdot 4 + 1 \cdot 1) + 2(2 \cdot 4 - 2 \cdot 0 + 1 \cdot 1)$$
$$+ 1 \cdot (2 \cdot 1 - 2 \cdot 0 + 1 \cdot 1)$$
$$= -4 + 1 + 2 \cdot 9 + 2 + 1 = 18.$$

Exercises

1. Evaluate the determinant obtained from that in the above example by interchanging rows and columns.

2. Find the value of a determinant for which $x_{ik} = 0$ for $i < k$.

3. Show that the value of a determinant remains unchanged if a multiple of one column is added to another column.

4. Prove that the value of a determinant changes its sign if two columns are interchanged.

5. Extend the definition of a shear to vector spaces of n dimensions and show that shears preserve spar products.

6. Show that n vectors $\mathbf{x}_1, \ldots, \mathbf{x}_n$ are linearly dependent, if and only if $[\mathbf{x}_1, \ldots, \mathbf{x}_n] = 0$.

4.10   Linear Mappings, Affine Mappings, and Matrices

Let $\mathbf{a}_1, \ldots, \mathbf{a}_n$ again be a basis of a vector space $\mathscr{V}$ over a field. We consider vectors $\mathbf{b}_1, \ldots, \mathbf{b}_n$ of this space:

$$\mathbf{b}_1 = a_{11}\mathbf{a}_1 + \cdots + a_{1n}\mathbf{a}_n$$
$$\vdots \tag{1}$$
$$\mathbf{b}_n = a_{n1}\mathbf{a}_1 + \cdots + a_{nn}\mathbf{a}_n.$$

They span a subspace $\mathscr{U}$ of the given vector space $\mathscr{V}$. If $\mathbf{b}_1, \ldots, \mathbf{b}_n$ are linearly independent, then $\mathscr{U} = \mathscr{V}$ (by Lemma 4.4.1). In this case, (1) can be interpreted as a *change of basis*.

Let us consider an arbitrary vector of $\mathscr{U}$

$$\mathbf{x}' = x_1\mathbf{b}_1 + \cdots + x_n\mathbf{b}_n.$$

In terms of the basis $\mathbf{a}_1, \ldots, \mathbf{a}_n$, we obtain for $\mathbf{x}'$

$$\mathbf{x}' = (a_{11}x_1 + \cdots + a_{n1}x_n)\mathbf{a}_1 + \cdots + (a_{1n}x_1 + \cdots + a_{nn}x_n)\mathbf{a}_n.$$

We denote the coefficients with respect to $\mathbf{a}_1, \ldots, \mathbf{a}_n$ by $x'_1, \ldots, x'_n$:

$$
\begin{aligned}
x'_1 &= a_{11}x_1 + \cdots + a_{n1}x_n \\
&\vdots \\
x'_n &= a_{1n}x_1 + \cdots + a_{nn}x_n.
\end{aligned} \tag{2}
$$

Now we give the system (2) a second interpretation: To every vector $\mathbf{x} = (x_1, \ldots, x_n)$ of $\mathscr{V}$ there is assigned a vector $\mathbf{x}' = (x'_1, \ldots, x'_n)$ of $\mathscr{U}$. In other words; $\mathscr{V}$ is mapped onto $\mathscr{U}$ by (2). We call such a mapping a **linear mapping** (or **linear transformation**). In case $\mathscr{U} = \mathscr{V}$ ($\mathbf{b}_1, \ldots, \mathbf{b}_n$ linearly independent), we can give (1) or (2) two interpretations:

1. Basis changed   Points of the space unchanged
2. Basis unchanged   Linear mapping of points.

If one rides on a merry-go-round, one may consider one's motion a rotation (special linear mapping) relative to the surrounding earth as a fixed frame of reference (basis unchanged); or one may insist that only the frame of reference changes continuously, whereas one is resting (basis changed). The latter interpretation corresponds to 1. The former corresponds to 2.
A useful way to write (2) is:

$$
\begin{pmatrix} x'_1 \\ \vdots \\ x'_n \end{pmatrix} = \begin{pmatrix} a_{11} & \cdots & a_{n1} \\ \vdots & & \vdots \\ a_{1n} & \cdots & a_{nn} \end{pmatrix} \begin{pmatrix} x_1 \\ \vdots \\ x_n \end{pmatrix}. \tag{2'}
$$

Here $\mathbf{x} = \begin{pmatrix} x_1 \\ \vdots \\ x_n \end{pmatrix}$ and $\mathbf{x}' = \begin{pmatrix} x'_1 \\ \vdots \\ x'_n \end{pmatrix}$ are **column vectors**.

For the sake of writing, it is, however, more convenient to use row vectors:

$$
\mathbf{x} = (x_1, \ldots, x_n) \qquad \mathbf{x}' = (x'_1, \ldots, x'_n).
$$

For reasons which will become apparent, in this case we write (2) as

$$
(x'_1, \ldots, x'_n) = (x_1, \ldots, x_n) \begin{pmatrix} a_{11} & \cdots & a_{1n} \\ \vdots & & \vdots \\ a_{n1} & \cdots & a_{nn} \end{pmatrix}. \tag{2''}
$$

Now we set

$$
\mathbf{A} = \begin{pmatrix} a_{11} & \cdots & a_{1n} \\ \vdots & & \vdots \\ a_{n1} & \cdots & a_{nn} \end{pmatrix}
$$

and express (2) in an even shorter form as

$$
\mathbf{x}' = \mathbf{xA}. \tag{2'''}
$$

$\mathbf{A}$ is called an **($n \times n$) matrix** or, briefly, a **matrix**. (2''') is a very suggestive notation, since it can be interpreted geometrically by considering $\mathbf{x}$ and $\mathbf{x}'$ as points of the space and $\mathbf{A}$ as a mapping. Furthermore, (2''') expresses "linearity," since it is analogous to an ordinary linear equation $x' = xa$ between scalars.

Examples Let $\mathscr{V}$ coordinatize ordinary Euclidean 3-space.

(a)

$$\mathbf{A} = \begin{pmatrix} 0 & -1 & 0 \\ -1 & 0 & 0 \\ 0 & 0 & 1 \end{pmatrix}$$

represents a reflection in a plane through the x_3-axis. The third coordinate remains unchanged, whereas the x_1- and x_2-coordinates are interchanged (Fig. 4.30).

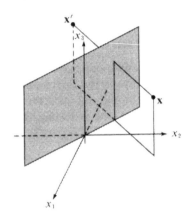

Figure 4.30

(b)

$$\mathbf{A} = \begin{pmatrix} d & 0 & 0 \\ 0 & d & 0 \\ 0 & 0 & d \end{pmatrix}; \quad d \neq 0,$$

represents a "dilatation" with center O. In case $d = -1$, this is a "central inversion" at 0: every vector is mapped onto its negative.

(c)

$$\mathbf{A} = \begin{pmatrix} 1 & 0 & 0 \\ 0 & 1 & 0 \\ 0 & 0 & 0 \end{pmatrix}$$

represents a projection of the whole space onto the (x_1, x_2)-plane. So this is a case in which $\mathscr{U} \neq \mathscr{V}$.

Now let two linear transformations be applied successively:

$$\mathbf{x}' = \mathbf{x}\mathbf{A}$$

and

$$\mathbf{x}'' = \mathbf{x}'\mathbf{B}.$$

By substituting, we obtain

$$\mathbf{x}'' = (\mathbf{x}\mathbf{A})\mathbf{B}. \tag{3}$$

If we substitute the corresponding linear equations, we see that (3) can be written as

$$\mathbf{x}'' = \mathbf{x}\mathbf{C},$$

where

$$\mathbf{C} = \begin{pmatrix} \sum\limits_{i=1}^{n} a_{1i}b_{i1} & \cdots & \sum\limits_{i=1}^{n} a_{1i}b_{in} \\ \vdots & & \vdots \\ \sum\limits_{i=1}^{n} a_{ni}b_{i1} & \cdots & \sum\limits_{i=1}^{n} a_{ni}b_{in} \end{pmatrix}, \; c_{ik} = \sum\limits_{j=1}^{n} a_{ij}b_{jk}.$$

We call $\mathbf{C}$ the **matrix product** $\mathbf{AB}$ of $\mathbf{A}$ and $\mathbf{B}$. Now we may write (3) also as

$$\mathbf{x}'' = \mathbf{x}(\mathbf{AB}). \tag{3'}$$

So the geometrical operation of applying two linear mappings successively is reflected in the algebraic operation of a matrix multiplication. (For a formal development of matrix calculus see Appendix 4.II.)

If $\mathscr{V}$ coordinatizes an affine space, we wish to consider not only mappings which leave the origin fixed—that is, mappings of the vector space—but also more general transformations which map lines onto lines. Therefore, we add, in affine spaces, a translation to a linear mapping of $\mathscr{V}$:

$$\mathbf{x}' = \mathbf{x}\mathbf{A} + \mathbf{c}, \tag{4}$$

where $\mathbf{c}$ is the translation vector. If (4) is a mapping of the affine space onto itself, we call it an **affine mapping**, or an **affinity**.

Examples

(d)
$$\mathbf{A} = \begin{pmatrix} 1 & 0 & 0 \\ 0 & 1 & 0 \\ 0 & 0 & 1 \end{pmatrix}, \quad \mathbf{c} = (c_1, c_2, c_3)$$

is an ordinary translation, since $\mathbf{x}\mathbf{A} = \mathbf{x}$.

(e)
$$\mathbf{A} = \begin{pmatrix} 1 & 1 & 0 \\ 0 & 1 & 0 \\ 0 & 0 & 1 \end{pmatrix}, \quad \mathbf{c} = (0, 0, 1)$$

is a shear with the (x_1, x_3)-plane as axis followed by a translation along the x_3 axis.

Exercises

1. Let $\mathscr{V}$ coordinatize an affine 3-space. Show explicitly that a linear mapping maps every plane of the affine space onto a plane, a line, or a point. Find in example (c) all planes that are mapped onto lines.

2. Describe the mapping $\mathbf{x}' = \mathbf{xA}$ where

$$A = \begin{pmatrix} 1 & 0 & 0 \\ 0 & 0 & 0 \\ 0 & 0 & 0 \end{pmatrix}.$$

In particular, find all planes that are mapped onto points.

3. Describe, in ordinary Euclidean 3-space, the mapping $\mathbf{x}' = \mathbf{Ax} + \mathbf{c}$ with

$$A = \begin{pmatrix} 0 & 0 & 1 \\ 0 & 1 & 0 \\ 1 & 0 & 0 \end{pmatrix}; \quad \mathbf{c} = (0, 1, 0).$$

Find the image of the triangle $\triangle PQR$ where $(O, P) = \mathbf{a}_1$, $(O, Q) = \mathbf{a}_2$, $(O, R) = \mathbf{a}_3$. Draw a figure.

4. Show that in a Pappian affine plane (coordinatized by a 2-dimensional vector space) every mapping $\mathbf{x}' = \mathbf{Ax}$ with

$$A = \begin{pmatrix} 1 & a \\ 0 & 1 \end{pmatrix}$$

represents a shear. Find the axis of this shear.

5. Express the skew reflection presented in Section 3.3, in matrix form.

6. Express the shears presented in Theorems 4.7.1 and 4.8.1 in matrix form.

4.11   Motions and Similarities

We are ready now to discuss special affine mappings

$$\mathbf{x}' = \mathbf{xA} + \mathbf{c},$$

where the skew field is assumed commutative. We have already considered translations

$$\mathbf{x}' = \mathbf{x} + \mathbf{c},$$

which we may write in the form $\mathbf{x}' = \mathbf{xI} + \mathbf{c}$ where $\mathbf{I}$ is the unit matrix. Also, the dilatations,

$$\mathbf{x}' = d\mathbf{x}$$

where d is a constant factor $\neq 0$, have been introduced. Writing formally, as in the first equation above, we may set $\mathbf{x}' = \mathbf{x}(d\mathbf{I})$, where $d\mathbf{I}$ is the matrix with d's as diagonal elements and zeros otherwise.

We shall now define motions for a general affine space. Compared to ordinary Euclidean 3-space, the generalization refers to the coordinate field as well as to the dimension. However, as with the general theory of vector spaces, a restriction is made: We assume that the *inner product* of two vectors $\mathbf{x}, \mathbf{y}$ is generally defined by

$$\mathbf{x} \cdot \mathbf{y} = x_1 y_1 + \cdots + x_n y_n$$

to satisfy the following condition: If $\mathbf{x} \cdot \mathbf{y} = 0$ for all vectors $\mathbf{y}$, then $\mathbf{x} = 0$

(for a more general treatment of inner products see Appendix 4.I). This is so in particular, for the field $\mathcal{R}$ of real numbers.

Two vectors $\mathbf{x}, \mathbf{y}$ are called **orthogonal** if $\mathbf{x} \cdot \mathbf{y} = 0$; they are called **orthonormal** if, in addition, $\mathbf{x} \cdot \mathbf{x} = 1$ and $\mathbf{y} \cdot \mathbf{y} = 1$. A basis of the vector space that defines the affine space is called **orthonormal** if each pair of its elements is orthonormal.

By the results of Section 4.6, we define a **motion** or **isometry** as an affine mapping that preserves the inner product: If $\mathbf{x}', \mathbf{y}'$ are the images of $\mathbf{x}, \mathbf{y}$, respectively, then

$$\mathbf{x}' \cdot \mathbf{y}' = \mathbf{x} \cdot \mathbf{y}.$$

Theorem 4.11.1 *Any translation*

$$\mathbf{x}' = \mathbf{x} + \mathbf{c}$$

is a motion.

Proof Let $\mathbf{x}_i = (O, X_i), \mathbf{y}_i = (O, Y_i), \mathbf{x}_i' = (O, X_i'), \mathbf{y}_i' = (O, Y_i'), (i = 1, 2)$. From

$$(\mathbf{x}_1' - \mathbf{y}_1') \cdot (\mathbf{x}_2' - \mathbf{y}_2') = (\mathbf{x}_1 + \mathbf{c} - \mathbf{y}_1 - \mathbf{c}) \cdot (\mathbf{x}_2 + \mathbf{c} - \mathbf{y}_2 - \mathbf{c})$$
$$= (\mathbf{x}_1 - \mathbf{y}_1) \cdot (\mathbf{x}_2 - \mathbf{y}_2),$$

the assertion follows. ◆

If we ask for conditions under which an affine mapping

$$\mathbf{x}' = \mathbf{x}\mathbf{A} + \mathbf{c}$$

is a motion, we may, because of Theorem 4.11.1, restrict ourselves to the case

$$\mathbf{x}' = \mathbf{x}\mathbf{A}. \tag{1}$$

Suppose (1) is a motion. Since all inner products remain unchanged, this applies, in particular, to the basis vectors $\mathbf{a}_1, \ldots, \mathbf{a}_n$. Substituting

$$\mathbf{a}_1 = (1, 0, \ldots, 0), \ldots, \mathbf{a}_n = (0, \ldots, 0, 1)$$

in (1) shows that the rows of $\mathbf{A}$ are just the images $\mathbf{a}_1', \ldots, \mathbf{a}_n'$ of $\mathbf{a}_1, \ldots, \mathbf{a}_n$ under (1):

$$\mathbf{A} = \begin{pmatrix} \mathbf{a}_1' \\ \vdots \\ \mathbf{a}_n' \end{pmatrix}.$$

Now we conclude from the assumption that $\mathbf{a}_1, \ldots, \mathbf{a}_n$ form an orthonormal basis and from the properties of a motion that

$$\mathbf{a}_1'^2 = \cdots = \mathbf{a}_1'^2 = \mathbf{a}_1^2 = \cdots = \mathbf{a}_n^2 = 1$$
$$\mathbf{a}_1' \cdot \mathbf{a}_2' = \mathbf{a}_1' \cdot \mathbf{a}_3' = \cdots = \mathbf{a}_{n-1}' \cdot \mathbf{a}_n'$$
$$= \mathbf{a}_1 \cdot \mathbf{a}_2 = \mathbf{a}_1 \cdot \mathbf{a}_3 = \cdots = \mathbf{a}_{n-1} \cdot \mathbf{a}_n = 0.$$

This can be expressed more simply by writing ($^T\mathbf{A}$ is the transposed matrix of $\mathbf{A}$; see Appendix 4.II)

$$^T\mathbf{A}\mathbf{A} = \begin{pmatrix} \mathbf{a}'_1 \\ \vdots \\ \mathbf{a}'_n \end{pmatrix} (\mathbf{a}'_1 \cdots \mathbf{a}'_n) = \begin{pmatrix} \mathbf{a}'^2_1 & \mathbf{a}'_1 \cdot \mathbf{a}'_2 & \cdots & \mathbf{a}'_1 \cdot \mathbf{a}'_n \\ \mathbf{a}'_2 \cdot \mathbf{a}'_1 & \mathbf{a}'^2_2 & & \\ \vdots & \vdots & & \vdots \\ \mathbf{a}'_n \cdot \mathbf{a}'_1 & \mathbf{a}'_n \cdot \mathbf{a}'_2 & \cdots & \mathbf{a}'^2_n \end{pmatrix}$$

$$= \begin{pmatrix} 1 & 0 & \cdots & 0 \\ 0 & & \ddots & \vdots \\ \vdots & & & 0 \\ 0 & \cdots & 0 & 1 \end{pmatrix}$$

or, shortly, by

$$\mathbf{A}^T\mathbf{A} = \mathbf{I}. \tag{2}$$

A matrix $\mathbf{A}$ satisfying (2) is called an **orthogonal matrix.**

Theorem 4.11.2 *An affine mapping*

$$\mathbf{x}' = \mathbf{x}\mathbf{A}$$

is a motion if and only if $\mathbf{A}$ is orthogonal.

Proof The necessity part has been shown above. To show sufficiency, assume $\mathbf{A}^T\mathbf{A} = \mathbf{I}$. Using the matrix rule $^T(\mathbf{x}\mathbf{A}) = {^T}\mathbf{A}\mathbf{x}$ (see appendix B), we find

$$\mathbf{x}' \cdot \mathbf{y}' = (\mathbf{x}\mathbf{A})^T(\mathbf{y}\mathbf{A}) = \mathbf{x}\mathbf{A}^T\mathbf{A}\mathbf{y}$$
$$= \mathbf{x}\mathbf{I}\mathbf{y} = \mathbf{x} \cdot \mathbf{y},$$

so that inner products are preserved and hence the affine mapping is a motion. ◆

Example

$$\mathbf{A} = \begin{pmatrix} 1 & 0 & 0 \\ 0 & \cos\theta & -\sin\theta \\ 0 & \sin\theta & \cos\theta \end{pmatrix}.$$

$\mathbf{A}^T\mathbf{A} = \mathbf{I}$ is easily verified. We shall show later that this matrix represents a rotation about the x_3 axis.

Theorem 4.11.3 *If $\mathbf{A}$, $\mathbf{B}$ are orthogonal, $\mathbf{A}\mathbf{B}$ is also orthogonal.*

Proof Since $\mathbf{A}$, $\mathbf{B}$ represent motions, $\mathbf{A}\mathbf{B}$ also represents a motion. Therefore, by Theorem 4.11.2, $\mathbf{A}\mathbf{B}$ is orthogonal. ◆

From Theorem 4.11.3 we can deduce the following: If $\mathbf{x}' = \mathbf{x}\mathbf{B}$ represents a motion and if $\mathbf{x} = \tilde{\mathbf{x}}\mathbf{A}$, then

$$\widetilde{\mathbf{x}'}\mathbf{A} = \tilde{\mathbf{x}}\mathbf{B}\mathbf{A},$$

or

$$\tilde{\mathbf{x}}' = \tilde{\mathbf{x}}\mathbf{ABA}^{-1}. \tag{1'}$$

By Theorem 4.11.3, $\mathbf{ABA}^{-1}$ is also orthogonal.

Theorem 4.11.4 *If we shift from one orthonormal basis to another, a motion* (1) *is expressed by* (1') *where* $\tilde{\mathbf{B}} = \mathbf{ABA}^{-1}$ *is an orthogonal matrix.*

By a ***similarity*** in Euclidean 3-space we mean an affine mapping that preserves ratios of distances between points. As the definition of trigonometric functions shows, this implies that similarities preserve angles. Conversely, if angles are preserved, trigonometric functions are also preserved, and by a simple argument, it follows that all ratios of segments remain invariant.

Given a similarity in Euclidean 3-space, we can show that there exists a constant $d \neq 0$ such that the distance of any pair of different points divided by the distance of the image points equals d. In fact, let

$$d = \frac{d(P_0, Q_0)}{d(P_0', Q_0')}$$

for two arbitrarily chosen points P_0, Q_0. For any other two different points P, Q we have, by assumption,

$$\frac{d(P_0, Q_0)}{d(P, Q)} = \frac{d(P_0', Q_0')}{d(P', Q')};$$

hence

$$\frac{d(P_0, Q_0)}{d(P_0', Q_0')} = \frac{d(P, Q)}{d(P', Q')} = d.$$

So there exists a universal "dilatation factor" d. If $|d| > 1$, then all figures are enlarged, whereas they are diminished if $|d| < 1$. For $d = +1$, the similarity is a motion.

Now we define in general a ***similarity*** as an affine map satisfying

$$\mathbf{x}' \cdot \mathbf{y}' = d^2 \mathbf{x} \cdot \mathbf{y}$$

for a constant $d \neq 0$. So, from Theorems 4.11.1 and 4.11.2, we derive the following theorem.

Theorem 4.11.5 *An affine mapping*

$$\mathbf{x}' = \mathbf{x}\mathbf{A} + \mathbf{c}$$

is a similarity if and only if

$$\mathbf{A}^{\mathrm{T}}\mathbf{A} = d^2\mathbf{I}$$

for a constant number $d \neq 0$.

Examples
(a) See Example (b) in Section 4.10.

(b)

$$A = \begin{pmatrix} 0 & 0 & 2 \\ 0 & 2 & 0 \\ 2 & 0 & 0 \end{pmatrix}$$

represents a reflection in a plane through the x_2 axis followed by a dilatation with dilatation factor 2. We may write

$$A = 2\begin{pmatrix} 0 & 0 & 1 \\ 0 & 1 & 0 \\ 1 & 0 & 0 \end{pmatrix} = \begin{pmatrix} 2 & 0 & 0 \\ 0 & 2 & 0 \\ 0 & 0 & 2 \end{pmatrix}\begin{pmatrix} 0 & 0 & 1 \\ 0 & 1 & 0 \\ 1 & 0 & 0 \end{pmatrix}.$$

Exercises

1. Find all values of s, t for which the following matrix A represents (a) a motion, (b) a similarity.

$$A = \begin{pmatrix} s \cdot t & 0 & 0 \\ 0 & 0 & s + t \\ 0 & s - t & 0 \end{pmatrix}$$

2. Show that every motion of an ordinary Euclidean plane that leaves the origin of a Cartesian coordinate system fixed can be represented by

$$A = \begin{pmatrix} \cos \theta & -\sin \theta \\ \varepsilon \sin \theta & \varepsilon \cos \theta \end{pmatrix}$$

for some angle θ, $0 \le \theta \le \pi$, where $\varepsilon = \pm 1$.

3. Show that every similarity is the product of a motion and a dilatation.

4. (a) Find a similarity that maps $(0, 0, 0)$, $(1, 0, 0)$, $(0, 1, 0)$ onto $(3, 0, 1)$, $(4, 0, 0)$, $(2, 0, 0)$, respectively. (b) Find all similarities in (a).

5. Show that any involutoric similarity is a motion.

6. Which results of this section would break down without the assumption made about inner products at the beginning of the section?

4.12   Equiaffine Mappings

Now we pick up again the study of volume-preserving maps, in terms of matrices. As in Section 4.8, we assume the coordinate skew field to be commutative. Also, we restrict ourselves again to considering the volume of spars, defined by spar products.

We call an affine map

$$\mathbf{x}' = \mathbf{x}\mathbf{A} + \mathbf{c}$$

volume preserving or *equiaffine* if it does not change the spar products except for the sign. In Section 4.8 we discussed two kinds of equiaffine maps: shears and translations. Since spar products are invariant under translations by their very definition, we can restrict our attention to affine maps

$$\mathbf{x}' = \mathbf{A}\mathbf{x}; \tag{1}$$

furthermore, we can assume that all spars considered have one vertex at O.

A spar of fundamental importance is that spanned by the given basis vectors $\mathbf{a}_1, \ldots, \mathbf{a}_n$. By the definition of spar products (Section 4.9),

$$[\mathbf{a}_1, \ldots, \mathbf{a}_n] = 1.$$

If (1) is applied to $\mathbf{a}_1, \ldots, \mathbf{a}_n$, we obtain

$$[\mathbf{a}_1', \ldots, \mathbf{a}_n']$$

as the volume of the image spar spanned by $\mathbf{a}_1', \ldots, \mathbf{a}_n'$. We call

$$[\mathbf{a}_1', \ldots, \mathbf{a}_n'] = [\mathbf{a}_1\mathbf{A}, \ldots, \mathbf{a}_n\mathbf{A}]$$

the determinant of $\mathbf{A}$ and denote it by $|\mathbf{A}|$:

$$|\mathbf{A}| = [\mathbf{a}_1', \ldots, \mathbf{a}_n'].$$

Theorem 4.12.1 *An affine mapping is equiaffine if and only if*

$$|\mathbf{A}| = \pm 1.$$

Now we submit $\mathbf{a}_1', \ldots, \mathbf{a}_n'$ to a second affine map

$$\mathbf{x}'' = \mathbf{x}'\mathbf{B}.$$

Relative to $\mathbf{a}_1', \ldots, \mathbf{a}_n'$ as basis, the volume of the spar spanned by $\mathbf{a}_1'', \ldots, \mathbf{a}_n''$ would be $|\mathbf{B}|$; in other words,

$$\frac{[\mathbf{a}_1'', \ldots, \mathbf{a}_n'']}{|\mathbf{A}|} = |\mathbf{B}|,$$

or

$$[\mathbf{a}_1'', \ldots, \mathbf{a}_n''] = |\mathbf{A}||\mathbf{B}|.$$

On the other hand, by joining the two affine maps together we find

$$[\mathbf{a}_1'', \ldots, \mathbf{a}_n''] = |\mathbf{A}\mathbf{B}|.$$

Now we can show an interesting law for determinants of matrices.

Theorem 4.12.2 $|\mathbf{B}\mathbf{A}| = |\mathbf{A}\mathbf{B}| = |\mathbf{A}||\mathbf{B}|.$

Proof Interchanging the roles of $\mathbf{A}$ and $\mathbf{B}$ and using $|\mathbf{A}||\mathbf{B}| = |\mathbf{B}||\mathbf{A}|$, we obtain

$$|\mathbf{A}\mathbf{B}| = |\mathbf{A}||\mathbf{B}| = |\mathbf{B}||\mathbf{A}| = |\mathbf{B}\mathbf{A}|.$$

The theorem is also true for $|A| = 0$ or $|B| = 0$, since, in this case, $a_1'', \ldots, a_n''$ are linearly dependent (by definition of the spar product, $[a_1'', \ldots, a_n''] = 0$, in this case). ◆

The next theorem readily follows from Theorem 4.12.2.

Theorem 4.12.3 *The family of all equiaffine mappings is a group under successive application.*

The following theorem on determinants can be verified by calculation.

Theorem 4.12.4 *The determinants of an $n \times n$ matrix A and its transposed matrix are equal:*

$$|{}^{T}A| = |A|.$$

With the aid of this theorem, we can show that every motion is an equiaffine map.

Theorem 4.12.5 *Every motion is an equiaffine map.*

Proof $1 = |I| = |A^{T}A| = |A||{}^{T}A| = |A|^2$, hence

$$|A| = \pm 1.$$ ◆

Theorem 4.12.6 *Every equiaffine similarity is a motion.*

Proof From ${}^{T}AA = d^2 I$ and $|A| = \pm 1$, we obtain

$$d^{2n} = |d^2 I| = |{}^{T}AA| = |{}^{T}A||A| = |A|^2 = 1;$$

hence,

$$A^{T}A = I.$$ ◆

Exercises

1. Carry out the calculation that proves Theorem 4.12.4.

2. Show explicitly that the determinant of the matrix of a shear in a 3-space (see Exercise 6 of Section 4.10) is 1.

3. Can every equiaffine map in 3-space be expressed as a product of shears?

4. Show that every equiaffine map in 3-space with determinant $+1$ can be expressed as a product of several shears. (*Hint*: Shift an arbitrarily given spar with one vertex at O by appropriate shears into the spar spanned by the basis vectors.)

5. Represent by matrices all shears in 3-space that have the (x_1, x_2)-plane as axis.

6. Show that any shear that is also a similarity is the identity map.

4.13 Fixed Points and Fixed Lines of Motions

In a further investigation of motions (see Section 4.11), fixed points and fixed lines play a vital role. In the case of a two-dimensional space, we observed this in Section 1.11; our approach was synthetic there. In this section we discuss a number of facts on fixed points and fixed lines analytically. We are given an arbitrary affine space, the coordinate field being commutative and of characteristic $\neq 2$. Furthermore, the inner product $\mathbf{x} \cdot \mathbf{y}$ possesses the property introduced at the beginning of Section 4.11. All assumptions are trivially true for the field $\mathscr{R}$ of real numbers. We shall find a complete classification of motions in ordinary Euclidean 3-space in the next section. If an arbitrary affine mapping

$$\mathbf{x}' = \mathbf{x}A + \mathbf{c} \tag{1}$$

possesses a fixed point $\mathbf{x}$, we obtain

$$\mathbf{x} = \mathbf{x}A + \mathbf{c}. \tag{2}$$

This can also be written (using $\mathbf{x} = \mathbf{x}I$)

$$\mathbf{x}(A - I) = -\mathbf{c}. \tag{3}$$

So, finding fixed points is equivalent to finding solutions of the system (3) of linear equations. Clearly, there exists a fixed point if $|A - I| \neq 0$, namely, the point $-\mathbf{c}(A - I)^{-1}$. If, however, $|A - I| = 0$, there need not exist any fixed point.

Theorem 4.13.1 *If $\mathbf{x}_1$, $\mathbf{x}_2$ are different fixed points of (1), the line spanned by $\mathbf{x}_1$, $\mathbf{x}_2$ consists of fixed points only.*

Proof The line spanned by $\mathbf{x}_1$, $\mathbf{x}_2$ has a representation (see Section 4.5)

$$\mathbf{x} = t(\mathbf{x}_2 - \mathbf{x}_1) + \mathbf{x}_1.$$

Substituting in (1), we obtain

$$\mathbf{x}A + \mathbf{c} = (1 - t)\mathbf{x}_1A + t\mathbf{x}_2A + \mathbf{c} = (1 - t)(\mathbf{x}_1A + \mathbf{c}) + t(\mathbf{x}_2A + \mathbf{c})$$
$$= (1 - t)\mathbf{x}_1 + t\mathbf{x}_2 = \mathbf{x}. \quad \blacklozenge$$

This theorem implies the following one.

Theorem 4.13.2 *If $\mathbf{x}_1, \mathbf{x}_2, \ldots, \mathbf{x}_r$ are fixed points of (1), the linear subspace of the affine space spanned by these points consists of fixed points only.*

It should be noted that Theorems 4.13.1 and 4.13.2 are true for arbitrary affine mappings. Now we turn to fixed lines. We restrict ourselves to the case $\mathbf{c} = \mathbf{0}$; that is,

$$\mathbf{x}' = \mathbf{x}A, \tag{4}$$

and we ask for fixed lines through O. If $\mathbf{u} \neq \mathbf{0}$, the line

$$\mathbf{x} = t\mathbf{u} \quad (t \text{ arbitrary})$$

represents a fixed line under (4) if $\mathbf{u}$ is mapped onto a multiple $\neq \mathbf{0}$ of itself:

$$\lambda\mathbf{u} = \mathbf{u}\mathbf{A}.$$

This can be written as

$$\mathbf{u}(\mathbf{A} - \lambda\mathbf{I}) = 0. \tag{5}$$

If $|\mathbf{A} - \lambda\mathbf{I}| \neq 0$, (5) has only $\mathbf{0}$ as a solution (see the preceding section). Therefore, $\mathbf{x} = t\mathbf{u}$ represents a fixed line only if

$$|\mathbf{A} - \lambda\mathbf{I}| = 0. \tag{6}$$

A value of λ for which (6) is true is called an *eigenvalue* of the matrix $\mathbf{A}$. A vector $\mathbf{u}$ different from the zero vector satisfying (5) is then called an *eigenvector* to the eigenvalue λ.

So far, all our considerations have been valid for arbitrary affinities. Let us now consider the special case of motions. Since, by definition of a motion, $(\mathbf{x}\mathbf{A})^2 = \mathbf{x}^2$, we obtain $\lambda^2 = 1$ and thus only two possibilities for λ:

$$\lambda = +1, \quad \text{or} \quad \lambda = -1.$$

In the first case, $\mathbf{u}$ is a fixed point of (4). Since O is another fixed point, we conclude in this case, from Theorem 4.13.1, that the line $\mathbf{x} = t\mathbf{u}$ remains pointwise fixed.

Theorem 4.13.3 *Every motion*

$$\mathbf{x}' = \mathbf{x}\mathbf{A}$$

in an odd-dimensional (in particular, three-dimensional) space possesses a fixed line.

This follows from the following statement (n the dimension of the given affine space):

Theorem 4.13.4 (a) *If n is odd and $|\mathbf{A}| = +1$ or if n is even and $|\mathbf{A}| = -1$, the orthogonal matrix $\mathbf{A}$ possesses 1 as an eigenvalue.*
(b) *If $|\mathbf{A}| = -1$, the orthogonal matrix $\mathbf{A}$ possesses -1 as an eigenvalue.*

Proof Since $^{\mathrm{T}}\mathbf{A}\mathbf{A} = \mathbf{I}$ (see Section 4.11) we find

$$(\mathbf{A} - \mathbf{I})^{\mathrm{T}}\mathbf{A} = -(^{\mathrm{T}}\mathbf{A} - \mathbf{I}) = -^{\mathrm{T}}(\mathbf{A} - \mathbf{I}).$$

Clearly $|-\mathbf{I}| = (-1)^n$ and thus, by Theorems 4.12.4 and 4.12.2,

$$|\mathbf{A} - \mathbf{I}| = \pm(-1)^n|\mathbf{A} - \mathbf{I}|,$$

where "$+$" is to be taken in case $|^{\mathrm{T}}\mathbf{A}| = |\mathbf{A}| = +1$, and "$-$" otherwise. If n is odd and $|\mathbf{A}| = +1$, we obtain

$$|\mathbf{A} - \mathbf{I}| = -|\mathbf{A} - \mathbf{I}|;$$

that is,

$$|\mathbf{A} - \mathbf{I}| = 0,$$

which expresses 1 as an eigenvalue. The same conclusion holds for n even and $|\mathbf{A}| = -1$. Similarly, we conclude from the equation

$$(\mathbf{A} + \mathbf{I})^{\mathrm{T}}\mathbf{A} = {}^{\mathrm{T}}\mathbf{A} + \mathbf{I} = {}^{\mathrm{T}}(\mathbf{A} + \mathbf{I}),$$

[4.13]

using Theorems 4.12.2 and 4.12.4, that

Affine Space and Vector Space

$$|\mathbf{A} + \mathbf{I}| = \pm |\mathbf{A} + \mathbf{I}|,$$

if the upper sign is taken for $|\mathbf{A}| = 1$. For $|\mathbf{A}| = -1$, this implies

$$|\mathbf{A} + \mathbf{I}| = 0,$$

that is, -1 is an eigenvalue of $\mathbf{A}$. ◆

Let s be the dimension of the linear space $\mathscr{F}$ of all fixed points of (4) (see Theorem 4.13.2). Let an orthonormal basis $\mathbf{a}_1, \ldots, \mathbf{a}_n$ of the space be given. We assume that $\mathbf{a}_1, \ldots, \mathbf{a}_s$ span $\mathscr{F}$. Then $\mathbf{a}_{s+1}, \ldots, \mathbf{a}_n$ span a subspace $\mathscr{F}^\perp$ consisting of vectors perpendicular to every vector of $\mathscr{F}$. Clearly, $\mathscr{F}^\perp$ is mapped onto itself under (4).

Theorem 4.13.5 $n - s$ *is even if* $|\mathbf{A}| = +1$ *and it is odd if* $|\mathbf{A}| = -1$.

Proof Since $\mathbf{a}_i\mathbf{A} = \mathbf{a}_i$ for $i = 1, \ldots, s$ and since $\mathscr{F}^\perp$ is mapped onto itself under (4), we obtain for $\mathbf{A}$:

$$\mathbf{A} = \left(\begin{array}{cc|c} 1 & & 0 \\ & \ddots & \\ & & 1 \\ \hline 0 & & \mathbf{B} \end{array} \right),$$

where the left upper square of the matrix is a unit matrix and the upper right and the lower left parts consist of zeros only. Clearly

$$|\mathbf{A}| = |\mathbf{B}|,$$

where $|\mathbf{A}|$ is a determinant of n rows and $|\mathbf{B}|$ is a determinant of $n - s$ rows. Since $\mathscr{F}^\perp$ contains no fixed point not equal to $\mathbf{0}$, $\mathbf{B}$ does not possess 1 as an eigenvalue. Hence, by (a) of Theorem 4.13.4, we conclude Theorem 4.13.5 ◆

Theorem 4.13.6 *If* 1 *and* -1 *are both eigenvalues of* $\mathbf{A}$ *and if* $\mathbf{u}$, $\mathbf{v}$ *are eigenvectors corresponding to* 1, -1, *respectively, then*

$$\mathbf{u} \cdot \mathbf{v} = 0.$$

Proof $\mathbf{u} \cdot \mathbf{v} = (\mathbf{u}\mathbf{A}) \cdot (\mathbf{v}\mathbf{A}) = \mathbf{u} \cdot (-\mathbf{v}) = -\mathbf{u}\mathbf{v}$; hence $\mathbf{u} \cdot \mathbf{v} = 0$. ◆

Exercises

1. Determine all fixed points and fixed lines of the mapping $\mathbf{x}' = \mathbf{x}\mathbf{A}$ where

$$\mathbf{A} = \begin{pmatrix} 0 & 1 & 0 & 0 \\ 1 & 0 & 0 & 0 \\ 0 & 0 & 0 & -1 \\ 0 & 0 & 1 & 0 \end{pmatrix}.$$

2. Let $\mathbf{A}$ be as in Exercise 1. Show that $\mathbf{x}' = \mathbf{x}\mathbf{A} + \mathbf{c}$ possesses a fixed point if $\mathbf{c} = (0, 0, 1, 1)$ and does not possess a fixed point if $\mathbf{c} = (1, 0, 0, 0)$.

3. Find the fixed point of $\mathbf{x}' = \mathbf{x}\mathbf{A} + \mathbf{c}$ where

$$\mathbf{A} = \begin{pmatrix} \dfrac{\sqrt{2}}{2} & -\dfrac{\sqrt{2}}{2} \\ \dfrac{\sqrt{2}}{2} & \dfrac{\sqrt{2}}{2} \end{pmatrix}, \qquad \mathbf{c} = (1, 3).$$

4. Show that if $\mathbf{A}$ possesses λ as an eigenvalue, then $\mathbf{A}^2$ possesses λ^2 as an eigenvalue.

5. Show that the matrices $\mathbf{A}$ and $\mathbf{B}^{-1}\mathbf{A}\mathbf{B}$ have the same eigenvalues (provided $|\mathbf{B}| \neq 0$).

6. Carry out the proof of Theorem 4.13.2 explicitly.

4.14 Classification of Motions in 3-Space

We are given the ordinary Euclidean 3-space ($\mathcal{K}$ = field of real numbers). A classification of planar motions presented in Section 1.18 states that a proper motion is a translation or a rotation; an improper motion is a glide reflection. In an analytic representation, a motion is proper if $|\mathbf{A}| = +1$, and improper if $|\mathbf{A}| = -1$.

Let, still in the planar case, $\mathbf{A}$ represent a rotation with center O. Since $a_{11}^2 + a_{12}^2 = 1$, we have $|a_{11}| \leq 1$ and we may set

$$a_{11} = \cos\theta.$$

Then, from $a_{11}^2 + a_{12}^2 = 1$, it follows that either $a_{12} = \sin\theta$ or $a_{12} = -\sin\theta$. Similarly (since $^{\mathrm{T}}\mathbf{A}$ is orthogonal if $\mathbf{A}$ is orthogonal), we obtain from $a_{11}^2 + a_{12}^2 = 1$:

$$a_{21} = \mp\sin\theta.$$

Again, $a_{22} = \pm \cos \theta$. Choosing $-\pi/2 \le \theta \le \pi/2$, we find from $|\mathbf{A}| = 1$
that $a_{22} = \cos \theta$. We may set $a_{12} = -\sin \theta$, $a_{21} = \sin \theta$ (otherwise replace
θ by $\theta' = -\theta$). So we find

$$\mathbf{A} = \begin{pmatrix} \cos \theta & -\sin \theta \\ \sin \theta & \cos \theta \end{pmatrix}.$$

Let

$$\mathbf{x'} = \mathbf{x}\mathbf{A} + \mathbf{c} \tag{1}$$

be a motion of the given 3-space. We assume first $|\mathbf{A}| = 1$, that is, the
motion is proper. Since $n = 3$, we conclude from Theorem 4.13.5 that
$s = 1$ or $s = 3$. In case $s = 3$, we obtain $\mathbf{A} = \mathbf{I}$ and (1) represents a translation. So let $s = 1$. In this case, (1) is not a translation. We choose an
orthonormal basis $\mathbf{b}_1, \mathbf{b}_2, \mathbf{b}_3$ in such a way that $\mathbf{b}_3 = \mathbf{b}_3\mathbf{A}$. In the (x_1, x_2)-
plane, (1) induces a rotation. Therefore we obtain

$$\mathbf{A} = \begin{pmatrix} \cos \theta & -\sin \theta & 0 \\ \sin \theta & \cos \theta & 0 \\ 0 & 0 & 1 \end{pmatrix}.$$

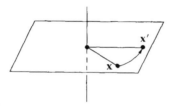

Figure 4.31

In every plane parallel to the (x_1, x_2)-plane, $\mathbf{x'} = \mathbf{x}\mathbf{A}$ induces a rotation, too
(Fig. 4.31). So $\mathbf{x'} = \mathbf{x}\mathbf{A}$ is what we call a *rotation* of the space. (1), written
explicitly, now reads

$$\begin{aligned}
x_1' &= x_1 \cos \theta + x_2 \sin \theta + c_1, \\
x_2' &= -x_1 \sin \theta + x_2 \cos \theta + c_2, \\
x_3' &= x_3 + c_3.
\end{aligned} \tag{1'}$$

The first two equations represent a proper motion of the (x_1, x_2)-plane, which
is not a translation (otherwise (1') would represent a translation) and hence
must be a rotation. By a shift of the origin into the (x_1, x_2)-plane, the origin
can be made the center of this rotation. (1') now reads

$$\begin{aligned}
x_1' &= x_1 \cos \theta + x_2 \sin \theta, \\
x_2' &= -x_1 \sin \theta + x_2 \cos \theta, \\
x_3' &= x_3 + c_3.
\end{aligned} \tag{1''}$$

This is a rotation about the x_3 axis followed by a translation in the direction
of the axis. We call it a *screw motion* (Fig. 4.32).

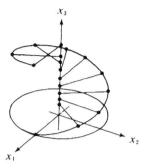

Figure 4.32

Let now the motion in (1) be improper, that is, $|\mathbf{A}| = -1$. By Theorem 4.13.5, we must distinguish $s = 0$ and $s = 2$. In both cases, -1 is an eigenvalue of $\mathbf{A}$, and thus there is a unit vector $\mathbf{b}_3$ satisfying $-\mathbf{b}_3 = \mathbf{b}_3\mathbf{A}$. Since $\mathbf{x} = t\mathbf{b}_3$ is a fixed line and o a fixed point under $\mathbf{x}' = \mathbf{xA}$, the plane perpendicular to this line and passing through o is mapped onto itself under $\mathbf{x}' = \mathbf{xA}$. Let $\mathbf{b}_1, \mathbf{b}_2$ be perpendicular unit vectors in this plane. Then $\mathbf{b}_1, \mathbf{b}_2, \mathbf{b}_3$ is a basis of the whole space.

Since fixed points belong to the eigenvalue $+1$, we conclude from Theorem 4.13.6 that all fixed points lie in a plane perpendicular to $\mathbf{b}_3$. Therefore, if $s = 2$, the plane consisting of fixed points is that spanned by $\mathbf{b}_1, \mathbf{b}_2$. Now (1) attains the form

$$
\begin{aligned}
x_1' &= x_1 + c_1 \\
x_2' &= x_2 + c_2 \\
x_3' &= -x_3 + c_3.
\end{aligned}
$$

By shifting the origin along the x_3 axis into $(0, 0, c_{3/2})$, we obtain (1) in the form

$$
\begin{aligned}
x_1' &= x_1 + c_1 \\
x_2' &= x_2 + c_2 \\
x_3' &= -x_3.
\end{aligned}
$$

This is a translation along the (x_1, x_2)-plane followed by a reflection in this plane. As in the planar case, we call it a **glide reflection** (Fig. 4.33).

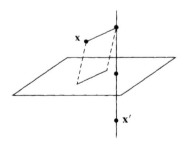

Figure 4.33

Let $s = 0$. With respect to the basis $\mathbf{b_1}, \mathbf{b_2}, \mathbf{b_3}$, $\mathbf{A}$ has the form

$$\mathbf{A} = \begin{pmatrix} & & 0 \\ & \mathbf{B} & 0 \\ 0 & 0 & -1 \end{pmatrix}$$

where $\mathbf{B}$ is a 2×2 matrix. Clearly,

$$|\mathbf{A}| = -|\mathbf{B}|;$$

hence $\mathbf{x}' = \mathbf{xA}$ induces a proper motion in the (x_1, x_2)-plane. Therefore, (1) attains the form

$$\begin{aligned} x_1' &= x_1 \cos \theta + x_2 \sin \theta + c_1, \\ x_2' &= -x_1 \sin \theta + x_2 \cos \theta + c_2, \\ x_3' &= -x_3 + c_3. \end{aligned} \tag{1''}$$

$\theta \neq 0$, since otherwise $s = 2$ would follow. So the first two equations represent a rotation with center (d_1, d_2). We shift the origin into $(d_1, d_2, c_3/2)$ and obtain, instead of (1''),

$$\begin{aligned} x_1' &= x_1 \cos \theta + x_2 \sin \theta \\ x_2' &= -x_1 \sin \theta + x_2 \cos \theta \\ x_3' &= -x_3. \end{aligned}$$

This is a rotation about the x_3 axis followed by a reflection in the (x_1, x_2)-plane. We call it a *rotary inversion* (Fig. 4.34). For $\theta = \pi$ the rotary inversion is said to be a *point symmetry*.

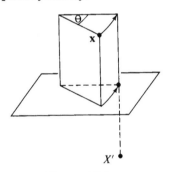

Figure 4.34

We now summarize the results of this section.

Theorem 4.14.1 Classification Theorem *A proper motion of Euclidean 3-space is a translation or a screw motion. An improper motion of Euclidean 3-space is a glide reflection or a rotary inversion.*

A screw motion is readily seen to possess no fixed point unless it is a rotation. Also a glide reflection does not possess a fixed point unless it is a reflection.

Theorem 4.14.2 *Every proper motion of Euclidean 3-space possessing a fixed point is a rotation. Every improper motion of Euclidean 3-space possessing a fixed point is a reflection or a rotary inversion.*

Exercises

1. Show explicitly that a screw motion with $c_3 \neq 0$ does not possess a fixed point.

2. Let $\mathbf{x}' = \mathbf{x}\mathbf{A}$ and $\mathbf{A} = \begin{pmatrix} \dfrac{\sqrt{2}}{2} & -\dfrac{1}{2} & -\dfrac{1}{2} \\[2mm] \dfrac{\sqrt{2}}{2} & \dfrac{1}{2} & \dfrac{1}{2} \\[2mm] 0 & \dfrac{\sqrt{2}}{2} & -\dfrac{\sqrt{2}}{2} \end{pmatrix}$.

   (a) Find all fixed points and fixed lines.
   (b) Determine what kind of motion this is.

3. A screw, turned by an angle of $2\pi/3$, is moved into a wall to a depth of 1 cm. Choose an appropriate coordinate system and determine the equations of this screw motion.

4. Show that the depth which the screw in Exercise 3 moves into the wall is proportional to the angle by which it is turned.

5. Show that all spatial rotations with center O form a group.

6. Show that the product of a glide reflection with itself is a translation.

4.15 The Platonic Solids and Their Symmetry Groups

We say that a figure has a center of symmetry 0 if it is mapped onto itself under the point symmetry at 0. If the figure is mapped onto itself under a reflection in a plane, we say that it has that plane as a plane of symmetry. A more general way to study the symmetry properties of some figure is that of asking for the set of all motions that map the figure onto itself. This set is clearly a group. We call it the **symmetry group** of the figure.

There are many sets that have only the identical motion as a symmetry group. Others, as, for example, a circle, have infinite symmetry groups. In the plane we find as a nontrivial finite symmetry group that of a regular n-gon—that is, a polygon $P_0 P_1 P_2 \cdots P_n$ inscribed in a circle—where $P_0 = P_n$ and $d(P_i, P_{i+1})$ is the same distance for $i = 0, \ldots, n - 1$. It is called a **dihedral group** $\mathscr{D}_n$ and consists of n rotations and n reflections. We are

from four points not in a plane: First, join any two of the four points by line segments and then join any two points of the set thus obtained (see Fig. 4.35).   The triangular regions with any three of the four points as vertices

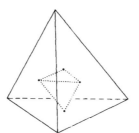

Figure 4.35

are called *faces* of the tetrahedron.   If Π is the union of all points belonging to tetrahedra $T_1, \ldots, T_k$, where T_i and T_{i+1} have a face in common, then Π is called a **polyhedron**.   We say Π is **convex** if it contains with any two points P, Q the line segment (P, Q).   Any tetrahedron clearly is convex.   If X lies in a convex polyhedron Π and is the end point of a ray not meeting Π, we call X a **boundary point** of Π.   If H is a plane that meets Π only in boundary points, we call the intersection $H \cap \Pi$ a *face* or an *edge* or a *vertex* of Π, depending on whether $H \cap \Pi$ contains three noncollinear points or two points or only one point.

A convex polyhedron is said to be *regular* if all its faces are regular k-gons with the same k.   It is easily seen that there exist only five regular convex polyhedra—the so-called *platonic solids*: The **regular tetrahedron**, the **cube**, the **regular octahedron**, the **dodecahedron**, and the **icosahedron** (see Fig. 4.36). If we join the midpoints of the faces of a cube as is indicated in Figure 4.36, we obtain a regular octahedron.   Analogously, we obtain a cube from an octahedron.   In the same way, the dodecahedron and the icosahedron are related to each other.   This leads to the striking fact that *the cube and the regular octahedron possess the same symmetry groups and the dodecahedron and the icosahedron possess the same symmetry groups.*

We study first the group $\mathscr{G}$ of all proper motions which leave a platonic solid fixed.   All vertices of a platonic solid lie on a sphere which is mapped onto itself if the solid is mapped onto itself.   The center of this sphere is a fixed point under any motion leaving the solid fixed (as a whole).   Therefore, by the results of Section 4.14, $\mathscr{G}$ consists only of rotations.   In the case of a cube, there are 13 possible axes for rotations $\neq$ I of $\mathscr{G}$:

> 3 joining the midpoints of opposite faces,
> 6 joining the midpoints of opposite edges,
> 4 joining opposite vertices.

The first three axes provide $3 \cdot 3 = 9$ rotations $\neq$ I, the second six provide 6,

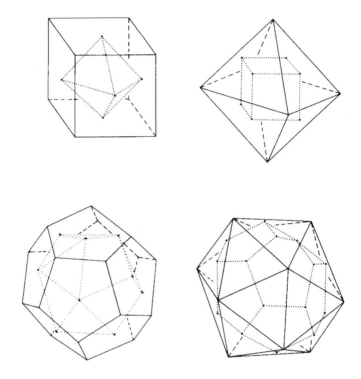

Figure 4.36

and the last four provide $4 \cdot 2 = 8$ rotations $\neq$ I. So altogether $\mathscr{G}$ possesses

$$9 + 6 + 8 + 1 = 24$$

elements; in other words, $\mathscr{G}$ has order 24.

In the case of a dodecahedron, we obtain 31 axes for rotations $\neq$ I of $\mathscr{G}$:

> 6 joining the midpoints of opposite faces,
> 15 joining the midpoints of opposite edges,
> 10 joining opposite vertices.

Altogether we obtain

$$6 \cdot 4 + 15 \cdot 1 + 10 \cdot 2 + 1 = 60$$

elements of $\mathscr{G}$. Clearly, in the case of a tetrahedron, $\mathscr{G}$ possesses 12 elements.

Now we ask for improper motions belonging to the symmetry group of a platonic solid. By Section 4.14, these can only be reflections in planes. If H is a symmetry plane, then a rotation of $\mathscr{G}$ shifts H into another symmetry plane of the solid, and any symmetry plane is obtained in this way. So the order of the full symmetry group is twice that of $\mathscr{G}$. We summarize these results in the following table.

Platonic solid	Proper symmetry groups		Order of full symmetry group
	symbol	order	
Tetrahedron	$\mathscr{A}_4$	12	24
Cube or Octahedron	$\mathscr{S}_4$	24	48
Dodecahedron or icosahedron	$\mathscr{A}_5$	60	120

Exercises

1. Find all subgroups of $\mathscr{A}_4$, $\mathscr{S}_4$, and $\mathscr{A}_5$ of order 2 and 3, respectively.

2. Find the proper symmetry group and the full symmetry group of (a) a "brick," (b) a regular prism, (c) a regular cone (basis a regular h-gon).

3. Show that $\mathscr{S}_4$ is abstractly the same group as that of all permutations of four objects. (*Hint*: Choose the space diagonals of a cube as such objects.)

4. Show that $\mathscr{A}_4$ is abstractly the same group as that of all even permutations of four objects ("alternating group").

5. Represent four of the 24 elements of $\mathscr{S}_4$ by matrices after introducing an appropriate coordinate system.

6. Show explicitly that other than the five platonic solids no regular convex polyhedron exists.

4.16 Finite Groups of Motions

We present now a full account of all finite groups of proper motions that exist in ordinary Euclidean 3-space. Let $\mathscr{G}$ be such a group. Clearly $\mathscr{G}$ does not contain a translation $\neq I$, since any such translation would generate an infinite (cyclic) subgroup of $\mathscr{G}$. In fact, $\mathscr{G}$ contains only rotations as is seen from the following theorem and Theorem 4.14.1.

Theorem 4.16.1 *Every finite group of motions possesses a fixed point.*

Proof Let P be some point and consider all images of P under the motions of the given group $\mathscr{G}$. We obtain a set Σ of finitely many points. We wish to show that there exists a sphere with a smallest possible diameter so that all points of Σ are inside S or on it. (A sphere clearly is a set of points with constant distance greater than 0 from a given point, the center of the sphere. A line segment through the center joining two points of the sphere is called a diameter. It is readily verified that four points not in a plane determine a unique sphere on which they lie; see Exercise 6 of this section.) This is clear if a limit argument ("topological" argument) is applied. However we can do without a limit argument. Among all distances between elements of Σ there is a largest one. Suppose P, P' have this distance.

Case 1 Let the sphere S' through P, P' with diameter $[PP']$ contain all points of Σ in it or on it. Then it is clearly a sphere S.

Case 2 If the sphere S' in Case 1 does not contain all points of Σ, consider all circumcenters of triangles formed by points of Σ. It can happen that a sphere with such a circumcenter C as its center and passing through the vertices of the respective triangle contains all points of Σ in it or on it. C is inside the triangle; so the circumscribed circle is the smallest circle containing the triangle. The sphere we have found is a sphere S.

Case 3 Suppose no circumcenter of a triangle formed by points of Σ is the center of a sphere S. We assert that there exists a sphere through four nonplanar points of Σ containing all other points of Σ in it or on it. *Proof*: Choose three noncollinear points P, P', P'' of Σ such that they determine a plane leaving all points of Σ on one side of the plane or inside the circumscribed circle of P, P', P'' or on that circle. Among all spheres determined by P, P', P'' and any of the other points Q of Σ, there is a largest one. Clearly it contains the points of Σ inside it or on its boundary. This proves the assertion. If there are several such spheres, choose one with center inside the tetrahedron $PP'P''Q$. It exists; otherwise, Case 2 would apply. It is readily seen that we obtain a sphere S.

S is mapped onto itself under all motions of $\mathcal{M}$, since, for an image $S' \neq S$ of S, the intersection $S \cap S'$ would also contain all the above points. $S \cap S'$, however, would lie in a sphere of smaller radius than that of S, contradicting the choice of S. The center of S is now a fixed point under all motions of $\mathcal{M}$. ◆

Let $\mathcal{G}$ be a group of proper motions, that is, a group of rotations. All axes of the rotations intersect in a common point 0. We call such an axis *k-gonal* if there are exactly $k - 1$ rotations $\neq I$ in $\mathcal{G}$ possessing this axis as the axis of rotation. Two k-gonal axes are called *equivalent* if there is a motion in $\mathcal{G}$ transforming one into the other. If a, a' are these axes and R is in $\mathcal{G}$, such that $a' = (a)R$, and if $R_1, \ldots, R_{k-1}$ possess a as their axis of rotation, then $R^{-1}R_1R, \ldots, R^{-1}R_{k-1}R$ possess a' as their axis of rotation. Clearly "equivalent" satisfies the rules of an equivalence relation.

Theorem 4.16.2 *Any equivalence class of k-gonal axes possesses n/k elements, where n is the number of elements (order) of $\mathcal{G}$.*

Proof Let P be a point not situated on any of the axes. If a is an axis belonging to a class of k-gonal axes, the images of P under the rotations of $\mathcal{G}$ with axis a form a k-gon. Applying all other rotations of $\mathcal{G}$ to this k-gon, we obtain a number l of k-gons. These can be considered to have no point in common if we choose P close enough to a. So $k \cdot l$ is the total number of images of P and thus is the order n of $\mathcal{G}: n = k \cdot l$. This proves $l = n/k$. ◆

If we consider a sphere with center O, every axis of rotation intersects this sphere in two different points.   We call such points **poles**, or, more precisely, **k-gonal poles** if they lie on a k-gonal axis.

There are $n - 1$ rotations $\neq I$ in $\mathscr{G}$.   These may be subdivided into rotations belonging to k-gonal points, $k - 1$ for each k-gonal axis.   Since every k-gonal axis determines two k-gonal poles, there are

$$\tfrac{1}{2}(k - 1)$$

rotations for every k-gonal pole or

$$\tfrac{1}{2}(k - 1)\frac{n}{k}$$

for any set of n/k equivalent poles.   This shows

$$n - 1 = \sum \frac{n}{2k}(k - 1) = \frac{n}{2}\sum\left(1 - \frac{1}{k}\right), \tag{1}$$

where the summation is taken over all classes of equivalent poles.   We may assume $n > 1$, since otherwise $\mathscr{G}$ consists of the identity alone.   We may write (1) as

$$2 - \frac{2}{n} = \sum\left(1 - \frac{1}{k}\right). \tag{2}$$

From

$$1 \leq 2 - \frac{2}{n} < 2,$$

we conclude that the sum on the right side of (2) possesses at least two and at most three terms.   So there are two or three equivalence classes of poles. If there are two equivalence classes we find

$$2 - \frac{2}{n} = 1 - \frac{1}{k_1} + 1 - \frac{1}{k_2},$$

or

$$2 = \frac{n}{k_1} + \frac{n}{k_2},$$

from which we can deduce $k_1 = k_2 = n$.   So there is only one axis, and $\mathscr{G} = C_n$ is the cyclic group of n elements.

If there are three equivalence classes, we obtain, similarly,

$$1 + \frac{2}{n} = \frac{1}{k_1} + \frac{1}{k_2} + \frac{1}{k_3}.$$

Since this term is greater than 1, at least one of the k_i, say k_1, is less than 3 and thus equals 2:

$$1 + \frac{2}{n} = \frac{1}{2} + \frac{1}{k_2} + \frac{1}{k_3};$$

hence

$$(k_2 - 2)(k_3 - 2) = 4\left(1 - \frac{k_2 k_3}{n}\right) < 4.$$

So there are four possibilities (assuming $k_2 \leq k_3$):

k_2	2	3	3	3
k_3	k	3	4	5
n	$2k$	12	24	60

From this and the results of the preceding section, we obtain the following theorem.

Theorem 4.16.3 *The only finite groups of rotations are*

> *cyclic groups (order k),*
> *dihedral groups (order $2k$),*
> *tetrahedral group (order 12),*
> *octahedral group (order 24),*
> *icosahedral group (order 60).*

It is also possible to determine all finite groups of proper and improper motions. (See Coxeter [1961], p. 276.)

Exercises

1. Determine by using matrices the symmetry group of a brick. Find the group table.

2. Find the lattice of all subgroups of the octahedral group.

3. Show that the symmetry group of the regular tetrahedrons is the group of all permutations of four objects.

4. Find the lattice of all subgroups of the group in Exercise 3.

5. Find the order of a group of motions generated by a finite group $\mathscr{G}$ of rotations in 3-space and a rotary inversion with O as center (O the common fixed point of the elements of $\mathscr{G}$).

6. Prove explicitly the properties of spheres stated (in parentheses) at the beginning of the proof of Theorem 4.16.1.

Chapter Five

Projective Geometry

If one sits in a room and looks at two pairs of parallel lines in which two walls meet the floor and ceiling, these lines seem to intersect in one point. This becomes even more apparent if we use a photograph and draw extensions of the four lines. As we pointed out early in Chapter 1, there is a certain freedom to decide in a geometrical theory whether parallel lines are to meet at "infinity" or not. In this chapter we assume throughout that they do intersect, as in the special case of an elliptic plane. The geometry thus obtained has a number of interesting and beautiful properties, in particular the so-called principle of duality, which is discussed in Section 5.3.

In the sense of Klein's Erlanger Programm, projective geometry is defined by the invariants of the group of all projective collineations (definition in Section 5.6). This group contains the group of affinities as a proper subgroup and thus adds a member to our hierarchy of geometries.

5.1   Affine Planes and Projective Planes

Suppose we are given two nonempty sets, the elements of the first being the **points** P, Q, ... and the elements of the second being the **lines** $a, b, \ldots$. There is an incidence relation defined for points and lines satisfying the following axioms (compare the axioms of Section 1.1).

Axiom P.1   *Two different points are incident with one and only one line.*

Axiom P.2   *Two different lines intersect in one and only one point.*

Axiom P.3   *There are at least three different points on every every line and at least three lines through every point.*

St. Jerome in His Study. Engraving, 1514, by Albrecht Dürer. Printed courtesy of National Museum of Prussian Art, West Berlin.

We call such a system of points and lines a ***projective plane.*** From a projective plane we obtain an affine plane as follows. Let u be an arbitrarily chosen line. We declare it to be the "line at infinity." We call two lines not equal to u ***parallel*** if they intersect on u. Now we take u away from the plane, together with all points lying on it. It is then readily seen that we obtain an affine plane. So there are as many affine planes "hidden" in a projective plane as there are lines in the projective plane.

The converse question arises: *Given an affine plane, does there exist a projective plane from which the affine plane is obtained by deleting a line?* This question can be answered in the affirmative: To prove this, we assign to every point the pencil of all lines passing through it. So we may describe the geometry of our affine plane in terms of "pencils" and "lines" (Fig. 5.1) instead of "points" and "lines." For example, instead of saying "two

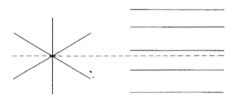

Figure 5.1

points are joined by a line," we say "two pencils have a line in common." Now we may add to the family of pencils all pencils of parallel lines (parallel pencils). The set of all parallel pencils we consider to be a further line ("line at infinity"). Any two pencils possess a unique common line, as is seen from the axioms of an affine plane (Section 3.1). Furthermore, any two lines are in a unique common pencil. Every pencil contains at least three lines, and every line is in at least three pencils. Translating this back into the language of points and lines, we obtain a projective plane with the desired properties.

We call a projective plane ***Desarguesian*** (see Fig. 5.2), if in it the following

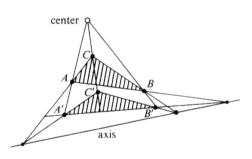

center

axis

Figure 5.2

projective version of Desargue's theorem is true (compare Section 3.9; $(XY)(VW)$, for example, denotes the point of intersection of two different lines XY, VW).

Desargues' Theorem *Let A, B, C and A', B', C' be two triplets of non-*
collinear points, all points different from each other. If the lines AA', BB'
and CC' intersect in a point called the center of the configuration, the points
(AB)(A'B'), (AC)(A'C'), and (BC)(B'C') are on a line, called the axis of
the configuration.

If the axis is chosen as the "line at infinity," we obtain the affine version
(D) of Desargues' theorem.

We say the projective plane is **Pappian** if in it the projective form of
Pappus' theorem is true.

Pappus' Theorem *Let A, B, C and A', B', C' be two triplets of collinear*
points, no point of one triplet being on the line joining the points of the other.
Then the points (AB')(A'B), (AC')(A'C), (BC')(B'C) are on a line.

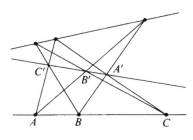

Figure 5.3

We obtain the affine version of Pappus' theorem by designating the line
joining $(AB')(A'B)$ and $(AC')(A'C)$ as the line at infinity (Fig. 5.3).

Exercises

1. Describe the minimal model of a projective plane (7 points and 7 lines).

2. Find all systems of points and lines satisfying P.1 and P.2 but not P.3.

3. Show that the axioms P.1–P.3 are equivalent to P.1, P.2, P.3', where P.3'
   states: On every line there is at least one point. There exist four points,
   no three of which are collinear.

4. State Desargues' little theorem (Section 3.9) in a projective version.

5. What is the difference between Desargues' theorem and its dual (Section
   3.10) in a projective plane?

6. Is Desargues' theorem true in the example of Exercise 1?

5.2   Projective Space

A nonempty system of **points** P, Q, ..., **lines** a, b, ..., and **planes** α, β, ... with an incidence relation is called a **projective 3-space**, or shortly, a **projective space** if it satisfies the following conditions.

Axiom PS.1   *Two points are joined by a unique line.*

Axiom PS.2   *Two planes meet in a unique line.*

Axiom PS.3   *A point and a line not passing through it lies on one and only one plane.*

Axiom PS.4   *A plane and a line not incident with it intersect in one and only one point.*

Axiom PS.5   *If a point is on a line and this line is on a plane, then the point lies on this plane.*

Axiom PS.6   *Every line is on at least three planes and contains at least three points; every plane contains at least three points not on a line; and every point is on at least three planes not intersecting in a common line.*

As in the case of a projective plane, we can proceed from a projective space to an affine space by deleting a plane ω and calling two planes that are not equal to ω or two lines in a plane not equal to ω **parallel** if their intersection lies in ω. The converse is also true. That is, every affine space can be considered as being obtained from a projective space by the deletion of a plane. We shall not go into the proof of this fact, because, the proof is simple, though rather lengthy, and we do not make use of the theorem.

Although a projective plane may, in general, be identified with the set of points, lines, and so forth lying on it, it is useful to make the following distinctions:

1. *Projective objects of order* 0. These are points, lines, and planes as introduced in the definition of a projective space (Fig. 5.4).

Figure 5.4

2. *Projective objects of order* 1. These are **pencils of planes** passing through a line and **linear ranges** consisting of all points on a line (Fig. 5.5).

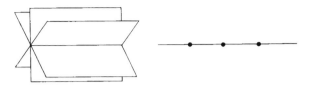

Figure 5.5

3. *Projective objects of order* 2. These are: **bundles of all planes** passing through one point;

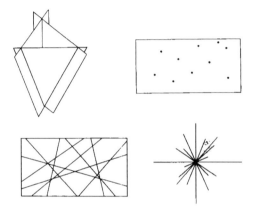

Figure 5.6

planar ranges of all points lying on one plane; **domains of all lines** incident with one plane; **bundles of all lines** passing through one point (Fig. 5.6).

A line, therefore, can be considered an object of order 0, a pencil of planes, or a linear range. Points and planes also possess three interpretations each (elements of order 0 and 2).

Theorem 5.2.1 *Every plane α in a projective space is a Desarguesian projective plane.*

Proof Extend the axis in the configuration of Desargues' Theorem to a plane β different from α and choose β as the plane at infinity. Then, by the results of Section 4.1, the theorem follows. ◆

Exercises

1. Find the minimal model of a projective space. How many points, lines, and planes does it possess?

2. Eliminate in Axioms PS.1–PS.6 those assumptions that can be proved (we stated them for reasons of symmetry).

3. Show explicitly that an affine space (see Section 4.1) is obtained from a projective space by deleting a plane.

4. Give an explicit proof of Theorem 5.2.1.

5. How are the projective objects of order 2 interrelated by the operations of intersection and central projection (defined as in Section 4.1, Exercise 6)?

6. Does the term "translation" make sense in either a projective space or a projective plane?

5.3   Principle of Duality

We have formulated the system of Axioms PS.1–PS.6 of a projective space in such a way that it remains unchanged if we

$$
\left.
\begin{array}{llll}
\text{replace "point"} & \text{by} & \text{"plane"} \\
\text{replace "plane"} & \text{by} & \text{"point"} \\
\text{leave "line"} & \text{as} & \text{"line"} \\
\text{leave "incident"} & \text{as} & \text{"incident."}
\end{array}
\right\} \tag{1}
$$

Hereby, on the left side "line" means "linear range" and on the right side it means "pencil of planes." "Incident" remains true, since it is a symmetric relation. This leads to one of the nicest observations in geometry, the so-called **principle of duality**.

From every true statement on points, lines, and planes there follows another true statement on points, lines, and planes if the substitutions of (1) *are applied.*

The same observation can be made in a projective plane. Here the principle of duality involves points and lines.

From every true statement on points and lines, there follows another true statement on points and lines if we

$$
\left.
\begin{array}{llll}
\text{replace "point"} & \text{by} & \text{"line"} \\
\text{replace "line"} & \text{by} & \text{"point"} \\
\text{and leave "incident"} & \text{as} & \text{"incident."}
\end{array}
\right\} \tag{2}
$$

As an example, take Desargues' theorem as given in Section 5.1. The theorem which we obtain from it by applying the principle of duality (in the plane) is the projective version of what we called in Section 3.10 the Dual Theorem of Desargues (this name now being justified). So the proof of the Dual Theorem of Desargues is accomplished simply by applying the duality principle.

Two geometrical figures are called *dual* if they can be described in "dual" terms, that is, by a substitution (1) or (2). Examples of dual figures are:

Linear range   ⟷ pencil of planes,
bundle of planes ⟷ planar range,
bundle of lines   ⟷ domain of lines,

and the two described in Figures 5.7 and 5.8.

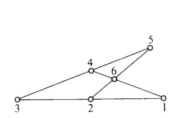

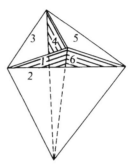

Figure 5.7   **Figure 5.8**

We have already met some pairs of dual figures in Euclidean geometry (Section 4.15):

tetrahedron   ⟷ tetrahedron,
cube   ⟷ octahedron,
dodecahedron ⟷ icosahedron.

In Section 5.12, we shall construct dualities explicitly.

Exercises

1. State the dual of Pappus' theorem in a projective plane.

2. State the dual of Desargues' little theorem.

3. Consider the figure of Desargues' theorem to be a figure in projective space. Draw a dual figure.

4. Find other objects besides the tetrahedron which are self-dual—that is, they correspond dually to the same kind of objects.

5. Find a principle of duality for the set of all subsets of a given set (for example, the set of all points in the plane). (*Hint*: Replace "dual" by "complementary," and "plane spanned by . . .," etc. by "set-theoretic union.")

6. Define the principle of duality in an abstract manner so that the principle in projective geometry and that of Exercise 5 are special cases of it.

5.4   Homogeneous Coordinates

Let a projective plane α in a projective space be given. We may describe α by the following "model": Suppose O is a point not on α. We join O to all points of α and obtain thus a bundle of lines. Every point in α corresponds to a unique line in the bundle, and conversely. So, by "projection," we assign to a domain of points a bundle of lines (both being projective elements of order 2). A line joining two points of α is then represented by the plane joining the corresponding lines through O.

Now we designate a plane different from α and not passing through O as a "plane at infinity." We obtain an affine 3-space (see Section 5.2). We choose O as the origin of a vector space which coordinatizes this affine space. The basis vectors $\mathbf{a}_1$, $\mathbf{a}_2$, $\mathbf{a}_3$ can be chosen such that the plane spanned by $\mathbf{a}_1$, $\mathbf{a}_2$ is parallel to α and such that $\mathbf{a}_3$ represents a point of α (Fig. 5.9).

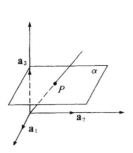

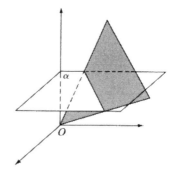

| **Figure 5.9** | **Figure 5.10** |

An arbitrary line through O is represented by all multiples $t\mathbf{p}$ of some vector $\mathbf{p} \neq \mathbf{0}$. We write shortly

$$\mathscr{F}\mathbf{p}$$

for the set of all $t\mathbf{p}$, where t is an element of the skew field $\mathscr{F}$, which is the skew field of scalars of the given vector space. Introducing coordinates with respect to $\mathbf{a}_1$, $\mathbf{a}_2$, $\mathbf{a}_3$, we write

$$\mathscr{F}\mathbf{p} = \mathscr{F}(p_1, p_2, p_3).$$

Any triplet

$$(x_1, x_2, x_3) = (tp_1, tp_2, tp_3), \quad t \neq 0,$$

can be considered a triplet of coordinates of the line $\mathscr{F}\mathbf{p}$ and, hence, of the point P of α represented by this line. We call these coordinates of P *homogeneous coordinates* and we set

$$P = \mathscr{F}\mathbf{p} = \mathscr{F}(p_1, p_2, p_3).$$

What is the advantage of introducing homogeneous coordinates? Mainly this: We threw out a plane w in order to get a vector space coordinatizing the given space. So, in particular, we lost a line of α. However, in our model of all lines through O, we have "saved" the points of this line; they are represented by those lines through O which are parallel to α (lying in the (x_1, x_2)-plane). So we have obtained coordinates of the *projective* plane α.

If we coordinatize the affine part of α by choosing basis vectors $\mathbf{a}_1', \mathbf{a}_2'$ parallel to $\mathbf{a}_1, \mathbf{a}_2$ (see Fig. 5.11), we obtain coordinates for P. Thus, in a

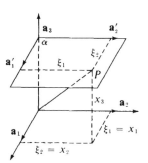

Figure 5.11

very formal way, we obtain homogeneous coordinates of α from inhomogeneous coordinates (ξ_1, ξ_2) (affine part of α) by setting

$$(x_1, x_2, x_3) = (t\,\xi_1, t\,\xi_2, t \cdot 1)$$

for affine points and

$$(x_1, x_2, x_3) = (t\,\xi_1, t\,\xi_2, 0)$$

for "points at infinity." ξ_1, ξ_2 can be calculated backward for affine points as follows:

$$\xi_1 = t^{-1}(t\xi_1) = x_3^{-1} \cdot x_1,$$
$$\xi_2 = t^{-1}(t\xi_2) = x_3^{-1} \cdot x_2.$$

In this formal introduction of homogeneous coordinates, it is not really necessary to introduce a 3-dimensional projective space first. However, the geometrical meaning of homogeneous coordinates remains obscure without the considerations discussed above.

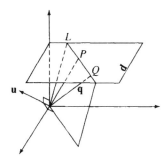

Figure 5.12

A line of α is represented by all nonzero linear combinations

$$s\mathbf{p} + t\mathbf{q},$$

where $\mathbf{p}$ and $\mathbf{q}$ are linearly independent (Figs. 5.10 and 5.12).

Assume an inner product to be given, for example,

$$\mathbf{x} \cdot \mathbf{y} = x_1 y_1 + x_2 y_2 + x_3 y_3$$

in ordinary Euclidean 3-space. Then a line of α is represented by a ***homogeneous*** equation

$$u_1 x_1 + u_2 x_2 + u_3 x_3 = 0,$$

or, briefly,

$$\mathbf{u} \cdot \mathbf{x} = 0,$$

where $\mathbf{u} = (u_1, u_2, u_3)$ is some fixed vector $\neq \mathbf{0}$. So every vector $\mathbf{u} \neq 0$ can also be considered a coordinate vector of a line l. Interpreted in the 3-space in which α is "embedded," $\mathbf{u}$ is a normal of the plane through O which joins l and $\mathbf{0}$. In particular,

$$\mathbf{u} = \mathbf{a}_3 = (0, 0, 1)$$

represents the "line at infinity" of α.

If in the equation

$$\mathbf{u} \cdot \mathbf{x} = 0$$

we consider $\mathbf{x}$ fixed ($\neq \mathbf{0}$) and $\mathbf{u}$ variable, we obtain the equation of a pencil of lines in α. So, in this equation, the principle of duality is reflected algebraically.

Exercises

1. Let $P = \mathscr{F}(0, 1, 1)$; $Q = \mathscr{F}(1, 1, 1)$; $R = \mathscr{F}(2, 2, 3)$; $S = \mathscr{F}(0, 2, 3)$. Find the point in which the lines PQ and RS meet.

2. Why does $K(0, 0, 0)$ not represent a point of the projective plane?

3. Introduce homogeneous coordinates in the plane of Exercise 1, Section 5.1 and set up the equations of all 7 lines (see Chapter 3).

4. Show that if the vectors $\mathbf{u}, \mathbf{v}, \mathbf{w}$ are linearly dependent and all $\neq 0$, the lines $\mathbf{ux} = 0$, $\mathbf{vx} = 0$, $\mathbf{wx} = 0$ have a point in common.

5. Show that if $\mathbf{u}, \mathbf{v}$ satisfy $\mathbf{uv} = 0$ ($\mathbf{u}, \mathbf{v}$ perpendicular in the vector space) they lie on the plane $\mathbf{vx} = 0$.

6. Let a Pappusian projective plane be coordinatized by homogeneous coordinates. Use determinants to set up a test whether three points (given by their coordinates) are on a line or three lines pass through a point.

5.5 Vector Space and Projective n-Space

Having explained geometrically the idea of projective coordinates for a projective plane (Section 5.4) we proceed now to *define*, conversely, an **n-dimensional projective space** $\mathscr{P}^n$ by assigning to every line

$$\mathscr{F}\mathbf{p} \quad (\mathbf{p} \neq 0)$$

of an $(n + 1)$-dimensional vector space $\mathscr{V}^{n+1}$ a point P. We again set $P = \mathscr{F}\mathbf{p}$. In this way every $(n + 1)$-dimensional vector space corresponds to an "n-dimensional" projective space $\mathscr{P}^n$, and conversely:

$$\boxed{\mathscr{V}^{n+1} \longleftrightarrow \mathscr{P}^n}$$

We may also introduce geometrical axioms of a projective n-space first and then show that it may be coordinatized by an $(n + 1)$-dimensional vector space. However, for simplicity's sake, we choose the algebraic definition, the geometrical procedure having been illustrated in the planar case.

The line joining two points P, Q of $\mathscr{P}^n$ is defined by the plane joining the lines of $\mathscr{V}^{n+1}$ that represent P, Q. *We will assume throughout this chapter that the coordinate skew field $\mathscr{F}$ is commutative—that is, a field $\mathscr{K}$.* If necessary, we assume, furthermore, that $\mathscr{K}$ has at least three elements, so that there are at least four points on every line. If $\mathscr{V}^{n+1}$ has a basis $\mathbf{a}_1, \ldots, \mathbf{a}_n, \mathbf{a}_{n+1}$, we introduce in the "hyperplane" α

$$x_{n+1} = 1$$

"inhomogeneous" coordinates $\xi_1, \ldots, \xi_n$, the first n coordinates of any point of the hyperplane with respect to $\mathbf{a}_1, \ldots, \mathbf{a}_{n+1}$. Now

$$(x_1, \ldots, x_{n+1}) = (t\xi_1, \ldots, t\xi_n, t \cdot 1), \quad t \neq 0,$$

are homogeneous coordinates of the "finite" points of α.

$$(x_1, \ldots, x_{n+1}) = (t\xi_1, \ldots, t\xi_n, 0)$$

represent the "points at infinity" of α. Calculating backward again, we obtain

$$\xi_1 = x_{n+1}^{-1}x_1, \ldots, \xi_n = x_{n+1}^{-1}x_n$$

for the "finite" points of α.

In the special case $n = 3$, we see, by analogy with the case $n = 2$ discussed in the preceding section, that the definition of a projective 3-space using $\mathscr{V}^4$ is equivalent to that given in Section 5.2.

Any linear subspace $\mathscr{U}$ of $\mathscr{V}^{n+1}$ of dimension $r + 1$ defines an r-dimensional **subspace** of $\mathscr{P}^n$. It is given by all linear combinations

$$t_1\mathbf{p}_1 + \cdots + t_r\mathbf{p}_r + t_{r+1}\mathbf{p}_{r+1}$$

of $r + 1$ linearly independent vectors of $\mathscr{V}^{n+1}$. We write for the set of these linear combinations

$$\mathscr{K}\mathbf{p}_1 + \cdots + \mathscr{K}\mathbf{p}_r + \mathscr{K}\mathbf{p}_{r+1}.$$

An $(n-1)$-dimensional subspace of $\mathscr{P}^n$ is called a **hyperplane.** Let

$$\mathbf{u} \cdot \mathbf{x} = u_1 x_1 + \cdots + u_{n+1} x_{n+1}$$

be an inner product in $\mathscr{V}^{n+1}$. Then $\mathscr{U}$ can also be represented by a set of equations

$$\mathbf{u}_1 \cdot \mathbf{x} = 0$$
$$\vdots$$
$$\mathbf{u}_{n-r} \cdot \mathbf{x} = 0.$$

Dually,

$$\mathbf{u} \cdot \mathbf{x}_1 = 0$$
$$\vdots \qquad (\mathbf{u} \text{ variable, } \mathbf{x}_i \text{ fixed})$$
$$\mathbf{u} \cdot \mathbf{x}_{r+1} = 0$$

denotes the set of hyperplanes common to a number of "bundles" of hyperplanes defined by

$$\mathbf{u} \cdot \mathbf{x}_i = 0 \quad (i = 1, \ldots, r+1).$$

Such a set of hyperplanes can be represented as

$$\mathscr{K} \mathbf{u}_1 + \cdots + \mathscr{K} \mathbf{u}_{n-r} = 0$$

where $\mathbf{u}_1, \ldots, \mathbf{u}_{n-r}$ are linearly independent. The projective objects of $\mathscr{P}^3$ can now be represented as follows:

Projective elements of order 0:

point	$\mathscr{K} \mathbf{p}$
plane	$\mathscr{K} \mathbf{u}$

Projective elements of order 1:

linear range	$\mathscr{K} \mathbf{p} + \mathscr{K} \mathbf{q}$
pencil of planes	$\mathscr{K} \mathbf{u} + \mathscr{K} \mathbf{v}$

Projective elements of order 2:

planar range	$\mathscr{K} \mathbf{p} + \mathscr{K} \mathbf{q} + \mathscr{K} \mathbf{r}$
bundle of planes	$\mathscr{K} \mathbf{u} + \mathscr{K} \mathbf{v} + \mathscr{K} \mathbf{w}$

Exercises

1. Given two planes,

$$3x_1 - x_2 + x_3 - 2x_4 = 0,$$
$$x_1 + x_2 - x_4 = 0,$$

in $\mathscr{P}^3$ ($\mathscr{K}$ = field of real numbers), find the plane passing through $(0, 1, 2, 1)$ and lying in the pencil spanned by the above planes.

2. Test whether or not the following four points of $\mathscr{P}^3$ are on a plane: $(1, 1, 1, 1)$, $(1, 0, 1, 1)$, $(0, 0, 1, 1)$, $(0, 1, 0, 0)$.

3. Find linear mappings of $\mathscr{V}^{n+1}$ that induce the identity map on $\mathscr{P}^n$.

4. Show explicitly that $\mathscr{P}^3$ is a projective space in the sense of Section 5.2.

5. Find the principle of duality for $\mathscr{P}^n$ (compare Section 5.3).

6. Find a condition for $\mathbf{u}_1, \ldots, \mathbf{u}_n$ so that the hyperplanes $\mathbf{u}_1\mathbf{x} = 0, \ldots, \mathbf{u}_n\mathbf{x} = 0$ of $\mathscr{P}^n$ have one and only one point in common.

7. Let $\mathbf{x}_1, \ldots, \mathbf{x}_{k+1}$ be linearly independent and let $\mathscr{U}$ be the subspace of $\mathscr{P}^n$ defined as the set of all $\mathscr{K}\mathbf{x}$, where $\mathbf{x}$ is a linear combination of $\mathbf{x}_1, \ldots, \mathbf{x}_{k+1}$. Any point of $\mathscr{U}$ can be obtained in r steps as follows: *Step 1*: Join every two of the points $P_1, \ldots, P_{n+1}$, by lines; *Step 2*: Join any two points obtained in Step 1 by lines; $\ldots$; *Step r*: Join any two points obtained in step $r - 1$ by lines. What is the smallest r needed?

8. Let $\mathscr{U} + \mathscr{W}$ denote the subspace of $\mathscr{P}^n$ consisting of the set of all linear combinations of points of the subspaces $\mathscr{U}$ and $\mathscr{W}$ of $\mathscr{P}^n$. The set-theoretic intersection $\mathscr{U} \cap \mathscr{W}$ is clearly again a subspace of $\mathscr{P}^n$. Define the dimension $d(\mathscr{U})$ for any subspace $\mathscr{U}$ of $\mathscr{P}^n$ and show:

$$d(\mathscr{U} + \mathscr{W}) + d(\mathscr{U} \cap \mathscr{W}) = d(\mathscr{U}) + d(\mathscr{W}).$$

5.6   Collineations

Let $\mathscr{P}^n$ and $\mathscr{V}^{n+1}$ be given as in Section 5.5. Clearly, any linear mapping of $\mathscr{V}^{n+1}$ onto itself induces in $\mathscr{P}^n$ a mapping that interchanges points and transforms lines into lines. So a mapping

$$\mathbf{x}' = \mathbf{x}\mathbf{A}, \quad |\mathbf{A}| \neq 0, \tag{1}$$

of $\mathscr{V}^{n+1}$ onto itself regarded as a mapping of $\mathscr{P}^n$ is called a *projective collineation*, in short a *collineation*, of $\mathscr{P}^n$. In coordinates,

$$(x'_1, \ldots, x'_n, x'_{n+1}) = (x_1, \ldots, x_n, x_{n+1}) \begin{pmatrix} a_{11} & \cdots & a_{1, n+1} \\ \vdots & & \vdots \\ a_{n+1, 1} & \cdots & a_{n+1, n+1} \end{pmatrix}. \tag{1'}$$

Theorem 5.6.1   *Any matrix $a\mathbf{A}, a \neq 0$, induces the same collineation in $\mathscr{P}^n$ as $\mathbf{A}$ does.*

If in $\mathscr{P}^n$ a hyperplane α is chosen as the hyperplane at infinity, we obtain an affine space $\mathscr{A}^n$. We shall now look at the relationship between collineations of $\mathscr{P}^n$ and affinities of $\mathscr{A}^n$. Let $x_{n+1} = 0$ be the equation of α. Affine mappings of $\mathscr{A}^n$ can be characterized as projective collineations of $\mathscr{P}^n$ under which α is mapped onto itself. So, in (1), $x'_{n+1} = 0$ whenever $x_{n+1} = 0$. Therefore,

$$x'_{n+1} = x_1 a_{1, n+1} + \cdots + x_n a_{n, n+1} + 0$$

vanishes for all $x_1, \ldots, x_n$. This implies

$$a_{1, n+1} = \cdots = a_{n, n+1} = 0.$$

Since $|\mathbf{A}| \neq 0$, $a_{n+1,n+1}$ cannot be 0 also. By Theorem 5.6.1, we may assume $a_{n+1,n+1} = 1$. Then

$$\mathbf{A} = \begin{pmatrix} a_{11} & \cdots & a_{1n} & 0 \\ \vdots & & \vdots & \vdots \\ a_{n1} & \cdots & a_{nn} & 0 \\ a_{n+1,1} & \cdots & a_{n+1,n} & 1 \end{pmatrix}$$

represents an affinity of $\mathscr{A}^n$. We may split $\mathbf{A}$ as follows:

$$\mathbf{A} = \begin{pmatrix} a_{11} & \cdots & a_{1n} & 0 \\ \vdots & & \vdots & \vdots \\ a_{n1} & \cdots & a_{nn} & 0 \\ 0 & \cdots & 0 & 1 \end{pmatrix} \begin{pmatrix} 1 & \cdots & 0 & 0 \\ \vdots & & \vdots & \vdots \\ 0 & \cdots & 1 & 0 \\ a_{n+1,1} & \cdots & a_{n+1,n} & 1 \end{pmatrix}. \tag{2}$$

We introduce (inhomogeneous) coordinates $\xi_1, \ldots \xi_n$ in $\mathscr{A}^n$ as is done in Section 5.5. Then

$$(\xi_1, \ldots, \xi_n, 1) \tag{3}$$

can be considered homogeneous coordinates of a point in $\mathscr{A}^n$. We write (2) briefly as

$$\mathbf{A} = \begin{pmatrix} \mathbf{A}_1 & 0 \\ 0 & 1 \end{pmatrix} \begin{pmatrix} \mathbf{I} & \mathbf{0} \\ \mathbf{c} & 1 \end{pmatrix}, \tag{4}$$

where

$$\mathbf{A}_1 = \begin{pmatrix} a_{11} & \cdots & a_{1n} \\ \vdots & & \vdots \\ a_{n1} & \cdots & a_{nn} \end{pmatrix} \quad \text{and} \quad \mathbf{c} = (a_{n+1,1} \ldots, a_{n+1,n}).$$

Furthermore, we set

$$\mathbf{x}_1 = (\xi_1, \ldots, \xi_n) \quad \text{and} \quad (\mathbf{x}_1, 1) = (\xi_1, \ldots, \xi_n, 1).$$

We find now

$$(\mathbf{x}_1, 1)\mathbf{A} = (\mathbf{x}_1\mathbf{A}_1, 1) \begin{pmatrix} \mathbf{I} & \mathbf{0} \\ \mathbf{c} & 1 \end{pmatrix} = (\mathbf{x}_1\mathbf{A}_1 + \mathbf{c}, 1).$$

Therefore, $\mathbf{A}$ induces in $\mathscr{A}^n$ the affinity

$$\mathbf{xA} = \mathbf{x}_1\mathbf{A}_1 + \mathbf{c}, \tag{5}$$

that is, a general affine mapping. The converse is also true.

Theorem 5.6.2 *If an affine space $\mathscr{A}^n$ is extended to a projective space $\mathscr{P}^n$, then any affinity (5) can be extended to a projective collineation (1) of $\mathscr{P}^n$ by setting*

$$\mathbf{A} = \begin{pmatrix} \mathbf{A}_1 & 0 \\ \mathbf{c} & 1 \end{pmatrix}.$$

Having explained the relationship between affinities and projective collineations, we turn now to the general problem of mapping $n + 2$ given points onto other $n + 2$ given points by a collineation. We call points $P_1, \ldots, P_r$ of a projective space $\mathscr{P}^n$ **projectively independent** if none of them lies in the subspace of $\mathscr{P}^n$ spanned by the others. Equivalently, we may postulate that vectors of $\mathscr{V}^{n+1}$ representing $P_1, \ldots, P_r$ are linearly independent.

Theorem 5.6.3 Fundamental theorem of projective geometry *Let $P_0, \ldots, P_{n+1}$ and $P'_0, \ldots, P'_{n+1}$ be two sets of $n + 2$ points in $\mathscr{P}^n$ such that each $n + 1$ of the first set and each $n + 1$ of the second set are projectively independent. Then there exists one and only one collineation mapping P_i onto P'_i ($i = 0, \ldots, n + 1$).*

Theorem 5.6.3 follows from the following theorem about vectors in $\mathscr{V}^{n+1}$.

Theorem 5.6.4 *Let $\mathbf{y}_0, \ldots, \mathbf{y}_{n+1}$ and $\mathbf{y}'_0, \ldots, \mathbf{y}'_{n+1}$ be two sets of vectors in $\mathscr{V}^{n+1}$ such that each $n + 1$ of them is linearly independent. Then there exists one and, up to a scalar multiple, only one linear mapping of $\mathscr{V}^{n+1}$ onto itself transforming $\mathbf{y}_i$ into a scalar multiple of $\mathbf{y}'_i$ ($i = 0, \ldots, n + 1$).*

Proof Since $\mathbf{y}_0, \ldots, \mathbf{y}_n$ are linearly independent vectors of $\mathscr{V}^{n+1}$, we may express $\mathbf{y}_{n+1}$ as

$$\mathbf{y}_{n+1} = a_0 \mathbf{y}_0 + \cdots + a_n \mathbf{y}_n. \tag{6}$$

All a_i are not equal to 0, since, otherwise, $n + 1$ of the vectors $\mathbf{y}_0, \ldots, \mathbf{y}_{n+1}$ would be linearly dependent. Correspondingly,

$$\mathbf{y}'_{n+1} = a'_0 \mathbf{y}'_0 + \cdots + a'_n \mathbf{y}'_n. \tag{7}$$

We wish to find $\mathbf{A}$ such that, for some $\lambda_i \neq 0$ ($i = 0, \ldots, n + 1$),

$$\lambda_i \mathbf{y}'_i = \mathbf{y}_i \mathbf{A}. \tag{8}$$

From (6) we obtain

$$\mathbf{y}_{n+1} \mathbf{A} = a_0 \mathbf{y}_0 \mathbf{A} + \cdots + a_n \mathbf{y}_n \mathbf{A}$$
$$= \lambda_{n+1} \mathbf{y}'_{n+1} = a_0 \lambda_0 \mathbf{y}'_0 + \cdots + a_n \lambda_n \mathbf{y}'_n.$$

Since, on the other hand,

$$\lambda_{n+1} \mathbf{y}'_{n+1} = \lambda_{n+1} a'_0 \mathbf{y}'_0 + \cdots + \lambda_{n+1} a'_n \mathbf{y}'_n,$$

we find

$$a'_i = \frac{\lambda_i}{\lambda_{n+1}} a_i \quad (i = 0, \ldots, n).$$

From (8) we can set up the matrix equation ($\mathbf{y}_i, \mathbf{y}_i'$ row vectors)

$$\begin{pmatrix} \mathbf{y}_0 \\ \vdots \\ \mathbf{y}_n \end{pmatrix} \mathbf{A} = \begin{pmatrix} \lambda_0 \mathbf{y}_0' \\ \vdots \\ \lambda_n \mathbf{y}_n' \end{pmatrix} = \lambda_{n+1} \begin{pmatrix} \dfrac{a_0'}{a_0} \mathbf{y}_0' \\ \vdots \\ \dfrac{a_n'}{a_n} \mathbf{y}_n' \end{pmatrix}$$

or

$$\mathbf{A} = \lambda_{n+1} \begin{pmatrix} \mathbf{y}_0 \\ \vdots \\ \mathbf{y}_n \end{pmatrix}^{-1} \begin{pmatrix} \dfrac{a_0'}{a_0} \mathbf{y}_0' \\ \vdots \\ \dfrac{a_n'}{a_n} \mathbf{y}_n' \end{pmatrix}. \tag{9}$$

So we have shown that if a matrix $\mathbf{A}$ as is required in Theorem 5.6.4 exists, it has the form (9). Hence, any two such matrices are distinct only by a factor λ_{n+1}. However, (9) does provide a matrix for which (8) is true. This proves Theorem 5.6.4 ◆

Exercises

1. Given two affinities of the kind in Equation (5), find the collineation corresponding to their product.

2. Find (by using Equations (6), (7), and (9)) a collineation mapping the points P_i in a projective plane onto the points P_i' where

   $$P_0 = \mathscr{K}(1, 0, 0); \quad P_1 = \mathscr{K}(1, 1, 0); \quad P_2 = \mathscr{K}(0, 1, -1); \quad P_3 = \mathscr{K}(1, 1, 1)$$

   and

   $$P_0' = \mathscr{K}(0, 0, 1); \quad P_1' = \mathscr{K}(0, 1, 0); \quad P_2' = \mathscr{K}(1, -1, 0); \quad P_3' = \mathscr{K}(1, 0, 1).$$

3. Let $\mathbf{u}_i \cdot \mathbf{x} = 0$ and $\mathbf{u}_i' \cdot \mathbf{x} = 0$; $i = 0, \ldots, 3$, be lines in a Pappian projective plane where $\mathbf{u}_0 = (0, 1, 0)$; $\mathbf{u}_1 = (0, 0, 2)$; $\mathbf{u}_2 = (1, 1, 1)$, $\mathbf{u}_3 = (-1, 0, 0)$; $\mathbf{u}_0' = (1, 1, 0)$; $\mathbf{u}_1' = (0, -1, -1)$; $\mathbf{u}_2' = (1, 0, 0)$; $\mathbf{u}_3' = (0, 2, -1)$. Find a collineation that maps the line $\mathbf{u}_i \cdot \mathbf{x} = 0$ onto the line $\mathbf{u}_i' \cdot \mathbf{x} = 0$, $i = 0, \ldots, 3$.

4. Show (by modifying (9)) that in a Pappian affine space $\mathscr{A}^n$ any $n + 1$ affinely independent points can be mapped onto any other affinely independent points by one and only one affinity.

5. Show that in a Desarguesian affine plane every affinity with two fixed points is a stretch or a shear.

6. Show that the group of all collineations of $\mathscr{P}^n$ leaving a hyperplane α pointwise fixed and possessing a 2-dimensional subspace, not contained in α, as fixed space, is isomorphic to the group of all dilatations and translations of an affine plane.

5.7   *Central Collineations*

By a *central collineation* we mean a collineation possessing all points
of a hyperplane c as fixed points and all hyperplanes passing through a
point C as fixed hyperplanes (notice Exercise 2 of this section!). C is called
the *center* and c is called the *axis* of the central collineation.   Briefly, we
call it a *(C, c)-collineation.*   We shall discuss only the case of a projective
plane.   Two kinds of central collineations are distinguished:

1.   *Elation,* defined by the assumption that C lies on c (Fig. 5.13).

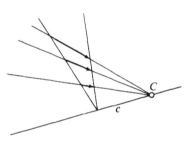

Figure 5.13

If c is considered a "line at infinity," *this is the case of a translation* (Fig.
5.14(a)) in an affine plane (see Section 3.2).   If C is at infinity but c is not
the line at infinity, *we obtain a shear* (Fig. 5.14(b)) in an affine plane.   *So
translations and shears are affine specializations of elations.*

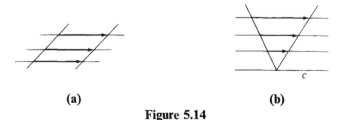

(a)   (b)

Figure 5.14

2. *Homology,* defined by assuming that C does not lie on c (Fig. 5.15).

Again we can subsume two affine possibilities: If c is the line at infinity,
the homology is a dilatation (Fig. 5.16(a)).   If C is at infinity (and hence
c is not), *we obtain a stretch* (See Fig. 5.16(b), compare Section 3.2).

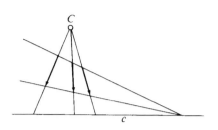

Figure 5.15

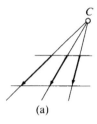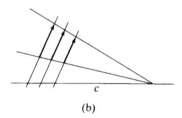

(a)   (b)

Figure 5.16

So we see that projective geometry is a natural frame for discussing the internal relationship between a number of affine transformations. Also the principle of duality becomes visible for those transformations that we saw in an incomplete setting in Chapter 3.

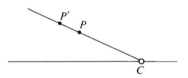

Figure 5.17

Theorem 5.7.1 (a) *If C, P, P' are different and collinear, P, P' not on c, there exists one and only one (C, c)-collineation mapping P onto P' (Fig. 5.17).*

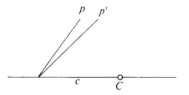

Figure 5.18

(b) *If c, p, p' are different and meet in a point, p, p' not containing C, there exists one and only one (C, c)-collineation mapping p onto p' (Fig. 5.18).*

From the results of Sections 1 of this chapter and 9 and 10 of Chapter 3 we conclude that Theorem 5.7.1(a) *is equivalent to Desargues' theorem and* 5.7.1(b) *is equivalent to the Dual Desargues' theorem.* Theorem 5.7.1 can also be expressed in group-theoretic terms (compare Section 2.4).

Theorem 5.7.2 *The set of all (C, c)-collineations $(C, c$ fixed) is a group under successive application which, in the case of a Desarguesian projective plane,*

(a) *is transitive on the set of all points $\neq C \not\in c$ of any line $\neq c$ through C,*
(b) *is transitive on the set of all lines $PC \neq c$ through point $P \neq C$ on c.*

We now prove a theorem on products of central collineations that may be considered an analog of the theorem on three reflections for Euclidean planes.

Theorem 5.7.3 *Let a Desarguesian projective plane be given. If three different lines a, b, c pass through a point P and if three different points A, B, C are on a line p, then any involutoric product of three involutoric homologies α, β, γ with centers, A, B, C and axes a, b, c respectively, is again a central collineation with its center on p and its axis through P.*

Proof It is readily seen that p is a fixed line under $\alpha\beta\gamma$, since A, B, C lie on p.

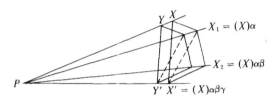

$X_1 = (X)\alpha$
$X_2 = (X)\alpha\beta$
$Y'\ X' = (X)\alpha\beta\gamma$

Figure 5.19

We show that there exists a fixed line $\neq p$ of $\alpha\beta\gamma$ not passing through P. Otherwise let $Q \neq P$ lie on a. Then $Q^{\alpha\beta\gamma}$ lies on a. Since $\alpha, \beta, \gamma, \alpha\beta\gamma$ involutoric, imply that $\alpha\beta\gamma = \gamma\beta\alpha$, we find $Q^{\beta\gamma} = Q^{\alpha\beta\gamma} = Q^{\gamma\beta\alpha}$; hence $Q^{\beta\gamma} = Q^{\gamma\beta}$ (points of a fixed under α). The line $u = QQ^{\beta} = Q^{\beta}Q = Q^{\beta}Q^{\beta^2} = u^{\beta}$ is fixed under β; also $u^{\gamma} = Q^{\gamma}Q^{\beta\gamma} = Q^{\gamma}Q^{\gamma\beta} = u^{\gamma\beta}$. Since $c, u,$ and u^{γ} intersect in a point, this point is B. Similarly it is seen that $v = QQ^{\gamma}, v^{\beta},$ and b intersect in c. Let V be the point in which v and c intersect. Clearly, $V \neq P$. Since $V^{\alpha\beta\gamma}$ lies on $CV^{\alpha} \neq c$, we find that $VV^{\alpha\beta\gamma}$ is a fixed line under $\alpha\beta\gamma$ not passing through P, contrary to our assumption. So there exists a fixed line q of $\alpha\beta\gamma$ not passing through P (Fig. 5.19). Let X lie on q and let $Y \neq X$ when P is a point on PX. Set $X_1 = (X)_{\alpha}, X_2 = (X)\alpha\beta\gamma, X' = (X)\alpha\beta$. Y_1, Y_2, Y' are defined correspondingly. By a three-fold application of Desargues' theorem (first to the triangles $XX_1X_2; YY_1Y_2$, then to $X_1X_2X'; Y_1Y_2Y'$, and finally to XX_1X'; YY_1Y') we see that XX' and YY' intersect in a point U of p. By varying one of the points X, Y, we obtain that every line through U is a fixed line. A straightforward argument shows now (see Exercise 2) that $\alpha\beta\gamma$ is a central collineation with center U on p and axis through p (p being a fixed point). ◆

Exercises

1. Show by an example that the product $\alpha\beta\gamma$ in Theorem 5.7.3 need not be involutoric if α, β, γ are involutoric.

2. Show that in a projective space any collineation leaving all points of a hyperplane fixed is a central collineation. State the dual theorem.

3. Can every collineation be expressed as a product of elations?

4. A *projectivity* is defined as the product of finitely many central collineations. Show that in a Pappian projective plane every collineation is a projectivity. (*Hint*: Use the fundamental theorem of projective geometry 5.6.3.)

5. In the 7-point projective plane (Section 5.1, Exercise 1) construct explicitly all central collineations.

6. Show that if in a finite projective plane there exists a *homology* of order 2, the number of points on every line is even.

5.8   Cross Ratios

According to Klein's Erlanger Programm, we may define projective geometry as *the theory of invariants under the group of all collineations*. One invariant of this kind is "collinearity" of points. We shall now study a numerical invariant.

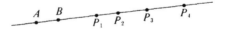

Figure 5.20

Let **a, b** be linearly independent and hence represent two different points A, B. On the line AB we choose four different points P_1, P_2, P_3, P_4 (Fig. 5.20). Their homogeneous coordinates can be written as

$$s_i\mathbf{a} + t_i\mathbf{b} \quad (i = 1, 2, 3, 4).\tag{1}$$

We set

$$(P_1P_2, P_3P_4) = \frac{\begin{vmatrix} s_1 & t_1 \\ s_3 & t_3 \end{vmatrix}}{\begin{vmatrix} s_2 & t_2 \\ s_3 & t_3 \end{vmatrix}} : \frac{\begin{vmatrix} s_1 & t_1 \\ s_4 & t_4 \end{vmatrix}}{\begin{vmatrix} s_2 & t_2 \\ s_4 & t_4 \end{vmatrix}}$$

and call this number the **cross ratio** of P_1, P_2, P_3, P_4. Since $\mathscr{K}\mathbf{a}$ and $\mathscr{K}\mathbf{b}$ may also be considered to represent hyperplanes a, b of the projective space, we may set for the hyperplanes h_1, h_2, h_3, h_4 defined by (1):

$$(h_1h_2, h_3h_4) = \frac{\begin{vmatrix} s_1 & t_1 \\ s_3 & t_3 \end{vmatrix}}{\begin{vmatrix} s_2 & t_2 \\ s_3 & t_3 \end{vmatrix}} : \frac{\begin{vmatrix} s_1 & t_1 \\ s_4 & t_4 \end{vmatrix}}{\begin{vmatrix} s_2 & t_2 \\ s_4 & t_4 \end{vmatrix}}$$

[5.8]

and call this number the **cross ratio** of the hyperplanes $h_1, \ldots, h_4$.

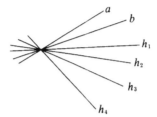

Figure 5.21

Theorem 5.8.1 *The cross ratio* (P_1P_2, P_3P_4) *(or* (h_1h_2, h_3h_4)*) is invariant under any collineation.*

Proof Let the linear mapping (of $\mathscr{V}^{n+1}$)

$$\mathbf{x}' = \mathbf{xA} \tag{2}$$

represent the given collineation. The image points P_i' of $P_i(i = 1, \ldots, 4)$ are represented by

$$s_i\mathbf{a}' + t_i\mathbf{b}' \quad (i = 1, \ldots, 4).$$

So the P_i' have the same "coordinates," (s_i, t_i), with respect to $\mathbf{a}', \mathbf{b}'$ as the P_i have with respect to $\mathbf{a}, \mathbf{b}$. Therefore, we need only show that the cross ratio (P_1P_2, P_3P_4) does not depend on the "basis" $\mathbf{a}, \mathbf{b}$ (or $\mathbf{a}', \mathbf{b}'$) to which the "coordinates" (s_i, t_i) refer. In order to do so, we may choose $\mathbf{a}, \mathbf{b}$ as basis vectors

$$\mathbf{a} = \mathbf{a}_1 = (1, 0), \qquad \mathbf{b} = \mathbf{a}_2 = (0, 1)$$

of a two-dimensional vector space (disregarding the remainder of $\mathscr{V}^{n+1}$). A change of basis in this space is given by

$$\mathbf{y}^* = \mathbf{yB}, \quad |\mathbf{B}| \neq 0, \tag{3}$$

where $\mathbf{B}$ is a 2×2 matrix. Letting $\mathbf{y}_i = (s_i, t_i)$ and

$$\mathbf{Y}_{(ik)} = \begin{pmatrix} s_i & t_i \\ s_k & t_k \end{pmatrix} = \begin{pmatrix} \mathbf{y}_i \\ \mathbf{y}_k \end{pmatrix};$$

$$\mathbf{Y}^*_{(ik)} \begin{pmatrix} \mathbf{y}_i^* \\ \mathbf{y}_k^* \end{pmatrix} = \begin{pmatrix} \mathbf{y}_i\mathbf{B} \\ \mathbf{y}_k\mathbf{B} \end{pmatrix} = \begin{pmatrix} \mathbf{y}_i \\ \mathbf{y}_k \end{pmatrix}\mathbf{B} = \mathbf{Y}_{(ik)}\mathbf{B},$$

we find for the images P_i^* of P_i under (3):

$$(P_1^* P_2^*, P_3^* P_4^*) = \frac{|\mathbf{Y}_{(13)}^*|}{|\mathbf{Y}_{(23)}^*|} : \frac{|\mathbf{Y}_{(14)}^*|}{|\mathbf{Y}_{(24)}^*|} = \frac{|\mathbf{Y}_{(13)}\mathbf{B}|}{|\mathbf{Y}_{(23)}\mathbf{B}|} : \frac{|\mathbf{Y}_{(14)}\mathbf{B}|}{|\mathbf{Y}_{(24)}\mathbf{B}|}$$

$$= \frac{|\mathbf{Y}_{(13)}|\,|\mathbf{B}|}{|\mathbf{Y}_{(23)}|\,|\mathbf{B}|} : \frac{|\mathbf{Y}_{(14)}|\,|\mathbf{B}|}{|\mathbf{Y}_{(24)}|\,|\mathbf{B}|} = (P_1 P_2, P_3 P_4). \quad \blacklozenge$$

A nice application of Theorem 5.8.1 is given in the following theorem.

Theorem 5.8.2 *If under a collineation (2) three different collinear points P_1, P_2, P_3 and their image points P_1', P_2', P_3' are prescribed, then the image X' of any point X on the line $P_1 P_2$ can be found from the equation*

$$(P_1 P_2, P_3 X) = (P_1' P_2', P_3' X').$$

Example In a projective plane, let the points $P_1 = \mathcal{K}(1, 2, -1)$, $P_2 = \mathcal{K}(1, 0, 1)$, $P_3 = \mathcal{K}(-1, 2, -3)$ be mapped under a projective collineation onto $P_1' = \mathcal{K}(0, 2, 2)$, $P_2' = \mathcal{K}(0, 1, 0)$, $P_3' = \mathcal{K}(0, 3, 2)$, respectively. Find the image of $X = (3, 4, -1)$.

Solution Since $P_3 = P_1 - 2P_2$; $X = 2P_1 + P_2$; $P_3' = P_1' + P_2'$, we obtain

$$(P_1 P_2, P_3 X) = \frac{\begin{vmatrix} 1 & 0 \\ 1 & -2 \end{vmatrix}}{\begin{vmatrix} 0 & 1 \\ 1 & -2 \end{vmatrix}} : \frac{\begin{vmatrix} 1 & 0 \\ 2 & 1 \end{vmatrix}}{\begin{vmatrix} 0 & 1 \\ 2 & 1 \end{vmatrix}} = \frac{\begin{vmatrix} 1 & 0 \\ 1 & 1 \end{vmatrix}}{\begin{vmatrix} 0 & 1 \\ 1 & 1 \end{vmatrix}} : \frac{\begin{vmatrix} 1 & 0 \\ s & t \end{vmatrix}}{\begin{vmatrix} 0 & 1 \\ s & t \end{vmatrix}}$$

$$= (P_1', P_2', P_3' X'),$$

from which $-4 = s/t$, or $s = -4t$, and thus $X' = \mathcal{K}(0, 7, 8)$ follows.

In Euclidean n-space, E^n, we may interpret cross ratios geometrically as follows: First consider E^n to be extended to a projective space $\mathcal{P}^n$ (by proceeding from $(\xi_1, \ldots, \xi_n)$ to $\mathcal{K}(\xi_1, \ldots, \xi_n, 1)$ and adding the points $\mathcal{K}(\xi_1, \ldots, \xi_n, 0)$). Let

$$A = \mathcal{K}\mathbf{a} = \mathcal{K}(a_1, \ldots, a_n, 1) \quad \text{and} \quad B = \mathcal{K}\mathbf{b} = \mathcal{K}(b_1, \ldots, b_n, 0).$$

We set

$$\tilde{\mathbf{a}} = (a_1, \ldots, a_n), \qquad \tilde{\mathbf{b}} = (b_1, \ldots, b_n).$$

In inhomogeneous coordinates, the line AB can be represented by

$$\tilde{\mathbf{x}} = \tilde{\mathbf{a}} + t\tilde{\mathbf{b}}.$$

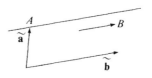

Figure 5.22

Assuming $|\bar{\mathbf{b}}| = 1$, the parameter t expresses, up to the sign, the distance of X and A. The cross ratio of four points P_1, P_2, P_3, P_4 of E^n reads now

$$(P_1P_2, P_3P_4) = \frac{\begin{vmatrix} 1 & t_1 \\ 1 & t_3 \end{vmatrix}}{\begin{vmatrix} 1 & t_2 \\ 1 & t_3 \end{vmatrix}} : \frac{\begin{vmatrix} 1 & t_1 \\ 1 & t_4 \end{vmatrix}}{\begin{vmatrix} 1 & t_2 \\ 1 & t_4 \end{vmatrix}} = \frac{t_3 - t_1}{t_3 - t_2} : \frac{t_4 - t_1}{t_4 - t_2}. \tag{4}$$

Let $\overline{P_iP_k}$ represent the **directed distance** of P_i, P_k, that is, the Euclidean distance of P_i, P_k times $+1$ or -1, depending on whether P_k is obtained from P_i by adding a positive multiple of $\mathbf{b}$, or a negative multiple, respectively. Then, clearly

$$(P_1P_2, P_3P_4) = \frac{\overline{P_1P_3}}{\overline{P_2P_3}} : \frac{\overline{P_1P_4}}{\overline{P_2P_4}}.$$

So, whereas in affine geometry ratios of lengths of line segments are invariants, in projective geometry *double ratios (cross ratios) of lengths of line segments are invariant*.

The following properties of cross ratios are readily verified.

Theorem 5.8.3 *If* $(P_1P_2, P_3P_4) = c$, *then also*

$$(P_2P_1, P_4P_3) = (P_3P_4, P_1P_2) = (P_4P_3, P_2P_1) = c$$

and

$$(P_2P_1, P_3P_4) = \frac{1}{c};$$
$$(P_1P_3, P_2P_4) = 1 - c;$$
$$(P_3P_2, P_1P_4) = \frac{c}{c - 1}.$$

Exercises

1. (a) Carry out the calculations that prove Theorem 5.8.3. (b) Apply those permutations of the P_i in (P_1P_2, P_3P_4) that do not occur in Theorem 5.8.3 and find the values of the corresponding cross ratios.

2. Extend as far as possible the definition of a cross ratio to the cases in which not all P_i are different.

3. State the dual theorems of Theorems 5.8.1 and 5.8.2.

4. In an affine plane, given three lines

$$\xi_1 + \xi_2 = 0, \qquad \xi_1 - \xi_2 = 0, \qquad \xi_1 - 2\xi_2 = 0$$

and their respective image lines

$$2\xi_1 - \xi_2 = 1, \qquad \xi_1 + \xi_2 = 2, \qquad 3\xi_1 - \xi_2 = 2$$

under an affinity, find by using cross ratios the image of the line $\xi_1 = 0$ under this affinity.

5. In the ordinary Euclidean metric plane find, by using cross ratios, a condition for three points PQR to satisfy the condition $[PQR]$ (i.e., Q between P and R).

6. In the ordinary Euclidean metric plane, find, by using cross ratios, a condition for two pairs of lines through a point to separate each other.

5.9   Perspectivities and Projectivities

Let l be a line in a projective space $\mathscr{P}^n$ and let P be a point not on l (Fig. 5.23). If P is joined to every point on l, the linear range of all points

Figure 5.23

on l is, in a natural way, mapped onto the pencil of all lines through P in the plane spanned by P and l. We call this assignment a **projection**.

Theorem 5.9.1 *If l_1, l_2, l_3, l_4 are obtained from P_1, P_2, P_3, P_4 by projection, then in the plane spanned by the l_i*

$$(P_1P_2, P_3P_4) = (l_1l_2, l_3l_4).$$

Proof It is sufficient to prove Theorem 5.9.1 for a projective plane $\mathscr{P}^2$. Furthermore, since cross ratios are invariant under projective collineations, we may choose the coordinate system in a convenient way. We let $P = \mathscr{K}(0, 0, 1)$ and let the line at infinity not pass through P_i ($i = 1, \ldots, 4$).

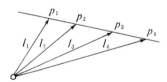

Figure 5.24

If, in inhomogeneous coordinates,

$$P_i = (p_1^{(i)}, p_2^{(i)}) = \mathbf{p}^{(i)} \quad (i = 1, \ldots, 4),$$

the line l_i is given by the equation

$$p_1^{(i)}\xi_1 + p_2^{(i)}\xi_2 = 0,$$

and hence it possesses the projective representation

$$l_i = \mathcal{K}(-p_2^{(i)}, p_1^{(i)}, 0) = \mathcal{K}\mathbf{q}^{(i)} \quad (i = 1, \ldots, 4).$$

From this it is readily seen that a linear relationship that holds between $\mathbf{p}^{(1)}, \ldots, \mathbf{p}^{(4)}$ also holds between $\mathbf{q}^{(1)}, \ldots, \mathbf{q}^{(4)}$. This proves Theorem 5.9.1.

$\blacklozenge$

Let m be a second line in the plane spanned by P, l that does not contain P (See Fig. 5.25). We assign to every point X on l the point X' in which m

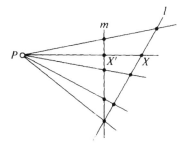

Figure 5.25

intersects PX and call this assignment a **perspectivity**. We set

$$X \overset{P}{\underset{\wedge}{=}} X'.$$

If points $X_1, \ldots, X_r$ are mapped onto points $X_1', \ldots, X_r'$ under this perspectivity, we set

$$X_1 X_2 \cdots X_r \overset{P}{\underset{\wedge}{=}} X_1' X_2' \cdots X_r'.$$

The inverse operation of a projection is called **intersection**. So we may define a perspectivity briefly as the product of a projection and an intersection. From Theorem 5.9.1, we immediately obtain the following theorem.

Theorem 5.9.2 *Perspectivities preserve cross ratios.*

The product of a finite number of perspectivities is called a **projectivity**. Theorem 5.9.2 implies the next one.

Theorem 5.9.3 *Projectivities preserve cross ratios.*

The fact that the converse of Theorem 5.9.3 is also true shows the strong position that cross ratios hold in projective geometry.

Theorem 5.9.4 *If a line a is mapped onto a line b by a one-to-one mapping that preserves cross ratios, this mapping is a projectivity.*

Proof We assume first $a \neq b$. Let the point of intersection $a \cdot b$ be different from P_i, P_i' $(i = 1, 2, 3)$. If $\tilde{P}_1, \tilde{P}_2, \tilde{P}_3$ are defined as indicated in Figure 5.26, we have

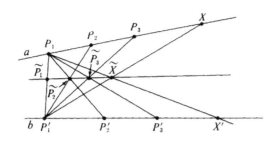

Figure 5.26

$$P_1P_2P_3 \underset{\wedge}{\overset{P_1'}{=}} \tilde{P}_1\tilde{P}_2\tilde{P}_3 \underset{\wedge}{\overset{P_1}{=}} P_1'P_2'P_3'.$$

Replacing P_3 by an arbitrary point X, we define $\tilde{X}$ as $(P_1'X)(\tilde{P}_1\tilde{P}_2)$ and then X' as $(P_1\tilde{X})b$. Clearly,

$$X \underset{\wedge}{\overset{P_1'}{=}} \tilde{X} \underset{\wedge}{\overset{P_1}{=}} X'.$$

Since cross ratios are preserved, we find $(X \neq P_1, P_2, P_3)$

$$(P_1P_2, P_3X) = (P_1'P_2', P_3'X').$$

By this equation, X' is determined uniquely if $P_1, P_2, P_3, P_1', P_2', P_3', X$ are given; this is seen directly from the definition of a cross ratio if we evaluate the determinants in it. So X' is the image of X under the given mapping.

If $a = b$, we introduce an auxiliary line $c \neq a$ and apply twice the above arguments.◆

The proof of Theorem 5.9.4 leads directly to the next theorem.

Theorem 5.9.5 *If P_1, P_2, P_3 and P_1', P_2', P_3' are two sets of different and collinear points, then there exists a unique projectivity of the line P_1P_2 onto the line $P_1'P_2'$ mapping P_i onto P_i' $(i = 1, 2, 3)$. It is given by*

$$(P_1P_2, P_3X) = (P_1'P_2', P_3'X').$$

Theorems 5.8.2 and 5.9.5 give us the following conclusion.

Theorem 5.9.6 *A projectivity of a line onto itself is a collineation of this line, and conversely.*

Exercises

1. Show that a projectivity between two different lines a, b is a perspectivity if and only if the point of intersection ab is mapped onto itself.

2. Show by using Pappus' theorem that the construction of X' in the proof of Theorem 5.9.4 can also be achieved by interchanging the roles of P_1, P_1' and P_2, P_2' or P_1, P_1' and P_3, P_3'.

3. Show that any projectivity between two different lines can be expressed as the product of at most two perspectivities.

4. Define collineations of $\mathscr{P}^1$ by using cross ratios.

5. Find by using Exercise 4 a collineation that maps $\mathscr{K}(0, 1)$, $\mathscr{K}(1, 0)$ onto $\mathscr{K}(1, 2)$, $\mathscr{K}(-2, 1)$, respectively.

6. Prove anew the fundamental theorem for $\mathscr{P}^1$ (Theorem 5.6.3) by using cross ratios.

5.10   *Harmonic Sets and Involutions*

By a complete quadrangle, we mean a set of four points no three of which are collinear (*vertices*) together with the six lines joining any two of them. The remaining three points in which these six lines intersect are called **diagonal points**. (See Fig. 5.27.) Let P, Q be two diagonal points and let

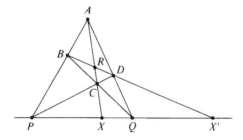

Figure 5.27

X, X' be the points in which PQ intersects the lines of the complete quadrangle not passing through P, Q. We call P, Q, X, X' a **harmonic set**, or we say that X' is the **fourth harmonic point** of P, Q, X.

Theorem 5.10.1 *Four points P, Q, X, X' form a harmonic set if and only if* $(PQ, XX') = -1$.

Proof Let P, Q, X, X' be a harmonic set. Using the notation of Figure 5.27, we have

$$PQXX' \stackrel{A}{\overline{\wedge}} BDRX' \stackrel{C}{\overline{\wedge}} QPXX'.$$

Let $(PQ, XX') = c$. By Theorem 5.8.3, $(QP, XX') = 1/c$; hence $c = 1/c$ and $c^2 = 1$. Suppose $c = +1$. In inhomogeneous coordinates, we may write (with obvious notation)

$$(PQ, XX') = \frac{x - p}{x - q} : \frac{x' - p}{x' - q} = +1,$$

from which

$$p(x - x') = q(x - x').$$

Since $x \neq x', p = q$ would follow, contradicting the definition of a complete quadrangle. Therefore, $c = -1$.

Let, conversely, $(PQ, XX') = -1$. We choose a point A not on PQ and a point B on AP different from P, A. Setting

$$C = (AX)(BQ),$$
$$D = (AQ)(PC),$$

and

$$X'' = (PQ)(BD),$$

we conclude from what we have shown above

$$(PQ, XX'') = -1.$$

Since also $(PQ, XX') = -1$, we find, by Theorem 5.9.5, $X' = X''$. ◆

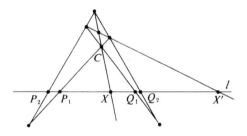

Figure 5.28

Now we intersect a complete quadrangle with a line l not passing through any of its points (possibly passing through one or two diagonal points). Let $P_1, P_2; Q_1, Q_2; X, X'$ be the three pairs of points in which l intersects opposite lines of the quadrangle (see Fig. 5.28). We call these six points a **quadrangular set** and denote by

$$Q(P_1 Q_1 X; P_2 Q_2 X')$$

the statement that $P_1, P_2, Q_1, Q_2, X, X'$ form a quadrangular set. Clearly, in a quadrangular set, $P_1 \neq Q_1, P_1 \neq Q_2, P_2 \neq Q_1, P_2 \neq Q_2$.

Theorem 5.10.2 *A quadrangular set is uniquely determined if five of its points are given.*

Proof Suppose $Q(P_1 Q_1 X; P_2 Q_2 X')$ and $Q(P_1 Q_1 X; P_2 Q_2 X'')$ are given (the proof in all other cases being analogous). Let A, B, C, D and A', B', C', D' be the vertices of the complete quadrangle constituting these quadrangular sets, respectively (see Fig. 5.29). Suppose first that $A \neq A'$, $B \neq B'$, and $AA' \neq BB'$. We set $Z = (AA')(BB')$ and $z = P_1 Q_1$. By Theorem 5.7.1, there exists a (Z, z)-collineation Γ mapping A onto A'.

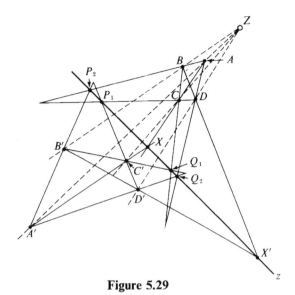

Figure 5.29

Since P_1, P_2, Q_1, Q_2 are fixed points of Γ, we conclude successively that Γ maps B onto B', C onto C', D onto D'. Therefore, the image of BD under Γ is $B'D'$, and hence

$$X' = (BD)z = (B'D')z = X''.$$

If $A = A'$ or $B = B'$ or $AA' = BB'$, we apply first for some Z^* a (Z^*, z)-collineation onto $ABCD$ which maps A, B onto points A'', B'' satisfying $A'' \neq A$; $A'' \neq A'$; $B'' \neq B$; $B'' \neq B'$; $AA'' \neq BB''$; $A'A'' \neq B'B''$ and apply the above argument twice. ◆

We consider the following projectivity with notation as in Figure 5.30:

$$P_1P_2XX' \overset{B}{\underset{\wedge}{=}} TAXR \overset{P_1}{\underset{\wedge}{=}} BSX'R \overset{A}{\underset{\wedge}{=}} P_2P_1X'X.$$

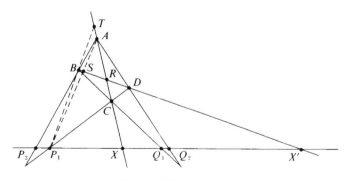

Figure 5.30

Since, by Theorem 5.9.6, every projectivity is a collineation, we conclude from the fundamental theorem (5.6.3) that there is a unique projectivity interchanging P_1 and P_2, and mapping X onto X'. This must be the projectivity just mentioned.

Theorem 5.10.3 *If a projectivity of a line onto itself interchanges one pair of points, it is involutoric.*

An involutoric projectivity is called an **involution**. As a special case of Theorem 5.10.3, we have $(P_1 = P_2; Q_1 = Q_2)$.

Theorem 5.10.4 *If P, Q are different points, then we obtain an involution of PQ onto itself by assigning to every point X the fourth harmonic point of P, Q, X.*

In the above discussion, the roles of P_i and Q_i can be interchanged, also the roles of X and X' (since an involution equals its inverse). Therefore, Q_1 and Q_2 are also interchanged under the involution of Theorem 5.10.3.

Theorem 5.10.5 *A line not passing through the vertices of a complete quadrangle intersects the three pairs of opposite lines of the quadrangle in three pairs of an involution.*

Theorem 5.10.5 provides a very simple method of constructing corresponding points under an involution if two pairs of corresponding points P_1, P_2 and Q_1, Q_2 are given. One simply draws lines through these points, no three of which have a point in common, and proceeds as indicated in Figure 5.28. So using Theorem 5.10.2, we make the following statement.

Theorem 5.10.6 *If two disjoint pairs P_1, P_2 and Q_1, Q_2 of points on a line are given (possibly $P_1 = P_2$ or $Q_1 = Q_2$ or both), then there exists one and only one involution interchanging P_1 and P_2 as well as Q_1 and Q_2.*

Exercises

1. State the dual of Theorem 5.10.1.

2. Show that a projectivity maps the fourth harmonic point of three points onto the fourth harmonic point of the images of the three points.

3. Show that the statement $Q(P_1 Q_1 X; P_2 Q_2 X')$ remains valid if P_1, Q_1, X are permuted or P_2, Q_2, X' are permuted. Show, furthermore, that $Q(P_1 Q_2 X'; P_2 Q_1 X)$, $Q(P_2 Q_1 X'; P_1 Q_2 X)$, $Q(P_2 Q_2 X; P_1 Q_1 X')$.

4. Show by using Pappus' theorem that $Q(P_1 Q_1 X; P_2 Q_2 X')$ implies $Q(P_2 Q_2 X'; P_1 Q_1 X)$.

5. In an ordinary Euclidean plane, find in terms of harmonic points a condition for a point to be the midpoint of two other points.

6. Show that in an ordinary Euclidean plane the fourth harmonic point X' can be obtained from P, Q, X as, indicated in Figure 5.31.

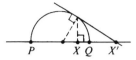

Figure 5.31

5.11   Intrinsic Geometry of the Projective Line

In the preceding two sections we have discussed some facts about the linear range of points on one line c (projectivities and involutions of c onto itself). This has been done mainly by considering c a line in a projective plane or a higher-dimensional projective space. We add, in this section, some observations that can be made without "leaving" the line c, that is, intrinsic properties of c. Clearly, these observations are made by using analytical tools, since there are no geometrical configurations on c except trivial ones. We restrict ourselves to the case where the coordinate field $\mathscr{K}$ is the field of real numbers. If homogeneous coordinates are introduced, then an arbitrary point of c can be expressed by

$$s\mathbf{p} + t\mathbf{q},$$

where $\mathbf{p}, \mathbf{q}$ are linearly independent vectors. If we choose $\mathbf{p}, \mathbf{q}$ as basis vectors $\mathbf{a}_1, \mathbf{a}_2$ of $\mathscr{V}^{n+1}$ and disregard all other basis vectors (that is, consider only a $\mathscr{V}^2$), we can let

$$s = x_1, \qquad t = x_2$$

and obtain the homogeneous coordinates (x_1, x_2) on the one-dimensional projective space $c = \mathscr{P}^1$, as introduced in Section 5.5.

A projective collineation Γ of $\mathscr{P}^1$ is given by a linear mapping

$$x' = \mathbf{x}\mathbf{A}, \quad |\mathbf{A}| \neq 0, \tag{1}$$

of $\mathscr{V}^2$: If we introduce inhomogeneous coordinates $(x_2 \neq 0)$

$$\xi = \frac{x_1}{x_2},$$

we obtain $\left(\mathbf{A} = \begin{pmatrix} a_{11} & a_{12} \\ a_{21} & a_{22} \end{pmatrix}\right)$

$$\xi' = \frac{x_1'}{x_2'} = \frac{a_{11}x_1 + a_{21}x_2}{a_{12}x_1 + a_{22}x_2} = \frac{a_{11}(x_1/x_2) + a_{21}}{a_{12}(x_1/x_2) + a_{22}}$$
$$= \frac{a_{11}\xi + a_{21}}{a_{12}\xi + a_{22}}.$$

Since $\mathbf{A}$ is arbitrary under the sole condition $|\mathbf{A}| \neq 0$, any rational equation

$$\xi' = \frac{a\xi + b}{c\xi + d}, \qquad ad - bc \neq 0, \tag{2}$$

defines a projective collineation (or projectivity; see Theorem 5.9.6). Formally, we may extend this mapping to the point at infinity U of $\mathscr{P}'$ by assigning to it "∞" as the value of ξ. So U is the image of the point with $\xi = -d/c$ and is mapped onto the point $\xi = a/c$. If P_1, P_2, P_3 are different points which are to be mapped onto the (different) points P_1', P_2', P_3' by a projective collineation, we obtain, in case all these points are not equal to U, as a consequence of Theorem 5.9.5 (p_i coordinates of P_i; p_i' coordinates of P_i'; ξ, ξ' coordinates of X, X'):

$$\frac{p_3 - p_1}{p_3 - p_2} \cdot \frac{\xi - p_1}{\xi - p_2} = \frac{p_3' - p_1'}{p_3' - p_2'} \cdot \frac{\xi' - p_1'}{\xi' - p_2'}.$$

This can be solved for ξ' and provides an equation of the form (2). In order to find the fixed points of Γ, one may either calculate the eigenvectors of $\mathbf{A}$ or set $\xi = \xi'$ in (2) and solve the quadratic equation

$$c\xi^2 + (d - a)\xi - b = 0. \tag{3}$$

Γ is called **hyperbolic, parabolic,** or **elliptic,** depending on whether it has two, one, or no fixed points, respectively.

If $c = 0$, and $a \neq d$, the point at infinity U is a fixed point and a second fixed point is obtained by its coordinate

$$\xi = \frac{b}{d - a}.$$

So Γ is hyperbolic.

If $c = 0$ and $a = d$, we conclude from (3) that there is no second fixed point unless $b = 0$. In the latter case, however, $\xi' = \xi$ and thus Γ is the identity map. Therefore, Γ is parabolic or the identity.

If $c \neq 0$, we have the following alternatives:

$$\Gamma \text{ is hyperbolic} \quad \text{if} \quad (d - a)^2 + 4bc > 0,$$
$$\Gamma \text{ is parabolic} \quad \text{if} \quad (d - a)^2 + 4bc = 0,$$
$$\Gamma \text{ is elliptic} \quad \text{if} \quad (d - a)^2 + 4bc < 0.$$

Now we investigate the special case of an involution. Since it equals its inverse, we can interchange in (2) the roles of ξ, ξ' and obtain

$$\xi' = \frac{-d\xi + b}{c\xi - a} = \frac{a\xi + b}{c\xi + d},$$

and thus

$$a = -d. \tag{4}$$

Conversely, (4) guarantees Γ to be involutoric, as is readily verified by calculation.

The fixed points of an involution are obtained by the solutions of

$$c\xi^2 - 2a\xi - b = 0.$$

Since $\qquad\qquad\qquad\qquad\qquad\qquad\qquad\qquad\qquad\qquad\qquad\qquad$

$$(d - a)^2 + 4bc = 4(a^2 + bc) = -4 \begin{vmatrix} a & b \\ c & d \end{vmatrix} \neq 0,$$

we conclude that *an involution is either hyperbolic or elliptic.*

Exercises

1. Find the fixed point of an involution by calculating the eigenvalues and the eigenvectors (see Section 4.13) of the matrix **A** in Equation (1).

2. Show that every collineation of $\mathscr{P}^1$ is the product of at most two involutions (therefore, the group of projectivities is generated by the set of all involutions).

3. Show that two involutions of $\mathscr{P}^1$ possess a common pair of corresponding points provided at least one of the involutions is elliptic.

4. State some properties of involutions if the coordinate field $\mathscr{K}$ of $\mathscr{P}^1$ is that of complex numbers, especially those which are not true for $\mathscr{K}$ real. Can an involution have more than two fixed points?

5. Show that (in an ordinary Euclidean plane) perpendicularity of lines in a pencil induces an involution in that pencil (which is clearly elliptic).

6. Is any map of $\mathscr{P}^1$ onto itself that preserves the relation "fourth harmonic point" a projectivity?

5.12   Correlations and Polarities

So far, most of the mappings considered have been mappings between objects of the same kind. One exception was that of a projection (see Section 5.9), in which we assigned lines to points. We begin now a systematic study of mappings of $\mathscr{P}^n$ in which points and hyperplanes are interchanged. This study will provide another example of the beauty unveiled in the principle of duality.

A mapping

$$\mathbf{x}' = \mathbf{x}\mathbf{A}, \quad |\mathbf{A}| \neq 0,$$

was interpreted in Section 5.6 as a mapping of points $\mathscr{K}\mathbf{x}$ onto points $\mathscr{K}\mathbf{x}'$ or of hyperplanes $\mathscr{K}\mathbf{x}$ onto hyperplanes $\mathscr{K}\mathbf{x}'$. However, we may also consider $\mathscr{K}\mathbf{x}$ to be a point, and $\mathscr{K}\mathbf{x}'$ to be a hyperplane. We denote hyperplanes by $\mathscr{K}\mathbf{u}$, etc., and set

$$\mathbf{u}' = \mathbf{x}\mathbf{A}; \quad |\mathbf{A}| \neq 0, \qquad\qquad (1)$$

considering this linear mapping of $\mathscr{V}^{n+1}$ onto itself a mapping of

points $\mathscr{K}\mathbf{x}$ onto hyperplanes $\mathscr{K}\mathbf{u}'(\mathbf{x} \neq 0; \mathbf{u}' \neq 0)$.

We call such a mapping a *correlation*. Naturally, we ask for a corresponding map

$$\mathbf{x}' = \mathbf{uB}, \tag{2}$$

which assigns to every hyperplane $\mathscr{K}\mathbf{u}$ a point $\mathscr{K}\mathbf{x}'$ in such a way that incidence is preserved; that is,

$$\mathbf{u} \cdot \mathbf{x} = 0 \quad \text{implies} \quad \mathbf{x}' \cdot \mathbf{u}' = 0.$$

Substituting (1), (2) in $\mathbf{x}' \cdot \mathbf{u}' = 0$ yields

$$\mathbf{uB} \cdot {}^{T}\mathbf{Ax} = 0$$

for all $\mathbf{x}, \mathbf{y}$. Since also $\mathbf{u} \cdot \mathbf{x} = 0$ for all $\mathbf{x}, \mathbf{y}$, we may set (up to a scalar multiple $\neq 0$)

$$\mathbf{B} \cdot {}^{T}\mathbf{A} = \mathbf{I} \quad \text{or} \quad \mathbf{B} = {}^{T}\mathbf{A}^{-1}.$$

So

$$\mathbf{x}' = \mathbf{u}^{T}\mathbf{A}^{-1}. \tag{2'}$$

This mapping is called the *contravariant* or *contragredient* mapping of (1). With respect to (2'), the mapping (1) is called *covariant* or *cogredient.*

So these terms "contravariant" and "covariant" are defined with respect to the inner product $\mathbf{u} \cdot \mathbf{x}$. Simultaneous linear transformations in both variables have to be such that vanishing of the inner product is preserved.

Example

$$\mathbf{A} = \begin{pmatrix} 1 & -1 & 0 \\ 0 & 1 & 0 \\ 0 & 0 & 2 \end{pmatrix}, \qquad {}^{T}\mathbf{A}^{-1} = \begin{pmatrix} 1 & 0 & 0 \\ 1 & 1 & 0 \\ 0 & 0 & \frac{1}{2} \end{pmatrix}.$$

The points $X_t = \mathscr{K}(t, 1, 1)$ are mapped onto the lines $X'_t = \mathscr{K}(t, 1 - t, 2)$. The lines $\mathscr{K}(t, 1 - t, 2)$ are mapped onto the points $X'' = \mathscr{K}(1, 1 - t, 1)$.

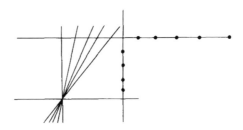

Figure 5.32

Clearly, any linear range of points

$$\mathscr{K}\mathbf{p} + \mathscr{K}\mathbf{q}$$

is mapped onto a pencil of lines

$$\mathcal{K}(\mathbf{pA}) + \mathcal{K}(\mathbf{qA}),$$

and, conversely (compare Figure 5.32), *any figure is mapped onto a dual figure* (see Section 5.3).

The product of two correlations is a collineation; the product of a correlation and a collineation is a correlation. Of a particular interest is the case in which the correlation is involutoric in the sense that multiplying the correlation by itself yields the identity map. We now determine under which condition for A this is so:

$$\mathcal{K}\mathbf{x}' = \mathcal{K}(\mathbf{u}^\mathrm{T}\mathbf{A}^{-1}) = \mathcal{K}(\mathbf{x}'\mathbf{A}^\mathrm{T}\mathbf{A}^{-1}) = \mathcal{K}\mathbf{x}'$$

for all $\mathbf{x}$; hence

$$\mathbf{A}^\mathrm{T}\mathbf{A}^{-1} = a\mathbf{I}, \quad a \neq 0,$$

or

$$\mathbf{A} = a^\mathrm{T}\mathbf{A}.$$

Applying transposition on both sides yields

$$^\mathrm{T}\mathbf{A} = a\mathbf{A};$$

hence

$$\mathbf{A} = a^2\mathbf{A},$$

and thus $a = \pm 1$, from which

$$^\mathrm{T}\mathbf{A} = \pm\,\mathbf{A}$$

follows as a necessary condition for $\mathbf{A}$ under which the correlation is involutoric. This condition is also sufficient, as can be verified directly.

If $^\mathrm{T}\mathbf{A} = +\mathbf{A}$, that is, if $\mathbf{A}$ is *symmetric*, we call the correlation a *polarity*. If $^\mathrm{T}\mathbf{A} = -\mathbf{A}$, that is, if $\mathbf{A}$ is *skew symmetric*, we call the correlation a *null system*. Clearly, the above example of $\mathbf{A}$ is neither symmetric nor skew symmetric, and the corresponding correlation is not involutoric.

We shall be mainly concerned with polarities from now on. If the hyperplane p is the image of a point P under a polarity, we call p the *polar* of P and P the *pole* of p. So we have the mutual correspondence

$$\text{pole} \longleftrightarrow \text{polar}.$$

Example of a polarity (Fig. 5.33):

$$\mathbf{A} = \begin{pmatrix} 4 & 0 & 0 \\ 0 & 1 & 0 \\ 1 & 0 & 1 \end{pmatrix}, \quad ^\mathrm{T}\mathbf{A}^{-1} = \begin{pmatrix} \tfrac{1}{4} & 0 & 0 \\ 0 & 1 & 0 \\ 0 & 0 & -1 \end{pmatrix}.$$

In Section 1.12, we introduced the terms "pole" and "polar lines." We shall see, in Chapter 7, that this is, in fact, a special case of our present definitions.

We call two points $X = \mathcal{K}\mathbf{x}$, $Y = \mathcal{K}\mathbf{y}$ *conjugate* if $\mathcal{K}\mathbf{y}$ lies on the polar of $\mathcal{K}\mathbf{x}$.

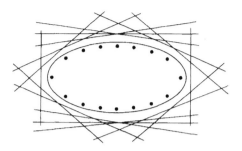

Figure 5.33

Theorem 5.12.1 $X = \mathcal{K}\mathbf{x}$ *is conjugate to* $Y = \mathcal{K}\mathbf{y}$ *if and only if*

$$\mathbf{x}A\mathbf{y} = 0. \qquad (3)$$

Proof I. Let $\mathcal{K}\mathbf{x}$ be conjugate to $\mathcal{K}\mathbf{y}$. If $\mathbf{x}' = \mathbf{x}A$ represents the polar of X, we find from Y lying on $\mathbf{x}' \cdot \mathbf{y} = 0$:

$$\mathbf{x}A\mathbf{y} = 0.$$

II. If, conversely, (3) is true, we conclude that $\mathcal{K}\mathbf{y}$ lies on the polar of $\mathcal{K}\mathbf{x}$. ◆

Theorem 5.12.2 x *is conjugate to* y *if and only if* y *is conjugate to* x.

Proof From (3) we find (the transpose of a matrix consisting of one element is the matrix itself $^T a = {}^T(a) = (a) = a$)

$$0 = {}^T0 = {}^T(\mathbf{x}A\mathbf{y}) = \mathbf{y}^T A \mathbf{x} = \mathbf{y}A\mathbf{x},$$

by using the symmetry of A. ◆

The symmetry $^T A = A$ of A carries over to the symmetry $^T A^{-1} = A^{-1}$ of A^{-1}, as is seen easily. We call a line $\mathcal{K}\mathbf{u}$ **conjugate** to a line $\mathcal{K}\mathbf{v}$ if $\mathcal{K}\mathbf{v}$ passes through the pole of $\mathcal{K}\mathbf{u}$. So, by applying the principle of duality to Theorems 5.12.1 and 5.12.2, we obtain the following two theorems.

Theorem 5.12.3 $u = \mathcal{K}\mathbf{u}$ *is conjugate to* $v = \mathcal{K}\mathbf{v}$ *if and only if* $\mathbf{u}A^{-1}\mathbf{v} = 0$.

Theorem 5.12.4 u *is conjugate to* v *if and only if* v *is conjugate to* u.

Exercises

1. Show that in a null system every point lies on its image hyperplane.
2. Show that in an even-dimensional projective space $\mathcal{P}^{2k}$ there exists no null system. (*Hint*: The determinant of A must be zero in this case.)

3. Let a null system be given by

$$A = \begin{pmatrix} 0 & 1 & 0 & 0 \\ -1 & 0 & 0 & 0 \\ 0 & 0 & 0 & -1 \\ 0 & 0 & 1 & 0 \end{pmatrix}.$$

Find the images of the four points

$$\mathscr{K}\,(\varepsilon(1, 1, 0, 1) + \varepsilon'(1, 1, 1, 1)), \varepsilon = \pm 1;\quad \varepsilon' = \pm 1.$$

Draw a picture.

4. If in (1) the condition $|A| \neq 0$ is replaced by $|A| = 0$, we call the "degenerate correlation" obtained in this way a quasi-correlation. Show that under a quasi-correlation the image hyperplanes of all points have a point in common.

5. What is the meaning of a correlation and that of a polarity in $\mathscr{P}^1$?

6. Suppose any point P on a subspace $\mathscr{U}$ of $\mathscr{P}^n$ has a polar p not containing $\mathscr{U}$. Define $p' = \mathscr{U} \cap p$. Is $P \to p'$ a polarity of $\mathscr{U}$?

5.13 Conics and Quadrics

In this section, the coordinate field is the field of real numbers. We define a *conic* (in $\mathscr{P}^2$) or a *quadric* (in $\mathscr{P}^3$) as the set of all points that, under a polarity, lie on their polar lines or polar planes, respectively. Generally, a *quadric* is the set of all points incident with their polars.

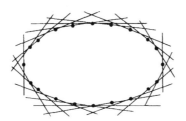

Figure 5.34

Analytically, a quadratic hypersurface is obtained from (1) of Section 5.12 by the condition $\mathbf{u}' \cdot \mathbf{x} = 0$ or

$$\mathbf{x}A\mathbf{x} = 0. \tag{1}$$

Written in coordinates, this is the quadratic equation

$$\sum_{\rho,\sigma=1}^{n+1} x_\rho a_{\rho\sigma} x_\sigma = 0. \tag{1'}$$

$$A = \begin{pmatrix} 1 & 0 & 0 \\ 0 & 4 & 0 \\ 0 & 0 & -4 \end{pmatrix},$$

$$xAx = x_1^2 + 4x_2^2 - 4x_3^2 = 0.$$

In inhomogeneous coordinates, we may write

$$\frac{\xi_1^2}{4} + \xi_2^2 = 1.$$

This is an *ellipse*, as Figure 5.34 indicates.

Although (1) was obtained from a polarity ($|A| \neq 0$), we may also, more generally, call the set of all points satisfying (1) a *quadric* in the case $|A| = 0$. We distinguish the case $|A| \neq 0$ by calling the conic or quadric in this case *regular*.

Let a regular quadric be given. Instead of (1) we can set up as a condition for points lying on their polar using (2′) in Section 5.12:

$$u^T A^{-1} u = 0. \tag{2}$$

(2) represents the totality of all hyperplanes incident with their poles, that is, the *envelope* of the hypersurface defined by (1) (compare Figure 5.34). We call this set of hyperplanes a *hypersurface of the second class.* Since $^T A^{-1}$ occurs in (2), we must assume $|A| \neq 0$. However, we may replace $^T A^{-1}$ in (2) by the *"adjoint matrix"* of $^T A$:

$$^T\tilde{A} = \begin{pmatrix} A_{11} & \cdots & A_{n+1,1} \\ \vdots & & \vdots \\ A_{1,n+1} & \cdots & A_{n+1,n+1} \end{pmatrix},$$

where A_{ik} is the determinant obtained from $|A|$ by deleting the ith row and the kth column. (Compare Appendix 4.III). In particular for $|A| \neq 0$,

$$^T A^{-1} = \frac{^T\tilde{A}}{|A|}$$

(2) is, for $|A| \neq 0$, equivalent to

$$u^T\tilde{A}u = 0. \tag{2′}$$

However, (2′) is also defined for $|A| = 0$; therefore, we can extend the definition of a hypersurface of second class to the general case.

Example With the above matrix A, we obtain

$$^T\tilde{A} = \tilde{A} = \begin{pmatrix} -16 & 0 & 0 \\ 0 & -4 & 0 \\ 0 & 0 & 4 \end{pmatrix};$$

hence

$$-16u_1^2 - 4u_2^2 + 4u_3^2 = 0,$$

or

$$-4u_1^2 - u_2^2 + u_3^2 = 0.$$

We shall now discuss a number of standard examples of conics in $\mathscr{P}^2$ and quadrics in $\mathscr{P}^3$. We remark that every conic or quadric can be transformed by a collineation into one of these standard examples, so that, according to Klein's Erlanger Programm, we can find in them all projective geometric properties of quadratic surfaces.

(a) *Conics in a projective plane* (Fig. 5.35):

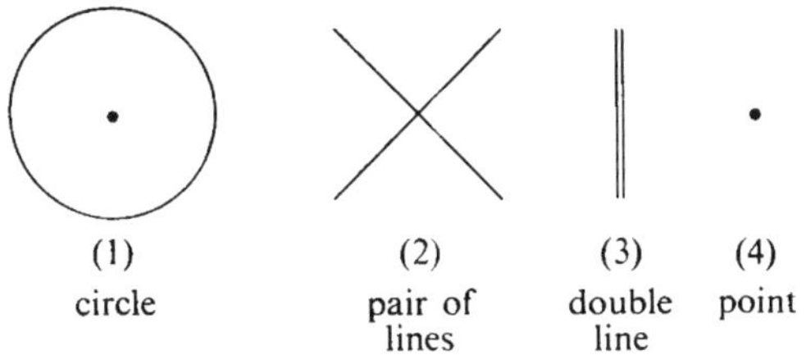

(1)	(2)	(3)	(4)
circle	pair of lines	double line	point

Figure 5.35

(1) $x_1^2 + x_2^2 - x_3^2 = 0$ circle (ellipse, hyperbola, parabola),
(2) $x_1^2 - x_2^2 \quad\quad = 0$ pair of lines,
(3) $x_1^2 \quad\quad\quad = 0$ double line,
(4) $x_1^2 + x_2^2 \quad = 0$ point,
(5) $x_1^2 + x_2^2 + x_3^2 = 0$ empty set.

We shall see in the next section how the circle, ellipse, and hyperbola are distinguished; this is not a projective but an affine distinction. If, in inhomogeneous coordinates,

$$\xi_1^2 - \xi_2^2 = 1$$

is the equation of a hyperbola, we obtain in homogeneous coordinates

$$x_1^2 - x_2^2 - x_3^2 = 0, \quad \text{or} \quad -x_1^2 + x_2^2 + x_3^2 = 0.$$

So, by the projective transformation

$$(x_1', x_2', x_3') = (x_3, x_2, x_1),$$

we obtain the normal form (1), hence a circle. Also, an ellipse of the form

$$\frac{\xi_1^2}{a^2} + \frac{\xi_2^2}{b^2} = 1$$

can be transformed projectively into (1): In homogeneous coordinates we obtain

$$\frac{x_1^2}{a^2} + \frac{x_2^2}{b^2} - x_3^2 = 0.$$

Letting

$$(x_1', x_2', x_3') = \left(\frac{1}{a}x_1, \frac{1}{b}x_2, x_3\right)$$

yields equation (1) (with x_i' instead of x_i).

(b) *Quadrics in a projective 3-space* (Figs. 5.36 and 5.37).

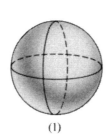 

(1)   (2)

Figure 5.36

(1) $x_1^2 + x_2^2 + x_3^2 - x_4^2 = 0$ sphere,
(2) $x_1^2 + x_2^2 - x_3^2 - x_4^2 = 0$ hyperboloid of one sheet,
(3) $x_1^2 + x_2^2 - x_3^2 \quad\;\; = 0$ circular cone,
(4) $x_1^2 - x_2^2 \qquad\qquad = 0$ pair of planes,
(5) $x_1^2 \qquad\qquad\qquad = 0$ double plane,
(6) $x_1^2 + x_2^2 \qquad\quad\; = 0$ line,
(7) $x_1^2 + x_2^2 + x_3^2 \quad\;\; = 0$ point,
(8) $x_1^2 + x_2^2 + x_3^2 + x_4^2 = 0$ empty set.

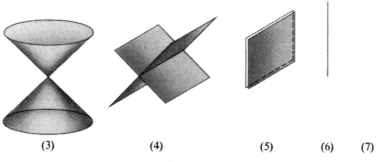

(3)   (4)   (5)   (6)   (7)

Figure 5.37

We shall see again in the following section how, in affine geometry, these types of quadrics are split into further types.

Exercises

1. Set up the matrices for all conics (1)–(5) and all quadrics (1)–(8). Which of these conics and quadrics stem from polarities?

2. Show why the equations in (4)–(8) of quadrics have the geometrical meaning expressed in their names.

3. Find a projectivity that transforms

$$x_1 x_2 - x_3 x_4 = 0$$

into one of the normal forms (1)–(8).

4. Investigate all intersections of the quadrics (2) and (3) by planes $x_2 = c$, and $x_3 = d$; c, d constant. Transform each conic obtained by a collineation into the plane $x_4 = 0$ such that it obtains one of the normal forms (1)–(4) of conics.

5. If we consider the field of complex numbers to be a coordinate field of $\mathscr{P}^3$, show which of the above normal forms of conics and quadrics are projectively equivalent—that is, which equations can be obtained from each other by (complex) collineations.

6. Find all lines on the hyperboloid of one sheet (2). (*Hint*: Intersect (2) by appropriate planes through $K(0, 0, 1, 0)$.)

7. Find all lines on the circular cone (3).

5.14 *Affine and Euclidean Distinction of Conics and Quadrics*

The set $\mathfrak{P}$ of all projective collineations of $\mathscr{P}^n$ is a group which we may call the **projective group** of $\mathscr{P}^n$. If we proceed from $\mathscr{P}^n$ to an affine space $\mathscr{A}^n$ by declaring a hyperplane ω to be a hyperplane at infinity, we obtain the **affine group** $\mathfrak{A}$ of $\mathscr{A}^n$ as the subgroup of $\mathfrak{P}$ consisting of all elements of $\mathfrak{P}$ that leave ω fixed.

$$\mathfrak{A} \subset \mathfrak{P}.$$

Since $\mathfrak{A}$ has fewer elements than $\mathfrak{P}$, there will be fewer geometrical figures that can be transformed into each other. For example, in the preceding section, we saw that in the extension of a Euclidean plane to a projective plane, a hyperbola can be transformed into an ellipse or even a circle. This ceases to be true in affine geometry, since a hyperbola "passes through infinity," but an ellipse does not (see Fig. 5.38). Generally we may define a

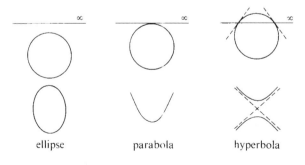

ellipse   parabola   hyperbola

Figure 5.38

nondegenerate (and nonempty) conic to be

an *ellipse* if it does not intersect ω,
a *parabola* if it intersects ω in one point,
a *hyperbola* if it intersect ω in two points.

Similarly, in $\mathscr{P}^3$ we obtain the following distinction: Given a (nonempty) quadric with $|A| \neq 0$, we generally call it a *sphere* if it does not contain lines, a *hyperboloid of one sheet* if it does contain lines (Fig. 5.39).

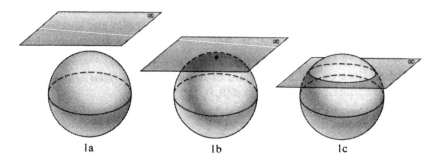

Figure 5.39

After introducing a plane ω at infinity, we define a sphere to be (Fig. 5.40):

(1-a) an *ellipsoid* if it does not intersect ω,
(1-b) an *elliptic paraboloid* if it intersects ω in one point,
(1-c) a *hyperboloid of 2 sheets* if it intersects ω in more than one point.

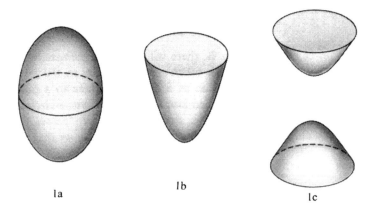

Figure 5.40

A hyperboloid is defined to be (Fig. 5.41):

(2-a) a *hyperboloid of one sheet* if it intersects ω in an ellipse (which is projectively equal to a hyperbola),
(2-b) a *hyperbolic paraboloid* (a "saddle") if it intersects ω in a pair of lines.

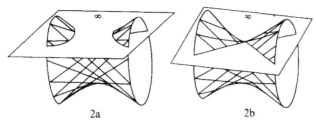

2a   2b

Figure 5.41

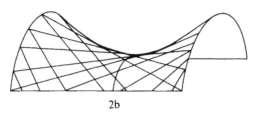

2b

Figure 5.42

So, instead of two types of regular quadrics in $\mathscr{P}^3$, we have found five types of regular quadrics in $\mathscr{A}^3$. An analogous discussion can be carried out for nonregular quadratic surfaces (see Exercise 1).

Finally, we consider the subgroup $\mathscr{M}$ of all motions in $\mathfrak{A}$, that is, the set of all affine mappings leaving the inner product $\mathbf{x} \cdot \mathbf{y}$ invariant.

$$\mathscr{M} \subset \mathfrak{A} \subset \mathfrak{P}.$$

Now we shall have an uncountable number of different classes of quadratic surfaces if equivalence under motions is considered. So, for example, any two ellipses

$$\frac{\xi_1^2}{a^2} + \frac{\xi_2^2}{b^2} = 1$$

that differ in a or b are not equivalent. In particular, for $a = b$ we obtain the case of a circle as distinct from any other ellipse. In Euclidean space, spheres are now special cases of general ellipsoids, circular cones are special cases of elliptic cones, and so on.

Exercises

1. Discuss the affine distinction of the nonregular quadrics (3)–(7) of Section 5.13. Introduce in particular the concept of a cylinder.

2. Using the above definitions, show that in inhomogeneous coordinates

   (a) $\xi_1^2 - \xi_2^2 - \xi_3^2 = 1$ represents a hyperboloid of two sheets,
   (b) $\xi_1^2 + \xi_2^2 - \xi_3 = 0$ represents an elliptic paraboloid.

3. Extend $\mathscr{A}^n$ to $\mathscr{P}^n$ and find a collineation mapping the hyperboloid (a) of two sheets in Exercise 2 onto the elliptic paraboloid (b).

4. Given the unit circle $\xi_1^2 + \xi_2^2 = 1$ in a Euclidean plane, consider this plane extended to a projective plane and find a projective collineation that maps the given circle onto a parabola passing through $(0, 0)$ and $(0, 1)$ and possessing the line $\xi_2 = 2\xi_1$ as axis (line of symmetry).

5. Discuss all plane sections of the hyperboloid $\xi_1^2 + \xi_2^2 = 1$ that are parallel to a coordinate plane.

6. In Euclidean 3-space, consider all rotations about an axis a. Show that all images of a line b not parallel to a lie on a hyperboloid of one sheet.

5.15   *Pencils of Conics*

This section on conics shows an interesting relationship among different kinds of conics. The results will also be needed in Section 6.6.

Suppose we are given an affine plane over a field $\mathscr{K}$. We might be given a projective plane as well; the following considerations can be changed easily into those of projective planes. Let

$$c(x_1, x_2) = rx_1^2 + sx_1x_2 + tx_2^2 + ux_1 + vx_2 + w = 0, \tag{1}$$

$$c'(x_1, x_2) = r'x_1^2 + s'x_1x_2 + t'x_2^2 + u'x_1 + v'x_2 + w' = 0 \tag{2}$$

be two different conics. Then, for arbitrary $p, q \in \mathscr{K}$, not both zero,

$$pc(x_1, x_2) + qc'(x_1, x_2) = (pr + qr')x_1^2 + \cdots + pw + qw' = 0 \tag{3}$$

is also a conic or empty. We call the totality of all conics (3) with p, q not both $=0$ a **pencil of conics** (Fig. 5.43). Any point common to the conics (1), (2) clearly belongs to the conic (3).

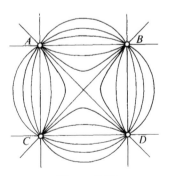

Figure 5.43

Now, let

$$r''x_1^2 + s''x_1x_2 + t''x_2^2 + u''x_1 + v''x_2 + w'' = 0 \tag{4}$$

be a further conic.

Theorem 5.15.1 *Three conics* (1), (2), *and* (4) *belong to a common pencil of conics if and only if the matrix*

$$\begin{pmatrix} r & s & t & u & v & w \\ r' & s' & t' & u' & v' & w' \\ r'' & s'' & t'' & u'' & v'' & w'' \end{pmatrix}$$

has rank at most two, that is, if all 3×3 *determinants consisting of columns of this matrix vanish.*

Proof In the language of vector algebra, the rank condition may be expressed as follows. The three row vectors of the matrix are linearly dependent (in a six-dimensional vector space). In other words, one is a linear combination of the other two, say the last one is a linear combination of the first two:

$$r'' = ar + br', s'' = as + bs', \ldots, w'' = aw + bw'.$$

This is clearly so if and only if all determinants formed by columns of the matrix vanish. ◆

Theorem 5.15.2 *Let four points* $A, B, C, D,$ *no three of which are collinear, be common to all conics of a pencil of conics. Then the pairs of lines* $AB, CD; AC, BD;$ *and* AD, BC *also belong to the pencil* (Fig. 5.43).

Proof Let $M = (m_1, m_2)$ be a point not equal to A, B on the line AB; (3) is the given pencil, and we let

$$pc(m_1, m_2) + qc'(m_1, m_2) = 0.$$

This is a linear equation in p, q which clearly has a solution $(p, q) \neq (0, 0)$; that is, in the pencil (3) there is a conic passing through M. Since A, B, M are collinear points on this conic, it is degenerate and so consists of two lines. AB is one of these lines; the other passes through C, D (since all conics (3) pass through A, B, C, D) and hence equals CD. The same argument applies for A, C or A, D instead of A, B. ◆

Figure 5.43 shows an example of a pencil of conics in the ordinary Euclidean plane. It is seen that the pairs AB, CD and AC, BD of lines occur as limiting positions of ellipses, whereas the pair AD, BC occurs as the limiting position of hyperbolas of the pencil.

Exercises

1. In the ordinary Euclidean plane, show that if (1) and (2) are circles, then (3) is a circle or a line or a point or is empty.

2. Investigate the pencil given by two circles (see Exercise 1) if the circles
   (a) intersect,
   (b) are tangent to each other,
   (c) do not intersect.

3. Investigate the pencil (3) if (1) and (2) are double lines.

4. Do the conics $x_1^2 + x_2^2 - x_3 + 1 = 0$, $x_1 - x_2^2 = 0$, $2x_1^2 - x_2^2 - 2x_3 + 3x_1 + 2 = 0$ belong to a common pencil of conics?

5. Find a line in the pencil spanned by $x_1^2 + x_2^2 = 1$ and $(x - 5)^2 + x_2^2 = 4$.

6. In the pencil (3) find a condition for the pencil to possess a conic tangent to the line $x_2 = 0$.

5.16   Polars, Tangents, and Asymptotes

In this section let the coordinate field of $\mathscr{P}^n$ be arbitrary (commutative) with the characteristic that it is not equal to 2. A regular quadric has been defined as the totality of points lying on their polars where a polarity is given. These lines or planes incident with their poles are called *tangents* to the surface.

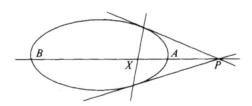

Figure 5.44

Let P be a pole. We seek a condition in which a point X lies on the polar of P. With $P = \mathscr{K}\mathbf{p}$; $X = \mathscr{K}\mathbf{x}$, we have

$$\mathbf{pAx} = 0, \tag{1}$$

which, for the variable $\mathbf{x}$, is the equation of the polar of P. $\mathbf{p}$ and $\mathbf{x}$ are also called *conjugate points* with respect to the conic or quadric surface

$$\mathbf{xAx} = 0.$$

Every point P of the quadric is conjugate to all points of the tangent at P. In particular, P is conjugate to itself, and the *quadric may be characterized as the set of all self-conjugate points under a polarity.* The relation "conjugate" is a symmetric one between points, since we obtain from (1) by applying a transposition on both sides and by using $^T\mathbf{A} = \mathbf{A}$:

$$^T(\mathbf{pAq}) = \mathbf{q}^T\mathbf{Ap} = \mathbf{qAp} = 0.$$

Theorem 5.16.1   *If Q lies on the polar of P, then P lies on the polar of Q. If, in particular, Q lies on the quadric, then P lies on the tangent to the quadric at Q.*

Let a be a line through P not tangent to the quadric. Then on this line there is one and only one conjugate point P' of P. So we obtain a mapping

$$P \longrightarrow P'$$

of a onto itself, which, by Theorem 5.16.1, is involutoric. It will turn out to be an involution in the sense of Section 5.10.

Let M be the pole of a and let M not lie on a. Any line through M is called a **conjugate line** of a. If P, P' are on a and if P is conjugate to P', then P is the pole of $P'M$ and P' is the pole of PM. We call $PP'M$ a **polar triangle**, since every vertex is the pole of the opposite side (Fig. 5.45). If MP and MP' both intersect the conic, we call them **conjugate diameters.** If

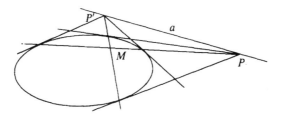

Figure 5.45

a is the line at infinity, we call M in the affine plane thus obtained the **center** of the quadric (compare Fig. 5.46).

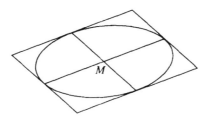

Figure 5.46

We shall restrict our discussion from now on to $\mathscr{P}^2$. It is convenient to introduce a coordinate system as follows. Let M be the point $\mathscr{K}(0, 0, 1)$; for some P_1 on a, let

$$P_1 = \mathscr{K}(1, 0, 0) \quad \text{and} \quad P'_1 = \mathscr{K}(0, 1, 0)$$

be conjugate points. If

$$\mathbf{u}' = \mathbf{x}\mathbf{A}$$

represents the given polarity in this coordinate system, we conclude from

$$0 = (1, 0, 0) \begin{pmatrix} a_{11} & a_{12} & a_{13} \\ a_{21} & a_{22} & a_{23} \\ a_{31} & a_{32} & a_{33} \end{pmatrix} \begin{pmatrix} 0 \\ 1 \\ 0 \end{pmatrix} = (a_{11} \quad a_{21} \quad a_{31}) \begin{pmatrix} 0 \\ 1 \\ 0 \end{pmatrix} = a_{21}$$

that $a_{21} = 0$. Since A is symmetric (that is, equal to ^TA), $a_{12} = 0$. Similarly since M, P and M, P' are both conjugate pairs, it follows that $a_{13} = a_{31} = 0$ and $a_{23} = a_{32} = 0$. So the equation $\mathbf{x}'\mathbf{Ax} = 0$ now reads

$$a_{11}x_1x_1' + a_{22}x_2x_2' + a_{33}x_3x_3' = 0.$$

a has the equation $x_3 = 0$. Hence, for the involutoric mapping $P \to P'$ introduced above on a, we obtain

$$a_{11}x_1x_1' + a_{22}x_2x_2' = 0 \tag{2}$$

as an equation or, setting

$$s = \frac{x_2}{x_1}, \qquad s' = \frac{x_2'}{x_1'},$$

we have

$$a_{11} + a_{22}ss' = 0. \tag{3}$$

We may assume $a_{11} \neq 0$ and $a_{22} \neq 0$, since, otherwise,

$$|\mathbf{A}| = a_{11}a_{22}a_{33} = 0.$$

Setting $k = -a_{11}/a_{22}$, we have

$$ss' = k. \tag{4}$$

Clearly, (4) represents an involution. k is called, in a Euclidean interpretation, an **orthogonality constant** and s, s' are denoted **slopes** of the lines MP, MP'. We now call these conjugate lines **perpendicular** or **orthogonal** (compare Appendix 3.II, the ordinary Euclidean perpendicularity constant is -1).

Theorem 5.16.2 *A polarity induces an involution on every line that is not tangent to the conic defined by the polarity. The involution is hyperbolic or elliptic, respectively, depending on whether the line does or does not meet the conic.*

Proof Only the second part of the theorem remains to be proved. If, in fact, the given line a meets the conic in A and B, these points are self-conjugate and hence $A' = A$; $B' = B$. Conversely, if $A' = A$ for a point on a, then A lies on its polar and hence on the conic. ◆

The following theorem is a conclusion from Theorem 5.16.2.

Theorem 5.16.3 *If a line a meets a conic in two different points A, B, then the conjugate point of a point P on a is the fourth harmonic point of A, B, P.*

___

Exercises

1. Show that for conjugate lines $\mathbf{u} \cdot \mathbf{x} = 0$ and $\mathbf{u}' \cdot \mathbf{x} = 0$, the relation

$$\mathbf{u}\mathbf{A}^{-1}\mathbf{u}' = 0$$

is true.

2. Show that a regular, nonempty conic possesses, up to projective collineation, the form (1) of conics in Section 5.13.

3. Extend Theorems 5.16.2 and 5.16.3 to $\mathscr{P}^n$.

4. Can the orthogonality constant k in (4) for an arbitrary field $\mathscr{K}$ be assumed to be -1 (after carrying out a collineation)?

5. Show that in ordinary Euclidean geometry one of two conjugate diameters of an ellipse bisects all chords parallel to the other.

6. Show that a correlation that induces elliptic or hyperbolic involutions on two different lines is a polarity.

5.17   *Projectivities and Conics*

Let a, b be two different lines in a projective plane and let these lines be related by a perspectivity from a point C (not on a or b).   From a point

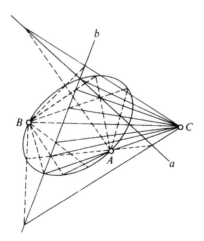

Figure 5.47

A not on a and a point B not on b, we project the point ranges on a and b, obtaining thus two projectively related pencils.   We intersect each two corresponding lines of the pencils.   This provides a point set in the plane which, according to Figure 5.47, looks like a conic.   We shall prove that it is indeed a conic.   Furthermore we shall show that any nondegenerate conic may be obtained by intersecting corresponding lines of projectively related pencils.

Theorem 5.17.1 *If two different pencils are projectively related, corresponding lines intersect in the points of a conic.*

Proof We may choose the coordinate system in such a way that the pencils are given by equations

$$x_2 = s x_1 \tag{1}$$

and

$$x_2 = s'(x_1 - x_3), \tag{2}$$

respectively, where the "slopes" s, s' may assume any value, including ∞. Since a linear relation between the line vectors $(s, -1, 0)$ of the first pencil holds, in particular for the first component s, and since the same is true for the s', a projective relation of the two pencils given by

$$(l_1\, l_2,\, l_3\, u) = (l'_1\, l'_2,\, l'_3\, u')$$

(l_i, u in the first pencil; l'_i, u' in the second pencil) may be expressed by a relation

$$s' = \frac{as + b}{cs + d}.$$

Substituting the fraction for s' in (2) and calculating s from (1) now yields as the equation for the points in which corresponding lines intersect:

$$x_2 = \frac{ax_2 + bx_1}{cx_2 + dx_1}(x_1 - x_3)$$

or

$$-bx_1^2 + cx_2^2 + (d - a)x_1x_2 + bx_1x_3 + ax_2x_3 = 0.$$

This is a quadratic equation and its solutions are nontrivial, since any two lines in a projective plane intersect.   So it is the equation of a conic. ◆

Before showing the converse of Theorem 5.17.1, we prove the following theorem.

Theorem 5.17.2 *If a quadrangle is inscribed in a nondegenerate conic, any diagonal point of this quadrangle is the pole of the line joining the other two diagonal points (under the polarity which defines the conic); and thus any two lines joining diagonal points are conjugate lines.*

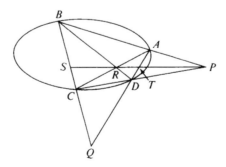

Figure 5.48

Proof Let $ABCD$ be the quadrangle, and let $P = (AB)(CD)$, $Q = (AD)(BC)$, $R = (AC)(BD)$ (Fig. 5.48). The line PR intersects BC and AD in points S and T, respectively. By the definition of harmonic points (Section 5.11), S is the fourth harmonic point of B, C, Q (consider the quadrangle $ARDP$). Similarly, T is the fourth harmonic point of A, D, Q. Therefore, by Theorem 5.16.3, the line $PR = ST$ joins two conjugate points of Q and is thus the polar of Q. Interchanging the roles of P, Q, R, we see also that P is the pole of QR and R is the pole of PQ. By definition, any two of the lines PQ, PR, QR are now conjugate. ◆

Theorem 5.17.3 *Let A, B be two different points on a nondegenerate conic and let X be a variable point on the conic. Then*

$$AX \longrightarrow BX$$

is a projectivity between the pencils with centers A and B.

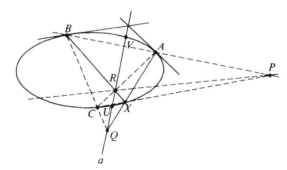

Figure 5.49

Proof Let a be a line through the pole of AB, not passing through A or B, such that a intersects the conic in two different points U, V (Fig. 5.49). We first assume X to be different from A, B, U, V. We set $Q = (AX)a$ and $R = (BX)a$. The line BQ meets the conic in a point $C \neq B$ (if C were $= B$, it would follow that Q is the pole of AB and hence $A = X$). Setting $P = (AB)(CX)$ and $R' = (AC)(BX)$, we conclude from Theorem 5.17.2 that QR' is the polar of P and hence passes through the pole of AB (P lying on AB). Therefore, $a = QR'$ and thus $R' = R$. Now Q, R are seen to be corresponding points under the hyperbolic involution on a with fixed points U, V. This proves that $AX = AQ$ and $BX = BR$ are corresponding lines under a (involutoric) projectivity. This correspondence is readily extended to the cases in which X coincides with A, B, U, or V. ◆

Exercises

1. (Seidewitz' theorem) Prove that if a triangle is inscribed in a conic, a line conjugate to one of the sides intersects the other two sides in conjugate points.

2. Does $x_1^2 + x_2^2 - x_3^2 = 0$ represent a conic for an arbitrary coordinate field $\mathscr{K}$?

3. Let $\mathscr{K}$ be the field with three elements. Show that, in the projective plane over $\mathscr{K}$, any set of four points, no three of which are collinear, is a conic.

4. Find a conic through the points $A = \mathscr{K}(0, 0, 1)$; $B = \mathscr{K}(1, 0, 1)$; $P_1 = \mathscr{K}(0, 1, 1)$; $P_2 = \mathscr{K}(1, 2, 1)$; $P_3 = \mathscr{K}(2, 1, 1)$ by using the fact that a projectivity between the pencils with centers A, B is determined by prescribing three lines and their images.

5. Show by analogy to Theorem 5.17.1 that a projectivity between two pencils of planes provides a hyperboloid of one sheet if the lines carrying the pencils do not intersect and if corresponding planes intersect.

6. Find edges of a cube in ordinary 3-space that lie on one hyperboloid of one sheet.

5.18   Theorems of Pascal and Brianchon

In closing this chapter we solve the problem of finding a conic through five given points. This leads to a most interesting theorem due to Pascal and the dual of this theorem due to Brianchon.

Theorem 5.18.1   *Let A, B, P_1, P_2, P_3 be five different points not situated on a pair of lines. There exists one and only one nondegenerate conic passing through these points.*

Proof   The proof provides, at the same time, an explicit pointwise construction of the conic. Let $C = (P_3 A)(P_1 B)$, as in Figure 5.50. We

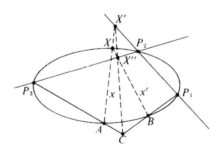

Figure 5.50

project the line $P_1 P_2$ by a perspectivity with center C onto $P_3 P_2$; join corresponding points X', X'' to A, B, respectively; and intersect the lines $x = AX'$ and $x' = BX''$. By Theorem 5.17.1, this point $X = xx'$ is the variable point of a conic. It is readily seen that A, B, P_1, P_2, P_3 lie on this conic. Suppose there were a second conic through A, B, P_1, P_2, P_3.

Then, by Theorem 5.17.3, for a variable point Y of this conic,

$$AY \longrightarrow BY$$

would be a projectivity. Since P_1, P_2, P_3 occur among the points Y, this projectivity is, by Theorems 5.9.1 and 5.9.5, the same as

$$x \longrightarrow x'.$$

Therefore, this conic would be the same as that constructed above, a contradiction. ◆

Theorem 5.18.2   Theorem of Pascal   *If a hexagon is inscribed in a conic, then opposite sides of the hexagon meet on a line* (Fig. 5.51).

Proof Denote the hexagon by $AXBP_1P_2P_3$. Then, according to Figure 5.50, the points $C = (P_3A)(P_1B)$, $X' = (AX)(P_1P_2)$, and $X'' = (BX)(P_2P_3)$ are on a line. ◆

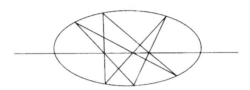

Figure 5.51

Dualizing the Theorem of Pascal yields Brianchon's theorem.

Theorem 5.18.3   Theorem of Brianchon   *If a hexagon is circumscribed about a nondegenerate conic, then opposite vertices of the hexagon are joined by lines through a point* (Fig. 5.52).

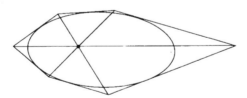

Figure 5.52

The Theorem of Pascal contains the Theorem of Pappus as a special case, namely for a conic consisting of a pair of lines.

Exercises

1. Carry out explicitly the dualization of Pascal's theorem.

2. State and prove the converse of Pascal's theorem.

3. State and prove the converse of Brianchon's theorem.

4. Show that a degenerate conic cannot be obtained by intersecting corresponding lines of projectively related pencils.

5. Find all nondegenerate conics in the projective plane consisting of seven points.

[5.18]

6. Construct a parabola analog to Figure 5.50.

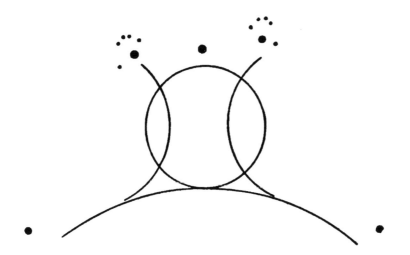

Chapter Six

Inversive Geometry

Inversive geometry is the geometry of circles, spheres, and hyperspheres. From a group-theoretic standpoint, it is the study of invariants under the group of all circle- or sphere-preserving mappings. This group can be generated by "reflections" or "inversions" in circles or spheres similarly to the way a Euclidean group of motions can be generated by reflections in lines.

Starting with a Euclidean plane, we know that the group of all similarities preserves the system of all circles and lines. If, in addition, inversions in circles are to be introduced, it is useful to add to the plane a single point U. This is analogous to adding a whole line to an affine plane when projectivities are to be discussed. Why we add only one point will become apparent in the first section. The extended plane is called the *inversive plane*.

Two parallels in the Euclidean plane will, by definition, intersect in the extended plane; that is, any line will pass through U. So the totality of all lines becomes a special case of a bundle of circles all passing through the same point (Fig. 6.1). This may be compared with the fact that a pencil of parallel lines in an affine plane becomes a special case of a pencil of lines through a point in the projective extension of the affine plane.

 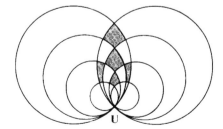

Figure 6.1

Kandinsky's "La Balleteuse." Reprinted from *Punkt und Linien zu Fläche*, courtesy of the publisher, Benteli Verlag Bern.

In the sense of Klein's *Erlanger Programm*, inversive geometry can be defined as a theory of invariants under the inversive group of all circle- or sphere-preserving mappings. This group contains the group of all similarities as a subgroup; it does not contain, however, the affine group or the projective group. A diagram of subgroups (or geometries) is shown in Figure 6.2.

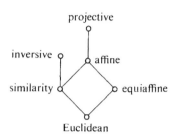

Figure 6.2

Inversive geometry is particularly interesting in the planar case because of its deep and nice relation to the geometrical study of complex numbers. So we shall dedicate several sections to this relation.

6.1 Inversions and Inversive Spaces

Let E^n be the ordinary Euclidean n-space, algebraically described by the set of all vectors $\mathbf{x} = (x_1, \ldots, x_n)$, where $x_1, \ldots, x_n$ are real numbers. The inner product is given by $\mathbf{x} \cdot \mathbf{y} = x_1 y_1 + \cdots + x_n y_n$ (see Section 4.6), and the distance between two points $X = \mathbf{x}$, $X' = \mathbf{x}'$ is

$$\overline{XX'} = |\mathbf{x} - \mathbf{x}'| = ((x_1 - x_1')^2 + \cdots + (x_n - x_n')^2)^{1/2}.$$

A *bounded hypersphere* is the set of all points that have equal distance from a point M. It can be characterized by the equation

$$(\mathbf{x} - \mathbf{m})^2 = r^2,$$

where $r > 0$ denotes the *radius* of the hypersphere and $\mathbf{m}$ the *midpoint* (or *center*), as the coordinate vector of M. If $n = 2$, we call the hyperspheres *circles*; if $n = 3$, we call them *spheres*.

Let X be an arbitrary point $\neq M$ (Fig. 6.3). We assign to X the point X' on the ray from M to X that satisfies

$$\overline{MX} \cdot \overline{MX'} = r^2. \tag{1}$$

The mapping $X \to X'$ is called an *inversion* or *reflection* in the hypersphere. It is readily seen that, in the terms of projective geometry, X and X' are conjugate points with respect to the hypersphere (see Exercise 1).

So far, the definition of an inversion is not defined for M. If X moves toward M, then X' moves toward "infinity." Conversely, if X moves, in

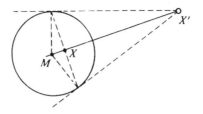

Figure 6.3

some direction, to "infinity," the image X' moves toward M. So it makes sense to enlarge E^n by one point U, which we consider to be incident with every line of E^n (and thus every linear subspace of dimension ≥ 1). The space

$$E^n \cup \{U\}$$

is called **inversive space**.

The above inversion is now extended to a mapping of the inversive space onto itself by assigning M to U and U to M. Again, we call this map an **inversion** (of the inversive space). In an inversive plane it is useful not to distinguish between lines and circles. We call every line that is augmented by U a **circle**. More generally, we denote a hyperplane of E^n augmented by U a (unbounded) **hypersphere**, in particular a **sphere** if $n = 3$.

It is possible to define an inversive space synthetically without distinguished elements as in the definition of a projective plane. We present such a definition only for inversive planes (see Section 6.4).

Theorem 6.1.1 *Any inversion maps hyperspheres onto hyperspheres.*

Proof We choose M as the origin of a Euclidean coordinate system. Then (1) can be written as

$$|\mathbf{x}| \cdot |\mathbf{x}'| = r^2. \tag{2}$$

Since $\mathbf{x}$ and $\mathbf{x}'$ are positive multiples of each other, we obtain from (2):

$$\mathbf{x}' = \frac{r^2}{\mathbf{x}^2} \mathbf{x}. \tag{3}$$

Since, by definition, an inversion is involutoric, we may replace (3) by

$$\mathbf{x} = \frac{r^2}{\mathbf{x}'^2} \mathbf{x}'. \tag{3'}$$

Suppose a hypersphere

$$(\mathbf{x} - \mathbf{p})^2 = s^2$$

is given. By (3'), we find

$$\left(\frac{r^2}{\mathbf{x}'^2} \mathbf{x}' - \mathbf{p} \right)^2 = s^2$$

or

$$(r^2\mathbf{x}' - \mathbf{p}\mathbf{x}'^2)^2 = s^2(\mathbf{x}'^2)^2.$$

By a simple calculation, we obtain

$$r^4 - 2r^2\mathbf{x}'\mathbf{p} + (\mathbf{p}^2 - s^2)\mathbf{x}'^2 = 0. \tag{4}$$

If $\mathbf{p}^2 - s^2 = 0$, that is, if the given hypersphere passes through O, its image (4) is a hyperplane augmented by U. If $\mathbf{p}^2 - s^2 \neq 0$, we set $c = \mathbf{p}^2 - s^2$ and find

$$\left(\mathbf{x}' - \frac{r^2}{c}\mathbf{p}\right)^2 = \left(\frac{r^2}{c}\right)^2 \mathbf{p}^2 - \frac{r^4}{c}. \tag{5}$$

Since

$$\frac{r^4}{c^2}\mathbf{p}^2 = \frac{r^4}{c}\frac{\mathbf{p}^2}{\mathbf{p}^2 - s^2} = \frac{r^4}{c}\frac{1}{1 - s^2/\mathbf{p}^2} > \frac{r^4}{c},$$

the right side of (5) is positive and (5) represents a hypersphere.

Finally, let a hyperplane be given. If it passes through O, it is clearly mapped onto itself. If it does not pass through O, we may write it in the form

$$\mathbf{x}\mathbf{p} = \frac{r^2}{2}.$$

Therefore, it occurs among the hyperplanes (4) for $\mathbf{p}^2 - s^2 = 0$. ◆

Another model of an inversive plane is that of all plane sections of a sphere in Euclidean 3-space. It can be obtained from the inversive plane introduced above by "stereographic projection" (see Fig. 6.4). To every point X of E^2

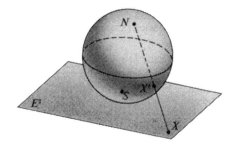

Figure 6.4

there is assigned a point X' on the sphere. All points on the sphere are obtained as images except its "north pole," N. We assign it to the "point at infinity" U of E^2. So we obtain a very suggestive way of closing (the image of) E^2 by only one point. We shall study the stereographic projection in Section 6.10.

Exercises

1. Show that the images X, X' under an inversion at a circle are conjugate with respect to this circle.

2. Find, in an inversive plane, all circles that are mapped onto themselves under a prescribed inversion.

3. (Special Appolonius problem) Given a point P and two circles c_1, c_2. Find (if possible) a circle passing through the point and tangent to c_1 and c_2. Discuss the cases in which this circle exists. (*Hint*: Apply an inversion in a circle with center P, find a common tangent of the image circles, and invert back.)

4. Find a method for constructing the image of a circle under an inversion by ruler and compass.

5. Show that a circle can be mapped onto any other circle by an inversion.

6. Show that the product of two inversions in circles with a common center is a dilatation.

6.2   *Complex Numbers and Circles*

Another way of studying inversive geometry uses complex numbers. We assign to any complex number

$$z = z_1 + z_2 i$$

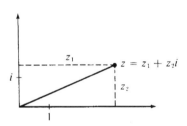

Figure 6.5

the point (z_1, z_2) of the Euclidean plane E^2 (Fig. 6.5). Furthermore, we add a "number ∞" to the system of complex numbers, to serve geometrically as point U which closes E^2 to an inversive plane. $E^2 \cup \{\infty\}$ is assumed to satisfy $c/\infty = 0$ and $c \cdot \infty = \infty$ for $c \neq 0$. A mapping of an inversive plane onto itself is called **circle-preserving** if circles are mapped onto circles. Examples:

1. *Translation:*

$$z' = z + c. \tag{1}$$

In E^2 this is clearly a translation

$$(z_1', z_2') = (z_1, z_2) + (c_1, c_2)$$

in the usual sense. It leaves ∞ fixed.

2. *Dilatation:*

$$z' = rz, \quad r \text{ is a real number} \neq 0. \tag{2}$$

This map can be expressed in Euclidean terms by $(z'_1, z'_2) = (rz, rz_2)$. It leaves 0 and ∞ fixed.

3. *Rotation:*

$$z' = az, \quad \text{where} \quad |a|^2 = a_1^2 + a_2^2 = 1. \tag{3}$$

Setting $a = \cos \theta + i \sin \theta$, we find

$$\begin{aligned}
z' = az &= (\cos \theta + i \sin \theta)(z_1 + iz_2) \\
&= z_1 \cos \theta - z_2 \sin \theta + i(z_1 \sin \theta + z_2 \cos \theta),
\end{aligned}$$

or, in matrix form for the two coordinates z'_1, z'_2:

$$(z'_1, z'_2) = (z_1, z_2) \begin{pmatrix} \cos \theta & \sin \theta \\ -\sin \theta & \cos \theta \end{pmatrix}.$$

This is a rotation about 0. Again, (3) leaves 0 and ∞ fixed. From (1), (2), (3) we readily conclude the following theorem.

Theorem 6.2.1 *Every similarity of E^2 can be represented by*

$$z' = az + c$$

where $a \neq 0$ and a and c are complex numbers.

4. *Reflection:*

$$z' = \bar{z} \quad \text{(conjugate complex number)} \tag{4}$$

It is the reflection $(z'_1, z'_2) = (z_1, -z_2)$ in the "real" axis.

5. *Inversion:*

$$z' = \frac{1}{z\bar{z}} z. \tag{5}$$

We may write (5) as

$$z' = \frac{1}{|z|^2} z.$$

This is equivalent to Equation (3) of Section 6.1 with $r = 1$. So (5) represents the inversion in the unit circle with center 0. It interchanges 0 and ∞.

Theorem 6.2.2 *Any rational mapping*

$$z' = \frac{az + b}{cz + d}, \quad \text{with } ad - bc \neq 0, \tag{6}$$

is a circle-preserving map of the inversive plane.

Proof We may split (6) into the following maps of the above types:

$$z^{(1)} = cz + d \qquad\qquad \text{(similarity)}$$

$$z^{(2)} = \frac{1}{\overline{z^{(1)}}} \qquad\qquad \text{(inversion)}$$

$$z^{(3)} = \overline{z^{(2)}} \qquad\qquad \text{(reflection)}$$

$$z' = -\frac{ad - bc}{c}\, z^{(3)} + \frac{a}{c} = \frac{az + b}{cz + d} \quad \text{(similarity)} \quad \blacklozenge$$

We obtain the following theorem if (6) is combined with a reflection (4).

Theorem 6.2.3 *Any mapping*

$$z' = \frac{a\bar{z} + b}{c\bar{z} + d}, \quad \text{with } ad - bc \neq 0, \tag{7}$$

is a circle-preserving map of the inversive plane.

We call (6) a **homography** and (7) an **anti-homography**. It should be noted that *from a projective point of view, the inversive plane can be considered to be a projective line over the field of complex numbers* as coordinate field. So it is "complex one-dimensional," but "real two-dimensional." In particular, cross ratios can be used for the study of the inversive plane.

As in Section 5.8, we define the cross ratio of four different complex numbers, a, b, c, d (points on the complex line), by

$$(ab, cd) = \frac{a - c}{b - c} : \frac{a - d}{b - d}.$$

If two of the four numbers coincide, we set

$$(aa, cd) = (ab, cc) = 1;$$
$$(ab, bd) = (ab, ca) = \infty;$$
$$(ab, ad) = (ab, cb) = 0.$$

In case one element is ∞, we set

$$(\infty b, cd) = \frac{b - d}{b - c},$$

and so forth. From Theorem 5.8.1, we have: *cross ratios are preserved under homographies* (6). (If we interpret the inversive plane as a projective line over the field of complex numbers, homographies are projective collineations.) Theorem 6.2.4 is implied by Theorem 5.8.2.

Theorem 6.2.4 *Three different points a, b, c can be mapped onto any other three different points by one and only one homography. This homography is given implicitly by*

$$(ab, cz) = (a'b', c'z'). \tag{8}$$

This can also be expressed as follows (concerning the terminology, see Section 2.4):

Theorem 6.2.4' *The group of all homographies is triply transitive on the set of all points of the inversive plane.*

Let, in particular, a', b', c' be real. Then in Equation (8) $(a'b', c'z')$ is real if and only if z' is real and, thus, if a, b, c, z lie on a circle. Therefore, we find the next theorem to be a nice test of whether or not four points a, b, c, z are on a circle.

Theorem 6.2.5 *Four different points a, b, c, z are on a circle if and only if their cross ratio (ab, cz) is a real number (∞ is considered to be real).*

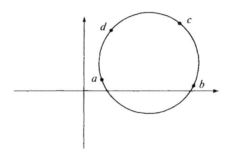

Figure 6.6

For a', b', c' real, Equation (8) can be considered a circle-preserving map of the real line onto the circle of all z for which (ab, cz) is real.

Exercises

1. Find a mapping (6) that transforms the "upper half-plane" $z_2 > 0$ (where $z = z_1 + iz_2$) onto the interior $|z| < 1$ of the unit circle with center O.

2. Find a mapping (6) that transforms the half-plane $z_2 > 0$ onto the interior of the circle $|z| < 1$.

3. Find by using complex numbers the product of a rotation about O through the angle $\pi/4$ and a rotation about i through the angle $\pi/3$.

4. Discuss all possibilities for which (6) is involutoric.

5. Show that $az\bar{z} + bz + \bar{b}\bar{z} + d = 0$; a, d, real, represents a circle if and only if $ad - b\bar{b} < 0$.

6. Use Exercise 5 to show anew that inversions map circles (lines included) onto circles.

6.3   Generalized Inversive Planes

In our axiomatic foundation of plane inversive geometry in the next sections we shall have to deal strongly with the fact that inversive geometry is "something quadratic." This fact becomes even clearer if, in the definition of an ordinary inversive plane, we leave out order and continuity properties. As in general affine space, we consider an arbitrary field $\mathcal{K}$ instead of the field $\mathcal{R}$ of real numbers. The field $\mathcal{C}$ of complex numbers, accordingly, will be replaced by a quadratic extension $\mathcal{K}(\sqrt{d})$ of $\mathcal{K}$ consisting of all elements

$$x = x_1 + x_2\sqrt{d}; \quad x_1, x_2 \text{ in } \mathcal{K}, d \text{ not a square in } \mathcal{K}.$$

(Compare Section 3.15). $\mathcal{C}$, of course, is the extension $\mathcal{C} = \mathcal{R}(\sqrt{-1})$ of $\mathcal{R}$. This leads us to the definition of a generalized inversive plane.

We assign to every number $x = x_1 + x_2\sqrt{d}$ the point (x_1, x_2) of an affine (Pappian) plane $\mathcal{A}^2$. This plane is again extended by a point ∞, which we consider to lie on every line. The extended lines are called **circles**. Furthermore, given any three numbers a, b, c not on a line, the totality of all x satisfying (ab, cx) in $\mathcal{K} \cup \{\infty\}$ (union of $\mathcal{K}$ and ∞) is called a **circle abc**. We call the totality of points and circles the **inversive plane over** $\mathcal{K}(\sqrt{d})$. Since in this section points are denoted by small italics (as elements of $\mathcal{K} \cup \{\infty\}$), we denote circles by script $\mathcal{C}$.

Theorem 6.3.1 *Three different points are on one and only one circle.*

Proof If the three given points are on a (extended) line, the uniqueness of a circle through them follows from the properties of an affine plane. Let a, b, c not be on an extended line. If $e \neq a$, b, c lies on the circle abc— that is, if (ab, ce) is in $\mathcal{K} \cup \{\infty\}$—then (ab, ec) is also in $\mathcal{K} \cup \{\infty\}$ (see Section 5.8) and so c lies on the circle abe. In this way a, b, c can be successively replaced by any other three different points e, f, g on the circle abc, and we see that $abc = efg$. So a, b, c lie on only one circle. The *existence* of a circle through a, b, c is guaranteed by definition. ◆

Theorem 6.3.2 *Every mapping*

$$x' = \frac{a + bx}{c + ex}; \quad \begin{vmatrix} a & b \\ c & e \end{vmatrix} \neq 0 \tag{1}$$

preserves cross ratios and thus is a circle-preserving map.

Proof Since the general inversive plane over $\mathcal{K}(\sqrt{d})$ can be considered as a projective line over $\mathcal{K}(\sqrt{d})$, Theorem 6.3.2 follows from Theorem 5.9.3. ◆

As in the preceding section, we find a circle-preserving map transforming p, q, r into p', q', r', respectively, by

$$(pa, rx) = (p'a', r'x'). \tag{2}$$

Theorem 6.3.3 *If p is a point on a circle $\mathcal{C}$ and if q is not on $\mathcal{C}$, then there exists one and only one circle $\mathcal{C}'$ passing through p, q and meeting $\mathcal{C}$ only in p* (Fig. 6.7).

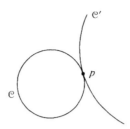

Figure 6.7

Proof By a circle-preserving map, we can move p to ∞. Theorem 6.3.3 now follows from the axiom of parallels in the affine plane $\mathscr{A}^2$. ◆

Theorem 6.3.4 *On every circle there are at least three points. There exist a point and a circle which are not incident.*

Proof Clearly 0, 1, and ∞ are different points and can be mapped by (2) onto every circle. $\mathscr{K} \cup \{\infty\}$ represents a circle that does not contain the point $\sqrt{d}$. ◆

The question arises whether the "circles" of the inversive plane over $\mathscr{K}(\sqrt{d})$ are conics in the (x_1, x_2)-plane. This is in fact so, as we are going to show now. Since lines may be considered as conics, we can restrict our attention to circles that do not pass through U. Let $\mathscr{C}$ be such a circle. If $\mathscr{C}$ is mapped onto $\mathscr{K} \cup \{\infty\}$ by a mapping (1), the points x of $\mathscr{C}$ are characterized by the condition that

$$\frac{a + bx}{c + ex}$$

is in $\mathscr{K}$. We set

$$a = a_1 + a_2\sqrt{d};$$
$$b = b_1 + b_2\sqrt{d};$$
$$c = c_1 + c_2\sqrt{d};$$
$$e = e_1 + e_2\sqrt{d};$$
$$x = x_1 + x_2\sqrt{d}.$$

By an obvious calculation, $(a + bx)/(c + ex)$ can be expressed by

$$\begin{aligned}
\frac{a + bx}{c + ex} = r \cdot s + r[&(e_1b_2 - e_2b_1)(x_1^2 - dx_2^2) \\
&+ (c_1b_2 - b_1c_2 + e_1a_2 - e_2a_1)x_1 \\
&+ (c_1b_1 - a_1e_1 + da_2e_2 - db_2c_2)x_2 \\
&+ c_1a_2 - a_1c_2]\sqrt{d},
\end{aligned}$$

where r and s are in $\mathscr{K}$ and $r \neq 0$. So $(a + bx)/(c + ex)$ is in $\mathscr{K}$ if and only if

$$(e_1b_2 - e_2b_1)(x_1^2 - dx_2^2) + (\cdots)x_1 + (\cdots)x_2 + c_1a_2 - a_1c_2 = 0.$$

If $e_1b_2 - e_2b_1 = 0$, this equation represents a line. If $e_1b_2 - e_2b_1 \neq 0$, we obtain a conic,

$$x_1^2 - dx_2^2 + ux_1 + vx_2 + w = 0. \tag{3}$$

In the example, $\mathscr{K}(\sqrt{d}) = \mathscr{R}(\sqrt{-1})$, that is, in the inversive plane over the field of complex numbers, this is the ordinary equation of a circle. So we have shown not only that every circle is a conic but also that all circles not containing U possess the same quadratic part, $x_1^2 - dx_2^2$, in their equation (3).

Every equation (3) that represents a circle possesses three noncollinear points as solutions. Conversely, we assert that if an equation (3) possesses three noncollinear points as solutions, it represents a circle. In fact, there exists a circle

$$x_1^2 - dx_2^2 + u'x_1 + v'x_2 + w' = 0 \tag{3'}$$

on which the given three points (solving (3)) lie. Subtracting (3) and (3′), we obtain

$$(u - u')x_1 + (v - v')x_2 + w - w' = 0.$$

Since this equation has three noncollinear solutions, we conclude

$$u - u' = v - v' = w - w' = 0,$$

and hence (3′) is identical with (3). The following theorem summarizes the foregoing discussion.

Theorem 6.3.5 *The inversive plane over $\mathscr{K}(\sqrt{d})$, where $\mathscr{K}$ is an arbitrary field containing a nonsquare d, can be described as follows: Points are all points of the affine plane over $\mathscr{K}$ together with an additional point U, circles are lines augmented by U, and conics are given by equations (3) with three noncollinear solutions.*

Exercises

1. Let $\mathscr{K}$ be the field of rational numbers and let $d = 2$. Describe the inversive plane over $\mathscr{K}(\sqrt{2})$ with the aid of the ordinary Euclidean plane.

2. Show that $x = x_1 + x_2\sqrt{d} \rightarrow \bar{x} = x_1 - x_2\sqrt{d}$ (x_1, x_2 in $\mathscr{K}$) is an automorphism of $\mathscr{K}(\sqrt{d})$ (that is, an isomorphism of the additive group of $\mathscr{K}(\sqrt{d})$ onto itself and an isomorphism of the multiplicative group of $\mathscr{K}$ onto itself).

3. Can $\mathscr{K}$ be one of the following fields? (a) the field of complex numbers, (b) the field consisting of 2 elements, (c) the field consisting of 3 elements.

4. Let the inversive plane over $\mathscr{K}(\sqrt{d})$ be given. Show that any two circles not passing through U (that is, not lines) can be mapped onto each other by the product of a dilatation and a translation of the affine (x_1, x_2)-plane,

provided each of the circles has a center, that is, can be mapped onto itself by an involutoric dilatation (half-turn).

5. Let $\mathscr{K}$ have characteristic $\neq 2$; that is, let $1 + 1 \neq 0$. Show that every circle can be represented by an equation $ax\bar{x} + bx + \bar{b}\bar{x} + c = 0$, where a and c are in $\mathscr{K}$.

6. Let $\mathscr{K}$ be the field with three elements. Find in the inversive plane over $\mathscr{K}(\sqrt{2})$ all points of the circle through $x = 0$; $y = 1 + \sqrt{2}$; $z = 1 + 2\sqrt{2}$.

6.4   *Axioms of Inversive Planes*

We present now a system of axioms by which the inversive planes over fields $\mathscr{K}(\sqrt{d})$ (see preceding section) are characterized, given the characteristic $\mathscr{K} \neq 2$ (that is, $1 + 1 \neq 0$). The characterization will be completed in Section 6.6.

Let a set of elements called *points*, P, Q, ..., and a set of elements called *circles*, $a, b, \ldots$, be given, for which an incidence relation is defined satisfying the following axioms:

Axiom C.1   *Three different points are on one and only one circle.*

Axiom C.2   *If P is a point on the circle a and if Q is not on a, there exists a unique circle b passing through P, Q and having only P in common with a.*

We say a and b are **tangent** at P.

Axiom C.3   *On each circle there are at least three points. There exist a point and a circle that are not incident.*

We call such a system of points and circles an **inversive plane**. The circle passing through three different points P, Q, R is denoted by

$$PQR.$$

The fact that four points P, Q, R, S lie on a common circle is expressed by the symbol

$$\mathscr{C}(PQRS).$$

If we say "*a **and** b **intersect in** P, Q*," we mean that P, Q are the only common points of a, b where $P = Q$ *may be possible* (a tangent to b).

We call an inversive plane *odd* (for a justification of this name see Theorem 6.4.2 below) if it satisfies the following postulate.

Axiom C.4   *There exist four circles each two of which are tangent such that all six points of tangency are different and such that no further circle through one of these points is tangent to three of the circles* (Fig. 6.8).

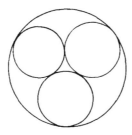

Figure 6.8

As in the case of an affine or a projective plane, we need a closure theorem in order to develop an algebraic theory of inversive planes.

Axiom C.5 (**Miquel's theorem**) *Let* P, P', Q, Q', R, R', S, S' *be different points with the exception that* P *may coincide with* P'. *If* $\mathscr{C}(PQRS)$, $\mathscr{C}(QQ'RR')$, $\mathscr{C}(RR'SS')$ *and if* PQQ', PSS' *intersect in* P, P', *then* $\mathscr{C}(P'Q'R'S')$.

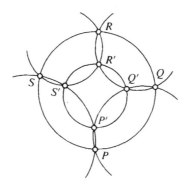

Figure 6.9

An inversive plane satisfying Axiom C.5 is called **Miquelian**.

By Theorems 6.3.1, 6.3.3, and 6.3.4, Axioms C.1–C.3 are true in every inversive plane over a field $\mathscr{K}(\sqrt{d})$. Before discussing Axiom C.4, we shall prove the following theorem.

Theorem 6.4.1 *If a point U is deleted from an inversive plane, the circles through U reduced by U form an affine plane.*

Proof Axiom C.1 guarantees the existence and uniqueness of a line through two different points. For $P = U$, C.2 is the axiom of parallels. By Axiom C.3, on every line there are at least two points, and there exists a point and a line that are not incident. ◆

We call this affine plane the **affine U-subplane** of the given inversive plane.

Theorem 6.4.2 *Given the characteristic $\mathscr{K} \neq 2$, the inversive plane over $\mathscr{K}(\sqrt{d})$ is odd (satisfies Axiom C.4).*

Proof If an arbitrary point of the inversive plane over $\mathcal{K}(\sqrt{d})$ is given, we map it, by a circle-preserving map, onto ∞. Now the affine ∞-subplane of the inversive plane over $\mathcal{K}(\sqrt{d})$ is the affine plane over $\mathcal{K}$. The circles

$$x_1^2 - d(x_2 + \tfrac{1}{2})^2 - \tfrac{1}{4} = 0,$$
$$x_1^2 - d(x_2 - \tfrac{1}{2})^2 - \tfrac{1}{4} = 0,$$
$$x_1 = \tfrac{1}{2},$$
$$x_1 = -\tfrac{1}{2},$$

satisfy the requirements of Axiom C.4. ◆

Finally, we show Axiom C.5 to be true.

Theorem 6.4.3 *In the inversive plane over $\mathcal{K}(\sqrt{d})$, Miquel's theorem holds.*

Proof We distinguish two cases: $P \neq P'$ and $P = P'$.

(a) $P \neq P'$: If every circle contains only three points, there is nothing to prove; Miquel's theorem follows simply from the fact that the assumptions are never fulfilled. So let every circle have at least four points. Then it is readily seen that there exists a point U different from P, P', Q, Q', R, R', S, S'. We choose it as point ∞ (applying a circle-preserving map). Axiom C.5 now follows from the following identity (which is easily verified), where the points are considered as numbers of $\mathcal{K}(\sqrt{d})$:

$$(Q'S', R'P') = (PR, QS)(RQ', QR')(RS', R'S)(PQ, P'Q)(PS', SP').$$

(b) $P = P'$: By a circle-preserving map, we can arrange it so that

$$P = \infty, \qquad S = 0, \qquad S' = 1.$$

Now we apply the identity

$$(P'R', S'Q') = (PR', S'Q')$$
$$= [(S'P, QS) - (S'P, Q'S)](PR, QS)(RQ', QR')(RS', R'S).$$

We have (note that $(X\infty, YZ) = (X - Y)/(X - Z)$)

$$(S'P, QS) = (S'P, Q'S) - \frac{1 - Q}{1 - 0} - \frac{1 - Q'}{1 - 0} = Q' - Q,$$

which is in $\mathcal{K}$, since Q, Q' lie on a line parallel to the line through 0, 1 (Fig. 6.10). Thus, the above identity shows Miquel's theorem to be true. ◆

Figure 6.10

If all points in Miquel's theorem are different ($P \neq P'$), we obtain a configuration of eight points and six circles where there are four points on each circle and three circles through each point. We call this an $(8_3, 6_4)$-*configuration*.

We remark that there exists a second $(8_3, 6_4)$-configuration which also expresses a closure theorem.

Theorem 6.4.4 (**bundle theorem**) *Let* P, P', Q, Q', R, R', S, S' *all be different and not lie on a common circle. If* $\mathscr{C}(PP'QQ')$, $\mathscr{C}(QQ'RR')$, $\mathscr{C}(RR'SS')$, $\mathscr{C}(PP'RR')$, $\mathscr{C}(QQ'SS')$, *then also* $\mathscr{C}(PP'SS')$.

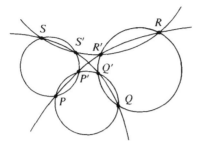

Figure 6.11

We shall prove the bundle theorem for a large class of inversive planes in Section 6.11. It is also shown there that in an inversive plane the bundle theorem does not imply Miquel's theorem. It can be shown, however, that the Bundle theorem is a consequence of Miquel's theorem. So we have an analogy to Pappus' and Desargues' theorems in affine or projective geometry.

Exercises

1. Discuss the minimal model of an inversive plane (five points and ten circles).

2. Show that in the axioms of an inversive plane, Axiom C.3 can be replaced by the weaker postulate, "On every circle there is at least one point. There exist four points that are not on a circle."

3. Show that two different circles PQR and $PQ'R'$ of the inversive plane over $\mathscr{K}(\sqrt{d})$ are tangent at P if and only if $(PQ, RQ') - (PQ, RR')$ is in $\mathscr{K}$.

4. If an inversive plane over a field $\mathscr{K}(\sqrt{d})$ contains only finitely many points, show that the inversive plane is odd if and only if every circle contains an even number of points; thus in every affine subplane each line contains an odd number of points each.

5. In an ordinary Euclidean plane E^2, let a closed strictly convex curve $\mathscr{C}$ be given—that is, a closed curve that is intersected by an arbitrary line in at most two points. Suppose $\mathscr{C}$ possesses no "corners," that is, it has a unique tangent at every point. Call every image of $\mathscr{C}$ under a product of a translation and a dilatation a circle. Furthermore, call every line of E^2 augmented by a symbol U a circle.

(a) Show that the points of $E^2 \cup \{U\}$ and the circles form an inversive plane.

(b) Show that $\mathscr{C}$ can be chosen such that Miquel's theorem is violated.

6. Consider the group $\mathscr{G}$ of all circle-preserving maps of an inversive plane. Is this group always transitive on the points of the given inversive plane? (Use Exercise 5.)

6.5   *Miquel and Pappus*

We wish to show that any odd inversive Miquelian plane is an inversive plane over a field $\mathscr{K}(\sqrt{d})$. Hereby Miquel's theorem has to perform two things: First, it has to make an affine U-subplane of the inversive plane a Pappian plane. Second, it has to make circles conics in this affine plane. This section accomplishes the first requirement.

If, in this discussion, we say that two of the points in Miquel's theorem coincide, we mean that the configuration circles passing through these points are tangent. We begin by proving some extensions of Miquel's theorem (Fig. 6.9 or Fig. 6.12).

Theorem 6.5.1 *Let the assumptions of Axiom C.5 be given except that $P' = Q'$ (or $P' = S'$). Then $Q'R'S'$ and PQQ' (or PSS') intersect in P' only (Fig. 6.13).*

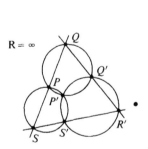

Figure 6.12

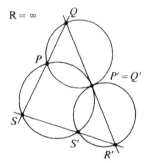

Figure 6.13

Proof $P' = Q'$ and $P' = S'$ cannot be true simultaneously, since in this case $Q' = S'$ would follow, and this condition is still excluded. It suffices to prove the theorem for $P' = Q'$ (that is, PQQ' tangent to $Q'R'S'$ in $P' = Q'$), since the assumptions of Axiom C.5 are symmetric for Q, S.

Let $Q'R'S'$ and PQQ' intersect in P'', Q'. If P'' coincides with one of the points P, Q, R, R', S, S', it would readily follow (consider Figure 6.13) that two of these six points coincide, contrary to our assumption. So, by Axiom C.5, $\mathscr{C}(PP''SS')$ and thus $P'' = P' = Q'$. ◆

Theorem 6.5.2 *Let the assumptions of Axiom C.5 be given except that* $Q' = R'$ *and that* P' *need not be different from* P, S'. *Then* PSS' *and the circle through* S', Q' *that is tangent to* $QQ'R$ *at* Q' *intersect in* P', S' (Fig. 6.14).

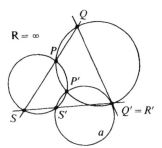

Figure 6.14

Proof If this is not so, the circle a through S', Q' that intersects PSS' in P', S' has a point $Q'' \neq Q'$ in common with $QQ'R$. We have $Q'' \neq Q$, since, otherwise, $a = PP'Q$, and thus $S' = P$ would follow. Then, by Axiom C.5 or Theorem 6.5.1 (in case $P' = S'$), we find that PQQ' passes through Q''; hence $PQQ' = RQQ'$ and so $P = Q$ or $P = R$, both of which are impossible by assumption. Therefore a is tangent to $QQ'R$ at Q'. ◆

The same technique can be used to prove the following theorem.

Theorem 6.5.3 *Let the assumptions of Axiom C.5 be given except that the following conditions need not all be true:*

$$P' \neq Q', S'; \qquad Q' \neq Q, R'; \qquad S' \neq S, R'; \qquad P \neq Q, S.$$

Then there are circles through P, Q, P', Q' *and* P', Q', R', S'; *respectively, which intersect in* P', Q'.

The proof is best achieved by first drawing all possible figures that may occur (starting with Figs. 6.12, 6.13, and 6.14) and then allowing successively an additional equality. We omit the details (compare Exercise 2).

Theorem 6.5.4 *Theorem 6.5.3 remains true if one of the following conditions is dropped:*

$$R \neq Q; \qquad R \neq R'; \qquad R \neq S.$$

Proof Let, for example, $R = Q$ (Fig. 6.15). We denote the circle through P, P', Q, Q' by a and the circle through P', R', S', which intersects the circle through R, R', S, S' in R', S', by b. If Theorem 6.5.4 is false, a and b intersect in P', Q'', where Q'' is not on the circle c through Q', R, R', which intersects b in R', Q'. The new point Q'' is different from R, R'. We replace c by the circle $RR'Q''$ and apply Theorem 6.5.3. Let Q^* be the second point of intersection of $RR'Q''$ and the circle d through P, R, S,

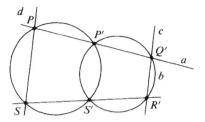

Figure 6.15

which intersects the circle through P, P', S, S' in P, S. It follows, by Theorem 6.5.3, that a, d intersect in P, Q^*, contrary to the assumption that $R \neq P$, Q^* also lies on a and d. So $Q'' = Q'$ and Theorem 6.5.4 is correct. The cases $R = R'$ and $R = S$ are treated in a similar way. ◆

We consider now the affine R-subplane of the given inversive plane (in Figs. 6.12–6.15, R has already served as the "point at infinity"). By a *quadrilateral of chords* $abcd$ in a circle k not passing through R we mean a set of different lines a, b, c, d for which points A, B, C, D on a circle k exist such that

k and a intersect in A, B;
k and b intersect in B, C;
k and c intersect in C, D;
k and d intersect in D, A.

It is readily seen that at most two of the points A, B, C, D may coincide.

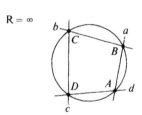

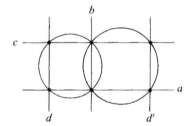

Figure 6.16   **Figure 6.17**

Theorem 6.5.5 *Let a, b, c, d be a quadrilateral of chords, and let d' be parallel to d such that a, b, c, d' do not pass through one point. Then $abcd'$ is again a quadrilateral of chords* (Fig. 6.17).

Proof If $d = d'$, there is nothing to prove. So let $d \neq d'$. Then Theorem 6.5.5 follows readily from Theorem 6.5.4. ◆

Theorem 6.5.6 *Let $abcd$ be a quadrilateral of chords and let a', b', c', d' be lines not passing through a common point and parallel to a, b, c, d, respectively. Then $a'b'c'd'$ is again a quadrilateral of chords* (Fig. 6.18).

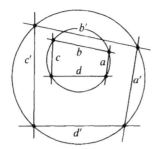

Figure 6.18

Proof Applying Theorem 6.5.5 four times, we can replace a, b, c, d by a', b', c', d', respectively. ◆

The converse of Theorem 6.5.6 is also true.

Theorem 6.5.7 *Let abcd and a'b'c'd' be quadrilaterals of chords and let a, b, c be parallel to a', b', c', respectively. Then d is parallel to d'.*

Proof Otherwise let d'' be a line through the intersection of a' and d' which is parallel to d. By Theorem 6.5.6, $a'b'c'd''$ is a quadrilateral of chords in a circle k''. If k' is the circle in which $a'b'c'd'$ is a quadrilateral of chords, we conclude with the aid of Axioms C.1 and C.2: $k' = k''$ and hence $d' = d''$. ◆

Theorem 6.5.7 includes the converse of Theorem 6.5.5.

Theorem 6.5.8 *In the affine R-subplane of the given inversive plane, Pappus' theorem (affine version) is true.*

Proof Let a, b, c, a', b', c' be the successive sides of a (nondegenerate) hexagon; let the points ac', bc, $a'b$ be different and lie on a line u; and let the points ab, $b'c'$, $a'c$ be different and lie on a line $v \neq u$ (Fig. 6.19). If a

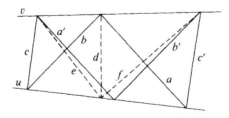

Figure 6.19

is parallel to a' and b is parallel to b' we must show that c is parallel to c'. Let d be the line through ab such that $cudv$ is a quadrilateral of chords.

Then *ud* is not on *v*, since otherwise *v* = *d* would follow.   Let *e* be the line joining *ud*, *cv*, and let *f* be the line joining *ud* and *c′v*.

Since *cudv* is a quadrilateral of chords, *buev* is a quadrilateral of chords, too (same vertices as *cudv*, etc.).   Since *b′* is parallel to *b*, we deduce from Theorem 6.5.5 that *b′uev*, and hence also *fua′v*, is a quadrilateral of chords.

Using Theorem 6.5.5 again we recognize *fuav*, and hence also *c′udv*, to be a quadrilateral of chords.   Now we apply Theorem 6.5.7 to *cudv* and *c′udv* and conclude that *c* is parallel to *c′*.   ◆

By Section 3.13 the affine *R*-subplane can now be coordinatized by a field $\mathcal{K}$.

Exercises

1. Let *a*, *b*, *c*, *d* be circles such that *a* is tangent to *b* at *A*; *b* is tangent to *c* at *B*; *c* is tangent to *d* at *C*; and *d* is tangent to *a* at *D*.   Show (by using a special case of Miquel's theorem) that $\mathscr{C}(ABCD)$.

2. Carry out in detail the proof of Theorem 6.5.3.

3. Carry out in detail the proof of Theorem 6.5.6.

4. Draw figures for all special cases of Miquel's theorem which may occur in Theorem 6.5.4.

5. In an inversive plane over $\mathcal{K}(\sqrt{d})$, let a circle $\mathscr{C}$ have a tangent line *a* at *A*, and let the line *b* be properly parallel to *a* (in the affine plane over $\mathcal{K}$).   Let the lines through *A* and two different points *B*, *C* of *b* intersect $\mathscr{C}$ in *D*, *E*, respectively.   Show that $\mathscr{C}(BCDE)$.

6. State and prove a theorem similar to Miquel's theorem which is suggested by Figure 6.20.   Assume all incidences of the figure to be true, except the one at *A*.   Show that the incidence at *A* is a consequence of the other incidences.   (*Hint*: Use Miquel's theorem twice; compare Figure 6.12.)

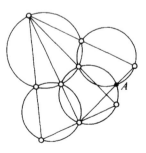

Figure 6.20

6.6   Foundation of Plane Inversive Geometry

We continue by showing that every circle not passing through R is a conic. For this purpose let the affine R-subplane be extended to a projective plane. We may characterize the line at infinity of this projective plane as the set of all pencils of R mutually tangent circles. We call the points at infinity **special points**. We set

$$\mathscr{S}(ABCD)$$

for special points A, B, C, D if there exists a quadrilateral of chords $abcd$ such that A, B, C, D lie on a, b, c, d, respectively. By Theorem 6.5.6, $\mathscr{S}(ABCD)$ does not depend on the choice of this quadrilateral of chords.

Theorem 6.6.1 *If $\mathscr{S}(AXCX')$ where A, C are fixed and X, X' are variable, the mapping $X \to X'$ ($X \neq A$) extended by $A \to C$ is an involution.*

Proof We must show that if $\mathscr{S}(AXCX')$ and $\mathscr{S}(AYCY')$, then A, C and X, X' and Y, Y' are pairs of an involution. In fact, let the line x pass through X and let P, Q be different points $\neq X$ on x (Fig. 6.21). We let

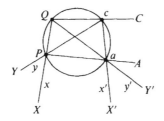

Figure 6.21

$a = PA$, $c = QC$, $y = PY$, $y' = QY'$. Since $\mathscr{S}(AYCY')$, the lines y, c and also the lines a, y' are not parallel. We join yc and ay' by the line x'. Since $\mathscr{S}(AYCY')$, the quadrilateral $aycy'$ is a quadrilateral of chords. Therefore, also $axcx'$ is a quadrilateral of chords, and from $\mathscr{S}(AXCX')$ it readily follows that X' is on x'. Since the pairs A, C; X, X'; Y, Y' are obtained by intersecting pairs of opposite sides of a complete quadrangle, they are, by Theorems 5.10.5 and 5.10.6, pairs of an involution. ◆

Theorem 6.6.2 *Every circle is a conic.*

Proof Lines are trivially conics. So let a circle k be given that does not pass through R. We choose two fixed points S_1, S_2 and a variable point X on k. We must show that the mapping between the pencils of lines through S_1 and S_2 defined by

$$S_1 X \longrightarrow S_2 X$$

is a projectivity. Let x_1 be the line $S_1 X$ or, if $S_1 = X$, the line tangent to k at S_1. Similarly let $x_2 = S_2 X$ or the tangent line of k in S_2.

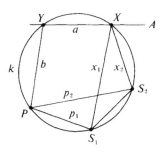

Figure 6.22

Let P be a fixed point $\neq S_1, S_2$ on k. We set $p_1 = S_1 P$, $p_2 = S_2 P$. Furthermore, let A be a fixed special point. We let $a = XA$. Let a intersect k in X, Y. If $P \neq Y$ let $b = PY$. Otherwise define b as the tangent of k at P. The special points of p_1, p_2, b, x_1, x_2 are called P_1, P_2, B, X_1, X_2, respectively.

Since, for $X \neq P$, $\mathscr{S}(P_1 B A X_1)$ and since P_1, A are fixed, the mapping

$$B \longrightarrow X_1$$

is, by Theorem 6.6.1, an involution. Similarly,

$$B \longrightarrow X_2$$

is an involution. Hence

$$X_1 \longrightarrow X_2$$

is a projectivity of the line at infinity, by Section 5.10. Therefore, $x_1 \to x_2$ is also a projectivity. ◆

We introduce in the affine R-subplane a coordinate system (see Section 3.7). Since k is a conic, it may be represented by a quadratic equation,

$$rx_1^2 + sx_1 x_2 + tx_2^2 + \cdots = 0, \tag{1}$$

where the linear part is not written out. In order to show that all other circles not passing through R may be represented by equations with the same quadratic part, we prove two lemmas.

Lemma 6.6.3 *Dilatations and translations of the affine R-subplane are circle-preserving maps.*

Proof With regard to Theorem 6.5.6, this follows from the fact that dilatations and translations map every line onto a parallel of itself (see Section 3.2). ◆

Lemma 6.6.4 *If $\mathcal{K}$ is the coordinate field of the affine R-subplane, then $\mathcal{K}$ has characteristic $\neq 2$ (that is, $1 + 1 \neq 0$).*

Proof By Section 3.7 it suffices to show that there exists a nonidentical translation T such that $T^2 \neq I$. We choose R such that it is one of the points of contact of the four circles a, b, c, d given by Axiom C.4. Let

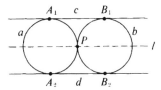

Figure 6.23

a, b not pass through R, and let P be their point of tangency. We draw a line l parallel to c (and d) through P. We denote by A_1, A_2 the points of contact of a and c, d, respectively, and by B_1, B_2 those of b and c, d, respectively. There exists a translation T mapping A_1 onto B_1 (see Theorem 3.5.1). Since c and d remain fixed under T, the image $(a)T$ of a is tangent to c at B_1 and tangent to d at a point B_2'. By Axiom C.4, $B_2' = B_2$. Since $b \neq a$, the image $(P)T$ of P does not lie on a. Clearly T^{-1} maps b onto a, and the image $(P)T^{-1}$ of P is not on b. Hence

$$(P)T \neq (P)T^{-1} \quad \text{or} \quad (P)T^2 \neq P$$

and thus $T^2 \neq I$. ◆

Theorem 6.6.5 *All circles that do not pass through R possess equations (1) with the same quadratic part*

$$rx_1^2 + sx_1x_2 + tx_2^2.$$

Proof Let $abcd$ be a quadrilateral of chords so that $\mathcal{S}(ABCD)$ for the corresponding special points. Let a, b, c, d be represented by equations

$$a_1x_1 + a_2x_2 + a_3 = 0,$$
$$b_1x_1 + b_2x_2 + b_3 = 0,$$
$$c_1x_1 + c_2x_2 + c_3 = 0,$$
$$d_1x_1 + d_2x_2 + d_3 = 0,$$

respectively. By Theorem 5.15.2, the given conic and the pairs of lines

$$(a_1x_1 + a_2x_2 + a_3)(c_1x_1 + c_2x_2 + c_3) = 0$$
$$(b_1x_1 + b_2x_2 + b_3)(d_1x_1 + d_2x_2 + d_3) = 0$$

are in a pencil of conics and so, by Theorem 5.15.1,

$$\begin{vmatrix} r & s & t \\ a_1c_1 & a_1c_2 + a_2c_1 & a_2c_2 \\ b_1d_1 & b_1d_2 + b_2d_1 & b_2d_2 \end{vmatrix} = 0. \tag{2}$$

This is so for any quadrilateral of chords *abcd*. Let, in particular, *a* and *c* satisfy

$$a_1 = 0, \qquad c_1 = 0.$$

Then $a_2 \neq 0$ and $c_2 \neq 0$ and, therefore, (2) implies

$$r(b_1 d_2 + b_2 d_1) - s b_1 d_1 = 0. \tag{3}$$

Similarly, for $a_2 = c_2 = 0$ or $a_1 = c_2 = 0$, we find equations

$$s b_2' d_2' - t(b_1' d_2' + b_2' d_1') = 0 \tag{4}$$

$$r b_2'' d_2'' - t b_1'' d_1'' = 0. \tag{5}$$

The matrix of the system (3), (4), (5) of linear equations in r, s, t is readily seen to have rank at least 2. So r, s, t are, up to a common factor, determined uniquely, independent of a_3, b_3, c_3, d_3. Therefore, they are the same (up to a common factor) for any circle not passing through R. ◆

We apply the transformation

$$x_1 = x_1' + (2t - s)x_2'$$
$$x_2 = -x_1' + (2r - s)x_2'.$$

In order to show that this is an affine transformation, we must prove that

$$\begin{vmatrix} 1 & 2t - s \\ -1 & 2r - s \end{vmatrix} = 2(r - s + t)$$

is not equal to 0. In fact: Suppose $r - s + t = 0$; the circle (1) intersects the line at infinity (given in homogeneous coordinates (x_1, x_2, x_3) by $x_3 = 0$) in $(-1, 1, 0)$, contrary to assumption. As an image of (1), we obtain

$$x_1'^2(r - s + t) + x_2'^2(r - s + t)(4tr - s^2) + \cdots = 0,$$

where the linear terms are not written out. Both coefficients are not equal to 0, since, otherwise, the circles would meet the line at infinity. Now we replace x_1', x_2' by x_1, x_2 again and set

$$d = s^2 - 4tr.$$

(1) thus obtains as its final form

$$x_1^2 - dx_2^2 + \cdots = 0.$$

Clearly d is not a square in $\mathscr{K}$; otherwise the circles would meet the line at infinity in points (written in homogeneous coordinates) $(1, \sqrt{d}, 0), (1, -\sqrt{d}, 0)$.
We conclude the section with a theorem derived from Theorem 6.3.5.

Theorem 6.6.6 Foundation theorem *Every odd Miquelian inversive plane is an inverse plane over a field $\mathscr{K}(\sqrt{d})$ with characteristic $\mathscr{K} \neq 2$, and conversely.*

Exercises

1. Find equations for the circles of the minimal model (see Exercise 1, Section 6.4) of an inversive plane ($\mathscr{K}$ consists of two elements; a quadratic form $x_1^2 + x_1 x_2 + x_2^2$ may be used).

2. Let an even (finite) inversive plane be given. Show that all tangent circles of a circle $\mathscr{C}$ passing through a point P have a second point Q in common.

3. Show that in an even Miquelian inversive plane there exists an inversion in every circle.

4. Are any two inversive planes over $\mathscr{K}(\sqrt{d_1})$ and $\mathscr{K}(\sqrt{d_2})$ isomorphic in the sense that the set of points of the first can be mapped onto the set of points of the second such that circles are transformed into circles and incidence is preserved?

5. Introduce axioms of order and continuity in one affine subplane of an inversive plane so that the subplane is turned into an ordinary Euclidean plane. Show that in this case all affine subplanes are ordinary Euclidean planes.

6. Show in Exercise 5 that for an appropriate coordinate system axiomatic circles become ordinary circles.

6.7   Conformal Mappings

We return now to the ordinary inversive plane (complex plane) as given by $\mathscr{R}(\sqrt{-1})$. Let ∞ be its point at infinity. Then $E^2 \cup \{\infty\}$ represents the set of all points of the inversive plane which may also be considered as the projective line over $\mathscr{R}(\sqrt{-1})$.

We wonder how the circle-preserving maps introduced in Section 6.2, that is, homographies and anti-homographies, affect "angles" between circles. We define the angle (mod π) between two circles that intersect in a point $P \neq \infty$ as the angle (mod π) between the tangents of the circles at P (Fig. 6.24). Clearly, two circles that are tangent have 0 as the angle between them. If we are given two pencils of circles mutually tangent at P, every circle of one pencil has the same angle with each circle of the other. So we may consider this angle also as the angle between the pencils.

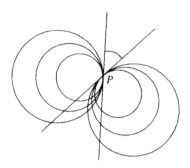

Figure 6.24

If two circles intersect in two points P, Q, the angles in P and Q are the same, by Section 2.18.   (See Fig. 6.25.)

The main result of this section is stated next.

Theorem 6.7.1 *Every homography*

$$z' = \frac{az + b}{cz + d}; \quad \begin{vmatrix} a & b \\ c & d \end{vmatrix} \neq 0, \tag{1}$$

and every anti-homography

$$z' = \frac{a\bar{z} + b}{c\bar{z} + d}; \quad \begin{vmatrix} a & b \\ c & d \end{vmatrix} \neq 0, \tag{2}$$

preserves angles between circles.

Proof (Note that from now on small italics will again represent circles.) As we have seen in the proof of Theorem 6.2.2, (1) and (2) can be broken down into products of similarities in E^2 and inversions.   Since similarities preserve angles, we need to show our theorem only for inversions. If a line l is inverted in a circle with center C, the image l' of l is a circle passing through C.   It is readily seen (compare Figure 6.26) that the

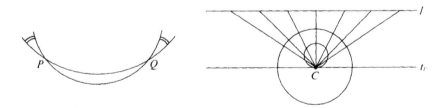

Figure 6.25   **Figure 6.26**

tangent $t_{l'}$ of l' at C (or l' itself if $l' = l$) is parallel to l.   So, if two lines l, m possess the angle α, their images possess as an angle the angle between $t_{l'}$ and $t_{m'}$, which equals α.   ◆

Because of Theorem 6.7.1, we may say that the mappings (1) and (2) preserve "form" locally. For this reason, these maps are called **conformal**. Of particular interest is the case in which two circles have an angle $\pi/2$. We call them **perpendicular** or **orthogonal**.

Theorem 6.7.2 *If a circle a intersects two circles through P, Q orthogonally, where P $\neq$ Q, then a intersects any circle through P, Q orthogonally* (Figs. 6.27 *and* 6.28).

Proof If $Q = U$, the center of a in E^2 is P, and Theorem 6.7.2 follows from the fact that every line through the center of a circle is perpendicular to this circle. If $Q \neq U$, we apply an inversion in a circle with center Q. This inversion maps Q onto U. Now Theorem 6.7.2 follows from Theorem 6.7.1 and from what we just have shown. ◆

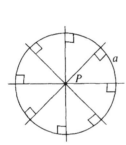 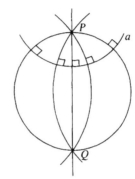

| Figure 6.27 | Figure 6.28 |

Theorem 6.7.3 *If P lies on a circle a and if Q $\neq$ P, there exists one and only one circle passing through P, Q and perpendicular to a.*

Proof Up to a homography we may assume that $P \neq \infty$. Let t be the tangent line of a at P (or a itself if a is a line), and let t' be the line perpendicular to t at P. There is a unique circle through P, Q tangent to t' and thus perpendicular to t and a. ◆

Theorem 6.7.4 *Given two different circles a, b and a point P not in the intersection of a, b, there exists a circle through P perpendicular to a and b.*

Proof Up to a conformal map we may assume that $P = \infty$. By assumption, a and b do not both pass through P; so let a be a Euclidean circle with center M. If b is also a Euclidean circle with center M', join M and M' by a line c. If b is a line, drop the perpendicular line c from M onto b. Now c is the circle we want. ◆

Theorem 6.7.5 *Under the inversion in a circle a, any circle perpendicular to a is mapped onto itself.*

Proof Let b be perpendicular to a and let a, b intersect in P, Q (Fig. 6.29). Since P, Q remain fixed under the inversion in a, and since angles (mod π)

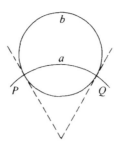

Figure 6.29

are preserved, the image b' of b passes through P, Q and is perpendicular to a. Hence, by Theorem 6.7.3, $b' = b$. ◆

We may employ the following two construction processes:

1. If a, b intersect in a point, find the second point of intersection.
2. Given a point Q on a circle a and a point P not on a. Drop the perpendicular (circle) through Q from P onto a.

Now the image of a point P under the inversion in a circle a can be found by choosing two points Q_1, Q_2 on a and applying constructions 2 and 1 (Fig. 6.30).

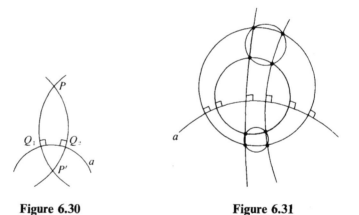

Figure 6.30   **Figure 6.31**

Figure 6.31 indicates a short way of finding the images of four points on a circle under an inversion. The fact that the image points do lie on a circle is directly expressed by Miquel's theorem (see Fig. 6.31).

Exercises

1. Map conformally the "upper half-plane" $x_2 > 0$ except the points $x_1^2 + x_2^2 \leq 1$ onto the intersection of the circular discs $x_1^2 + x_2^2 < 4$ and $(x_1 - 2)^2 + (x_2 - 2)^2 < 4$.

2. Show that any two circles of $E^2 \cup \{\infty\}$ that do not intersect can be mapped conformally onto a pair of concentric circles of E^2.

3. Let three circles be given, two of which do not intersect. Prove (by using Exercise 2) that there exists a circle perpendicular to all three of them.

4. Let two points P, Q and a circle c be given. Show (using a conformal map) that there exists a circle through P, Q perpendicular to c.

5. Let in E^2 the points P, Q both lie outside or both lie inside a circle c. Show that there exist two circles through P, Q that are tangent to c.

6. Show that in terms of cross ratios the angle α between two circles c, c' can be expressed by

$$\alpha = \frac{-1}{2i} \log (PQ, AB)$$

if c, c' intersect in P, Q, and if A is on c but not on c', furthermore, if B is on c' but not on c.

6.8 Subgroups of the Inversive Group

The product of two homographies is a homography; the product of a homography and an anti-homography is an anti-homography and the product of two anti-homographies is a homography. This is readily seen by combining Equations (1) and (2) of Section 6.7. (See also Exercise 7 below.) So the set consisting of all homographies and anti-homographies is a group under successive application. We call it the *inversive group*. From Theorem 6.7.1 we see that angles (mod π) are invariants of the inversive group.

The similarities of the Euclidean plane E^2 can be extended to homographies or anti-homographies, respectively, by defining $\infty \to \infty$. So the group of all similarities of E^2 can be considered a subgroup of the inversive group (compare the diagram in the introduction of this chapter). Now we are going to study further subgroups of the inversive group. First we need some preparations. If A denotes a homography or an anti-homography we write, instead of (1) or (2) in Section 6.7:

$$z' = (z)\text{A}.$$

For the image of a circle k under A we write $(k)\text{A}$ (considered as the set of all $(z)\text{A}$ with z on k).

Theorem 6.8.1 *If* A, B *are elements of the inversive group and if* A *possesses* z_0 *as a fixed point or* k *as a fixed circle, then* B^{-1}AB *possesses* $z'_0 = (z_0)\text{B}$ *as a fixed point or* $k' = (k)\text{B}$ *as a fixed circle, respectively.*

Proof $(z'_0)\text{B}^{-1}\text{AB} = (z_0)\text{BB}^{-1}\text{AB} = (z_0)\text{AB} = (z_0)\text{B} = z'_0,$
$(k')\text{B}^{-1}\text{AB} = (k)\text{BB}^{-1}\text{AB} = (k)\text{AB} = (k)\text{B} = k'.$ ◆

We call two elements A, A' of the inversive group *similar* if there exists an element B of the inversive group such that

$$A' = B^{-1}AB.$$

We consider now the group of all homographies which leave two different points U, V fixed. First let $V = \infty$ and $U = 0$. Given a pair of points

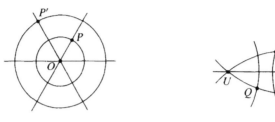

Figure 6.32   **Figure 6.33**

P, P' both $\neq 0$, ∞ we can find a rotation R that moves the line OP onto the line OP' (Figs. 6.32 and 6.33). The image $(P)R$ of P can now be mapped onto P' by a dilatation D with center 0. Thus $(P)RD = P'$. Clearly

$$RD = DR. \tag{1}$$

Now let U, V be arbitrary. We may find, by Section 6.2, a homography B such that $(O)B = U$ and $(\infty)B = V$. If Q, Q' are points $\neq U$, V, we let $P = (Q)B^{-1}$; $P' = (Q')B^{-1}$; and we map P onto P' by a product RD as defined above. This implies

$$(Q)B^{-1}RDB = (P)RDB = (P')B = Q'.$$

We write the left side of this equation somewhat differently as

$$(Q)(B^{-1}RB)(B^{-1}DB).$$

Let us investigate $D_0 = B^{-1}DB$. If k is any circle passing through U, V then

$$(k)D_0 = (k)B^{-1}DB = ((k)B^{-1})DB = ((k)B^{-1})B = k,$$

since $(k)B^{-1}$ is a line through O. So D_0 *possesses every circle through U, V as a fixed circle.*

Consequently, by Section 6.7, the circles perpendicular to the circles through U, V (see Theorem 6.7.2) are interchanged by D_0. The dilatation D induces, on every line through O augmented by ∞, a projectivity (see Section 5.11) with two different fixed points—that is, a hyperbolic projectivity. So we call the similar map D_0 of D a *hyperbolic* homography.

Setting $R_0 = B^{-1}RB$ we find, correspondingly, that R_0 interchanges the circles through U, V and leaves every circle perpendicular to the circles through U, V fixed. If such a fixed circle is considered to be a projective line, R_0 induces on it a projectivity without a fixed point (unless $R_0 = I$). So we call R_0 an *elliptic* homography.

If $R_0 \neq I$ and $D_0 \neq I$, the composed homography $R_0 D_0$ is called *loxodromic*. From (1) we find:

$$R_0 D_0 = B^{-1} R B B^{-1} D B = B^{-1} R D B = B^{-1} D R B = D_0 R_0.$$

By Theorem 6.2.4, the only homography that leaves U, V fixed and maps Q onto Q' is $R_0 D_0$. So we have proved the following theorem.

Theorem 6.8.2 *The set of all homographies leaving two different points U, V fixed is a commutative group consisting of elliptic, hyperbolic, and loxodromic homographies.*

We call the totality of all circles passing through two different points an *elliptic pencil*: The set of all circles perpendicular to those of an elliptic pencil is said to be the *hyperbolic pencil* orthogonal to the elliptic pencil. Finally, the set of all circles tangent to a circle c at a point P together with c is called a *parabolic pencil*. So a hyperbolic homography has all circles of an elliptic pencil as fixed circles; and an elliptic homography has all circles of a hyperbolic pencil as fixed circles.

Elliptic and hyperbolic homographies have a widespread application in physics. For example, U can be considered a source and V a sink. If water is flowing from U to V (under ideal conditions), this flow can be considered a "continuous chain" of hyperbolic homographies. Every particle moves on a circle of the elliptic pencil determined by U, V. Those particles which appear in U at the same time will be distributed on circles of the orthogonal hyperbolic pencil (Fig. 6.34).

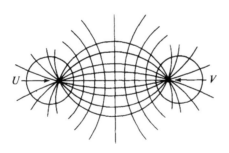

Figure 6.34

If $\mathscr{L}$ is the group of all homographies leaving U, V fixed, then all elliptic elements in $\mathscr{L}$ form a subgroup $\mathscr{E}$ and all hyperbolic elements form a subgroup $\mathscr{H}$. Clearly, $\mathscr{E}$ and $\mathscr{H}$ have only the identity map in common.

We consider now inversions that interchange U and V. Clearly any such inversion leaves every circle through U, V fixed.

Theorem 6.8.3 *Let U, V, P, P' be points on a circle c and let P be different from U, V. There exists one and only one inversion $\mathscr{I}$ that interchanges U, V and also interchanges P, P'.*

Proof With the aid of Theorem 6.8.1 it is readily seen that a similar image of an inversion is also an inversion. So we can restrict ourselves (by

Theorem 6.2.4) to the case $U = 0$, $V = \infty$, P real $= p$. Then P' is also a real number p'. We set $r^2 = pp'$. Then

$$z' = \frac{r^2}{|z|^2} z$$

is, by equation (3) of Section 6.1, an inversion $\mathscr{I}$. It has the desired properties. Suppose $\mathscr{I}^*$ is a second inversion of this kind. Then $\mathscr{I}\mathscr{I}^*$ is a homography and it leaves U, V, P, P' fixed; hence, by Theorem 6.2.4, $\mathscr{I}\mathscr{I}^* = \mathrm{I}$ or $\mathscr{I}^* = \mathscr{I}^{-1} = \mathscr{I}$. So $\mathscr{I}$ is uniquely determined. ◆

Theorem 6.8.4 *If* H *is a hyperbolic homography with fixed points* U, V, *and if* $\mathscr{I}$ *is an inversion interchanging* U, V *then* H$\mathscr{I}$ *and* $\mathscr{I}$H *are inversions interchanging* U, V.

Proof H$\mathscr{I}$ possesses all circles through U, V as fixed circles. Let $P \neq U$, V and set $P' = (P)$H$\mathscr{I}$. By Theorem 6.8.3, there exists an inversion $\mathscr{I}_1$, interchanging U, V and interchanging P, P'. The product H$\mathscr{I}\mathscr{I}_1$ is a homography possessing U, V, P as fixed points. Therefore, by Theorem 6.2.4, H$\mathscr{I}\mathscr{I}_1 = \mathrm{I}$ or H$\mathscr{I} = \mathscr{I}_1^{-1} = \mathscr{I}_1$. An analogous proof is established for $\mathscr{I}$H instead of H$\mathscr{I}$. ◆

Theorem 6.8.5 *If we add to the group* $\mathscr{H}$ *of hyperbolic homographies with fixed points* U, V *all inversions which interchange* U, V, *we obtain a group* $\hat{\mathscr{H}}$, *which is generated by its inversions.*

Proof The group properties of $\hat{\mathscr{H}}$ are deduced from Theorem 6.8.4 and the fact that the product of two inversions which both interchange U, V is a homography that leaves all circles through U, V fixed and, therefore, belongs to $\mathscr{H}$. Let H be an element of $\mathscr{H}$, and let for some $P \neq U$, V the

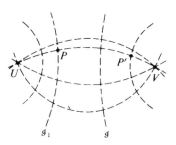

Figure 6.35

image (P)H be denoted by P'. By Theorem 6.8.3 there exists an inversion $\mathscr{I}$ interchanging U, V and interchanging P, P'. Furthermore, there exists an inversion $\mathscr{I}_1$ interchanging U, V and leaving P fixed (Theorem 6.8.3 applied for $P = P'$). $\mathscr{I}_1\mathscr{I}$ is a homography with fixed points U, V and $(P)\mathscr{I}_1\mathscr{I} = P'$. By Theorem 6.2.4,

$$\mathrm{H} = \mathscr{I}_1\mathscr{I}.$$

Therefore, every element of $\mathscr{H}$ is a product of inversions of $\hat{\mathscr{H}}$. ◆

We now extend $\mathscr{E}$ analogously to a set $\hat{\mathscr{E}}$ by adding all inversions in circles passing through U, V. We prove three theorems that correspond to the last three.

Theorem 6.8.6 *Let P, P' be two points on a circle c perpendicular to the hyperbolic pencil determined by two points U, V. There exists one and only one inversion leaving U, V fixed and interchanging P, P'.*

Proof By Theorem 6.8.3 (switching the roles of U, V and P, P'), there exists an inversion $\mathscr{I}$ in a circle d that interchanges P, P' and leaves U fixed. Clearly d is perpendicular to c and intersects c in a point P_0 (Fig. 6.36).

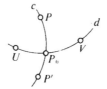

Figure 6.36

By assumption, any circle through U, V is perpendicular to c, in particular UVP_0. By Theorem 6.7.3, $d = UVP_0$, and thus V is also a fixed point of $\mathscr{I}$. ◆

Theorem 6.8.7 *If E is an elliptic homography with fixed points U, V and if $\mathscr{I}$ is an inversion in a circle through U, V, then $E\mathscr{I}$ and $\mathscr{I}E$ are inversions in circles through U, V.*

The proof proceeds in strict analogy to that of Theorem 6.8.4.

Theorem 6.8.8 *If we add to the group $\mathscr{E}$ of all elliptic homographies with fixed points U, V all inversions in circles through U, V, we obtain a group $\hat{\mathscr{E}}$, which is generated by its inversions.*

Again we may argue analogously to the proof of Theorem 6.8.5.

If we are given a parabolic pencil of circles, the totality of all circles perpendicular to it is again a parabolic pencil. Intuitively speaking, this is the limiting case of a hyperbolic pencil where U and V approach each other, in which case the orthogonal elliptic pencil is also "pushed together" ending up in a parabolic pencil. If we wish to study **parabolic** homographies defined as homographies that possess all circles of a parabolic pencil as fixed circles, we may refer again to the special case where the point of tangency of all circles is ∞. The group which corresponds to the above group $\mathscr{L}$ of hyperbolic, elliptic, and loxodromic homographies is now simply the group of all translations of E^2. Also inversions have become ordinary Euclidean reflections. (See Figs. 6.37 and 6.38.)

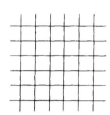

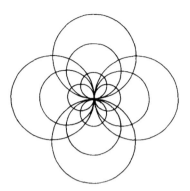

Figure 6.37   **Figure 6.38**

As a last subgroup of the inversive group, we consider the group of all homographies with a given circle c as fixed circle. Up to a homography we may assume c to be the real axis.

Theorem 6.8.9 *The group of all homographies that have the real line as a fixed line is given by all mappings*

$$z' = \frac{rz + s}{tz + u}; \quad \begin{vmatrix} r & s \\ t & u \end{vmatrix} \neq 0, \tag{2}$$

where r, s, t, u are real numbers.

Proof Clearly, every map (2) leaves the real line fixed. Conversely, since any three real numbers can be transformed into any other three real numbers by a mapping (2), we conclude from Theorem 6.2.4 that the homographies (2) are the only ones leaving the real line fixed. ◆

Exercises

1. Let $\mathscr{L}$ be the subgroup of the inversive group consisting of all elements leaving two points U, V fixed. Find the diagram of subgroups by which $\mathscr{L}$ and the groups $\mathscr{E}$, $\hat{\mathscr{E}}$, $\mathscr{H}$, $\hat{\mathscr{H}}$, $\mathscr{L}$ are related.

2. Is either of the groups $\hat{\mathscr{E}}$, $\hat{\mathscr{H}}$ commutative?

3. Find the homography which leaves $U = 0$ and $V = 1 + i$ fixed and maps $-i$ onto 2.

4. Show that a homography (2) maps any pair of conjugate complex numbers onto another pair of conjugate complex numbers.

5. Carry out explicitly the proofs of (a) Theorem 6.8.7, (b) Theorem 6.8.8.

6. Find the theorems for parabolic homographies that correspond to Theorems 6.8.3–6.8.8.

7. If two homographies

$$z' = (z)A = \frac{az + b}{cz + d} \quad \text{and} \quad z'' = (z')B = \frac{a'z' + b'}{c'z' + d'}$$

are combined, find explicitly the map $z'' = (z)AB$ (in the form (1) of Section 6.7). Show that AB corresponds to the matrix product **BA** where

$$\mathbf{A} = \begin{pmatrix} a & b \\ c & d \end{pmatrix}; \quad \mathbf{B} = \begin{pmatrix} a' & b' \\ c' & d' \end{pmatrix}.$$

If, however, we start from

$$z' = (z)A = \frac{a + bz}{c + dz},$$

etc., show that we find the corresponding matrix products without switching the factors.

8. Show that the group of all homographies leaving a given circle fixed is isomorphic to the group of all 2×2 matrices **A** (see Exercise 7) with real entries and a/c, b/d, different constants.

9. Prove that two different circles lie either in one and only one hyperbolic pencil or in one and only one elliptic pencil or in one and only one parabolic pencil.

6.9 *Poincaré's Model of the Hyperbolic Plane*

We return now to the model of a hyperbolic plane mentioned in Section 1.6. Consider in the inversive plane a circle c. If c does not contain ∞, it lies in E^2 and the interior of c is well defined (Fig. 6.39). If c passes through ∞, we may choose one of the two half-planes of E^2 bounded by c as "interior" of c (Fig. 6.40). In the following we assume c to lie in E^2.

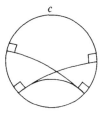

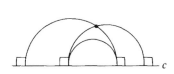

Figure 6.39   **Figure 6.40**

We choose the set Π of all points in the interior of c as points of a geometry (the points of c are excluded). We call them *h-points*; *h-lines* are defined as intersections with Π of circles perpendicular to c. Incidence and perpendicularity are taken over in a natural way from inversive geometry.

Theorem 6.9.1 *The incidence axioms of a metric plane are satisfied* (I.1–I.3 *of Section* 1.1) *for h-points and h-lines.*

Proof Let P, Q be two different h-points. The circles through P that are perpendicular to c meet again in a point P' of the inversive plane (not in Π). Intersecting PQP' with Π, we obtain an h-line through P, Q. It is uniquely determined, since P, P', Q determine a circle. So I.1 is satisfied. I.2 and I.3 are trivial from our definitions. ◆

Theorem 6.9.2 *The perpendicularity axioms of a metric plane* (P.1–P.3 *in Section* 1.2) *are satisfied for h-points and h-lines.*

Proof Since perpendicularity of Euclidean lines is symmetric, perpendicularity of circles, and thus of h-lines, is also symmetric. Hence Axiom P.1 is true.

Perpendicular circles intersect, by definition, in two points. Let a, b be perpendicular h-lines, obtained from circles a^*, b^*, respectively, by intersection with Π. Let P, Q be the points of intersection of a^*, b^*. Then P, Q are interchanged under the inversion in c. So either P or Q lies in Π. This implies Axiom P.3.

Let P be an h-point and a an h-line; a^* and c intersect in points U, V. By Theorem 6.8.3 (applied for $P = P'$), there exists a unique inversion in a circle b^* leaving P fixed and interchanging U, V. Intersecting b^* with Π we obtain an h-line b through P perpendicular to a. Any circle through P perpendicular to a, c provides an inversion interchanging U, V and hence coincides with b^*. So b is uniquely determined. This shows Axiom P.2. ◆

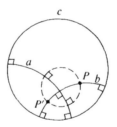

Figure 6.41

Before defining reflections we prove a lemma:

Lemma 6.9.3 *Let a be an h-line, $a = a^* \cap \Pi$, and let P be an h-point. The image of P under the inversion in a^* is again an h-point.*

Proof Let b be the perpendicular to a through P, $b = b^* \cap \Pi$. The circle d through P (dotted line in Fig. 6.41), which is perpendicular to a and b, does not intersect c; if it did intersect c in Q, then c, d would be circles through Q determining inversions under which the points of intersection of a^*, b^* would be interchanged, contrary to Theorem 6.8.3. So it readily follows that d lies completely in Π. In particular, the second point, P', in which b, d intersect, is an h-point. P', however, is the image of P under the inversion in a^*. ◆

If $a = a^* \cap \Pi$, the restriction of the inversion in a^* to Π provides, by
Lemma 6.9.3, a mapping of Π onto itself. We call it the **reflection** in a.
Clearly this definition fulfills the requirements (1)–(4) of a reflection introduced in Section 1.3. Also Axiom M.1 is true. Furthermore, it is involutoric, so that Axiom M.2 is satisfied.

In order to prove Axiom M.3, we proceed as follows: Let the h-lines a, a'
intersect in the h-point R, and let P be an h-point $\neq R$ on a. By Theorem

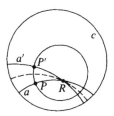

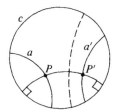

Figure 6.42　　　　　　　　　**Figure 6.43**

6.7.4, there is a circle d through P perpendicular to a, a'; and d intersects a'
in an h-point P' (Fig. 6.42). Using Theorem 6.8.6, we find a reflection R
in an h-line through R mapping P onto P'. If R_1, R_2 are two reflections in
h-lines through R, then $R_1 R_2$ is a rotation about R. Suppose $(a)R_1 R_2 = a'$.
Then $(P)R_1 R_2 = P'$, since both R_1, R_2 leave d fixed. Similarly, $(X)R_1 R_2 =
(X)R$ for any X on a. This shows Axiom M.3.

The proof of Axiom M.4 is shown analogously, using Theorem 6.8.3
instead of Theorem 6.8.6 (compare Fig. 6.43).

Two h-lines are said to be *parallel* if they have neither an h-point nor a
perpendicular h-line in common. This is so if they "meet" on a point of c
(compare Figs. 6.39 and 6.40). This clearly implies the hyperbolic Axiom
Hyp. The following theorem states what we have shown.

Theorem 6.9.4　*h-points and h-lines form a hyperbolic metric plane. We call
it Poincaré's model of the hyperbolic plane.*

It can now be shown that the axioms of order and continuity hold and that,
therefore, *Poincaré's model is isomorphic to the ordinary hyperbolic plane as
introduced axiomatically in Section 2.10.*

Exercises

1. Show that the group of proper motions of Poincaré's model is isomorphic
   to the group of all homographies of the inversive plane leaving c fixed.

2. Show that horocycles and hypercycles (see Section 2.11) are circles or parts of circles.

3. Prove that in Poincaré's model the axioms of order and continuity (Sections 2.2 and 2.9) are fulfilled.

4. Prove Axiom M.4 explicitly.

5. Prove that any three noncollinear points are each on a side of an asymptotic triangle (see Section 2.11).

6.10   Stereographic Projection

We shall now discuss in more detail the stereographic projection mentioned in Section 6.1. In ordinary Euclidean 3-space we introduce a Cartesian coordinate system so that points are represented by vectors (x_1, x_2, x_3). The sphere S,

$$x_1^2 + x_2^2 + (x_3 - \tfrac{1}{2})^2 = \tfrac{1}{4},$$

is tangent to the (x_1, x_2)-plane E in $O = (0, 0, 0)$ and has diameter 1. Its "north pole" $(0, 0, 1)$ may be denoted by N. If $P = (\xi_1, \xi_2, 0)$ is any point in E, we wish to find its image P' on S under a stereographic projection, that is, the point $\neq N$ in which the line NP and the sphere S meet (Fig. 6.44).

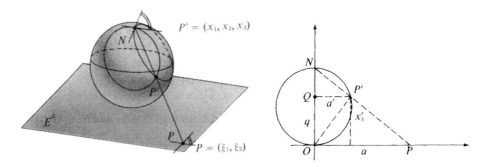

Figure 6.44   **Figure 6.45**

We drop a perpendicular line from P' onto the x_3 axis, meeting this axis in a point $Q = (O, O, q)$ (Fig. 6.45). Let $a = d(O, P)$ and $a' = d(Q, P')$ be the distances from O to P and from Q to P', respectively. The triangles $\triangle OPN$ and $\triangle OQP'$ are similar, and

$$\frac{q}{a'} = \frac{a}{1}, \qquad \frac{a'}{1-q} = \frac{a}{1};$$

hence

$$a' = \frac{a}{1 + a^2}, \qquad q = \frac{a^2}{1 + a^2}.$$

Let θ be the angle between the positive x_1 axis and the ray from O to P.
Then

$$x_1' = a' \cos \theta = \frac{a \cos \theta}{1 + a^2} = \frac{\xi_1}{1 + \xi_1^2 + \xi_2^2},$$

$$x_2' = a' \sin \theta = \frac{a \sin \theta}{1 + a^2} = \frac{\xi_2}{1 + \xi_1^2 + \xi_2^2},$$

$$x_3' = q = \frac{a^2}{1 + a^2} = \frac{\xi_1^2 + \xi_2^2}{1 + \xi_1^2 + \xi_2^2}.$$

Therefore,

$$(\xi_1, \xi_2, 0) \longrightarrow \frac{1}{1 + \xi_1^2 + \xi_2^2}(\xi_1, \xi_2, \xi_1^2 + \xi_2^2)$$

represents the stereographic projection. The inverse map (of S diminished by N) is

$$\xi_1 = \frac{x_1'}{1 - x_3'}; \qquad \xi_2 = \frac{x_2'}{1 - x_3'}.$$

Theorem 6.10.1 *The stereographic projection is a circle-preserving map.*
Proof Let

$$a(\xi_1^2 + \xi_2^2) + b\xi_1 + c\xi_2 + d = 0 \qquad (1)$$

be a circle or a line ($a = 0$) in E^2(a, b, c, d are real numbers). Its image on S satisfies the equation

$$ax_3' + bx_1' + cx_2' + d(1 - x_3') = 0.$$

This is the equation of a plane. So the image of the circle (1) is a plane section of S, that is, a circle. ◆

If E is extended by ∞ to an inversive plane, this point ∞ can be, in a natural way, mapped onto N.

Two circles on the sphere S that intersect in a point R include, as angle (mod π), the angle between the tangents of the circles at R. If R^* is the second point in which the circles intersect, the angle at R^* is the same as that at R; this is readily seen by applying a reflection in a plane through the center M of S and perpendicular to the line RR^* ($\triangle MRR^*$ is an isosceles triangle).

Theorem 6.10.2 *The stereographic projection is conformal—that is, it preserves angles between circles.*

Proof Every line l of E is mapped onto a circle l' passing through N. Since the tangent plane ($x_3 = 1$) of S at N is parallel to E, the tangent of l' at N is parallel to l. Therefore, the angle between two lines remains unchanged under a stereographic projection. The same applies for circles, since the angle between two circles is defined by means of tangent lines. ◆

If c is a circle on S, we denote the plane which cuts c out of S by $\Pi(c)$. In Section 5.16 we called two planes conjugate with respect to S if one passes through the pole (with respect to S) of the other.

Theorem 6.10.3 *Two circles c_1, c_2 on S are perpendicular if and only if the planes $\Pi(c_1)$ and $\Pi(c_2)$ are conjugate with respect to S.*

Proof Let $\tilde{c}_1$, $\tilde{c}_2$ be the circles of $E \cup \{\infty\}$, whose images are $\tilde{c}_1$, $\tilde{c}_2$. Up to a homography (which is conformal) we may assume that c_1, c_2 are lines through O. Then $\tilde{c}_1$, $\tilde{c}_2$ are circles through O, N, and M which are cut out from S by planes perpendicular to each other. Those planes clearly are conjugate with respect to S. ◆

A wide class of inversive planes can be obtained if we replace the sphere S by an egg-shaped surface which, for convenience, we define as follows:

An ***egg-shaped surface*** in Euclidean 3-space is a nonempty point set Π with the following two properties:

Figure 6.46

1. Through every point P of the set Π there passes one and only one plane $t(P)$ intersecting Π only in P, called the ***tangent plane*** of Π at P.
2. If a line a intersects Π in precisely one point P, then a lies in $t(P)$.

It is easy to prove that the plane sections of Π that contain more than one point form an inversive plane. This plane may also be projected onto a tangent plane of Π by the inverse process of a stereographic projection. We obtain then as "circles" a system of planar curves.

Theorem 6.10.4 *The inversive plane defined by the plane sections of an egg-shaped surface satisfies the bundle theorem (see Section 6.4).*

Proof Using the notation of the bundle theorem in Section 6.4, we deduce from the assumptions of that theorem that the lines PP', QQ', RR', SS' all meet in a point or are all parallel, and hence that P, P', S, S' lie in a plane. ◆

We remark that the egg-shaped surface can be chosen in such a way that Miquel's theorem does not hold (see Exercise 8). So in an inversive plane Miquel's theorem cannot be proved from the bundle theorem.

Exercises

1. Show that the unit circle of E is mapped under a stereographic projection onto the "equator" of S (intersection with $x_3 = \frac{1}{2}$).

2. Describe elliptic, hyperbolic, and parabolic pencils of circles (Section 6.8) in terms of the planes by which they are cut out of S.

3. Show that under a stereographic projection any great circle of S (radius $\frac{1}{2}$) is obtained as the image of a circle passing through diametrically opposite points of the unit circle (center O) of E.

4. (a) Consider the "lower hemisphere" H of S, that is, all points of S satisfying $x_3 < \frac{1}{2}$. Intersect H by planes perpendicular to E and show that in this way a hyperbolic metric plane (isomorphic to Poincaré's model) is obtained.

   (b) Prove anew the incidence and perpendicularity axioms for this plane by means of spatial projective geometry (using Theorem 6.10.3).

5. Prove the following closure theorem involving incidence and perpendicularity (see Fig. 6.47). Let $a \cap b = \{P, Q\}$, $c \cap d = \{R, S\}$, $a' \cap b' = \{P', Q'\}$, $c' \cap d' = \{R', S'\}$; let a, b both be perpendicular to a', b';

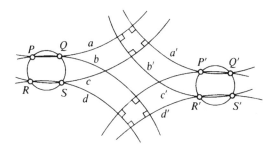

Figure 6.47

and let c, d both be perpendicular to c', d', where $P, Q, R, S, P', Q', R', S'$ are all different points. If $\mathscr{C}(PQRS)$ then also $\mathscr{C}(P'Q'R'S')$.

6. Prove anew Miquel's theorem by considering the plane sections of a sphere.

7. Carry out the proof that the plane sections of an egg-shaped surface form an inversive plane.

8. Find an egg-shaped surface such that its plane sections do not satisfy Miquel's theorem.

6.11   Klein's Model of the Elliptic Plane

A metric plane has been called elliptic (see Section 1.5) if any two of its lines intersect. So an elliptic plane is always a projective plane. Here, we shall consider the real projective plane—that is, the projective plane obtained as an extension of the ordinary Euclidean plane E^2. We may represent it in Euclidean 3-space by considering all lines passing through O (compare Section 5.4) as "points" and considering planes through O as "lines." These lines and planes through O may, in turn, be intersected with the unit sphere S, center O. Then we obtain another model of the projective plane: "Points" of the projective plane are pairs of diametrically opposite points of S; we call them *e-points*. "Lines" of the projective plane are great circles of S—that is, circles cut out of S by planes through O; we denote them *e-lines* (see Fig. 6.48).

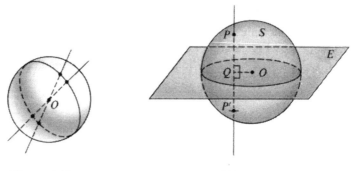

Figure 6.48   Figure 6.49

This new model is of particular advantage if the projective plane is to be an elliptic metric plane. We say two *e*-lines are perpendicular if they are perpendicular as Euclidean circles. If an *e*-line a lies in the plane E, we intersect the perpendicular line of E through O with S, obtaining an *e*-point P. We call P the **pole** of a and a the **polar** of P. Clearly, two *e*-lines are perpendicular if and only if one passes through the pole of the other.

The incidence axioms of a metric plane are satisfied, since, as we just have seen, the *e*-points and *e*-lines form a projective plane. To show the perpendicularity axioms, we proceed as follows: The symmetry of the relation "perpendicular" is guaranteed by definition (P.1). Perpendicular lines intersect, since any two lines intersect (P.3). If a is a line and Q is any point, join Q and the pole P of a by a line. This line is a perpendicular of a. If Q lies on a, there cannot be two perpendiculars of a at Q, since, otherwise, the Euclidean tangent of the circle a at Q would have two Euclidean perpendiculars, both lying in the tangent plane of S at Q, which is impossible (P.2).

We apply a Euclidean reflection of E^3 in a plane E through O. This reflection maps the sphere S onto itself, the image P' of any point P being obtained by dropping the perpendicular line from P onto E and intersecting again by S (in Fig. 6.49, clearly $\triangle OPQ$ and $\triangle OP'Q$ are congruent, since P and P' have equal distance from O). Every plane through O is mapped onto another plane through O; hence *e*-lines are mapped onto *e*-lines, and also *e*-points are mapped onto *e*-points. Those *e*-points which lie in E remain fixed. Since Euclidean perpendicularity is preserved under a reflection, perpendicularity of *e*-lines is preserved, too. So the Euclidean reflection in E induces a reflection R on the system of *e*-points and *e*-lines. It satisfies Axioms M.1 and M.2, by definition. Axiom M.4 is trivially satisfied, since there are no lines which don't intersect. To show Axiom M.3, we reflect a line successively in two e-lines through an e-point R, obtaining an e-line a'. The product of these reflections stems from the product of two Euclidean reflections in planes through the Euclidean line l determined by R, hence from a rotation about l.

If $P \neq R$ is mapped onto P' under this rotation, we reflect P in the plane through l perpendicular to the Euclidean line PP'. This reflection induces in the plane of e-points and e-lines a reflection which satisfies the requirements of Axiom M.3. Thus, we have proved the following theorem.

Theorem 6.11.1 *The e-points and e-lines form an elliptic metric plane. Furthermore, under a Euclidean reflection in a plane E through O, plane sections of S are mapped onto plane sections of S, that is, circles onto circles.*

Theorem 6.11.2 *The group of motions of the elliptic metric plane defined above is a subgroup of the inversive group of the ordinary inversive plane.*

In closing, let us compare Klein's model of an elliptic plane with hyperbolic and Euclidean planes that can be defined on the sphere S.

In Section 6.9, a hyperbolic metric plane was defined by circular arcs of an inversive plane perpendicular to a particular circle c. If we transfer the inversive plane onto S by a stereographic projection under which c is mapped onto a great circle of S, the hyperbolic plane is carried by a hemisphere S_1 and the circular arcs may be considered to be cut out of S_1 by planes perpendicular to the plane E in which c lies. By analogy to Klein's model of an elliptic plane, we may choose as **h-points** pairs of different points obtained by (Euclidean) reflection at E, and as **h-lines** circles on S perpendicular to c. In other words, we intersect S by lines perpendicular to E. If we obtain different points of intersection with S, we choose this pair of points as an h-point, etc.

If the euclidean space E^3 is considered to be extended to a projective space, all lines perpendicular to E meet in a point M. So the h-points are obtained by intersecting S with lines through a point M, just as e-points have been obtained by intersecting S with lines through O.

Finally, let M be a point on S. All points not equal to M and all circles through M form a Euclidean plane, as can be seen by applying a stereographic projection with M as "north pole". Again, we may consider all pairs (M, P) of points, $P \neq M$, as points of the Euclidean plane, instead of the points P alone. Then we obtain the "points" of the Euclidean plane by intersecting S with all lines through M.

It can easily be seen that, instead of O, any interior point of S can be chosen for the definition of an elliptic plane, and any exterior point M of S can be chosen for the definition of a hyperbolic plane. So we find the following general method for constructing metric (Euclidean and non-Euclidean) planes on a sphere S (see Figs. 6.50–6.52): *Take a point M in 3-space, intersect S by all lines and planes through M. Choose all pairs of different points obtained hereby as "points" and all circles as "lines" of the metric plane. In each case, the group of motions can be considered to be a subgroup of the (ordinary) inversive group.*

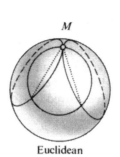

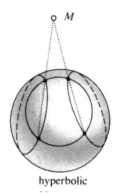

elliptic

Euclidean

hyperbolic

Figure 6.50 **Figure 6.51** **Figure 6.52**

Exercises

1. Find the group of all homographies that induce the identity map on Klein's model.

2. Prove that the group of all motions in Klein's model is isomorphic to the group of all rotations of E^3 with center O.

3. Describe the perpendicularity of lines in Klein's model in terms of the inner product given on the vector space of all vectors emanating from O.

4. Show explicitly that for M on S an ordinary Euclidean plane is obtained.

5. By comparing the above models of elliptic, Euclidean, and hyperbolic planes, find a reason why there are dilatations other than motions in the Euclidean case but not in the elliptic and the hyperbolic cases.

6. Describe the common part of the elliptic, Euclidean, and hyperbolic geometries given by the above models in projective terms (replace the sphere by an ellipsoid).

Chapter Seven

Foundation of Euclidean and Non-Euclidean Geometry

In this last chapter we return to the discussion of metric planes initiated in Chapter 1. In other chapters, we have developed several powerful tools for this discussion: vector calculus (Chapter 4), projective geometry (Chapter 5), and inversive geometry (Chapter 6). It turns out, that every metric plane, Euclidean or non-Euclidean, may be considered as part of a projective plane. Perpendicularity of the metric plane is then described by a polarity. This is the essential content of the fundamental theorem of metric planes (Section 7.11). The present chapter is primarily dedicated to proving this fundamental theorem. Nevertheless, on the way to reaching this goal, we shall meet a number of other nice geometrical facts that are common to Euclidean and non-Euclidean geometry.

Throughout this chapter we shall use the following convention: We use

$$a \text{ instead of } R_a,$$
$$P \text{ instead of } H_P.$$

In other words, we shall use the same letters for lines and points as for the reflections and half-turns determined by them. So, concerning products of motions,

$$ab \text{ replaces } R_a R_b,$$
$$aP \text{ replaces } R_a H_P,$$
$$PQ \text{ replaces } H_P H_Q,$$

and so on. This notation provides a very helpful simplification. It is based on the one-to-one-correspondence between reflections and lines as well as that between half-turns and points (see Section 1.15).

7.1 Definition of Snail Maps

We consider a metric plane as introduced in Sections 1.1–1.4. The next problem we wish to solve is that of "embedding" the

---

Figure reprinted from H. S. M. Coxeter, *Introduction to Geometry*, courtesy of John Wiley & Sons, Inc

metric plane into a projective plane, in the following sense: We wish to find a projective plane such that points and lines of the metric plane are points and lines in this projective plane, and such that a point lies on a line in the metric plane if and only if it lies on this line in the projective plane.

If the metric plane happens to be elliptic, there is nothing to be done: Since, by assumption, in an elliptic plane every two lines intersect, the plane is already a projective plane.

In case the metric plane is Euclidean, it is an affine plane (see Section 3.1) and its extension to a projective plane was achieved in Section 5.1. Nevertheless we shall carry out the embedding for arbitrary metric planes. The underlying idea is similar to that used in the extension of affine planes to projective ones: We define pencils of lines in order to assign points to them. Generally these will be pencils of the first, the second, and the third kind (see Section 1.16). In Euclidean planes we have only pencils of the first and of the second kind. Therefore, we may assign to the set of all pencils of the second kind ("parallel pencils") a single line, the "line at infinity." In general, however, we shall have to define many new lines. This is done by using a nice idea due to the Danish geometer J. Hjelmslev: By certain mappings which we are going to call "snail maps," pencils of the second and third kind are "pulled in," that is, mapped onto pencils of the first kind. The lines joining pencils of the first kind may then by reverse maps be "pushed out" and so join pencils of the second and of the third kind.

Let O be a point of a metric plane and let the lines u, v pass through O so that the product uv (of reflections) is not involutoric—that is, $(uv)^2 \neq I$ (or: the line u is not perpendicular to the line v; see Section 1.15).

If w is an arbitrary line through O we assign to w the line

$$w^* = wuv.$$

In an ordinary Euclidean plane this means that we rotate w into w^* about an angle θ $(0 < \theta < \pi)$ by a rotation which also moves u into v (Fig. 7.1).

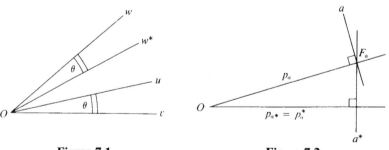

Figure 7.1   **Figure 7.2**

Now assume that a does not pass through O and (in case of an elliptic plane) is not a polar line of O. The perpendicular line p_a of a through O (by Section 1.12 it is uniquely determined) intersects a in F_a $(=ap_a)$. (See Fig. 7.2.) If p_a^* is the image of p_a as defined above, we assign to a the perpendicular a^* of p_a^* through F_a (a^* is uniquely determined, since, otherwise, by

Section 1.12, $p_a^* \perp p_a$ and thus $v \perp u$ would follow, contrary to assumption). Finally, we assign, in case of an elliptic plane, the polar line of O to itself. We call the mapping of the set of all lines into itself thus defined a ***snail map***. It induces a mapping of the set of all points F_a into itself by setting

$$F_a^* = a^* p_a^* \quad \text{and} \quad O^* = O.$$

O is called the ***center*** of the snail map.

The name snail map may be justified as follows: If we apply a snail map again and again in an ordinary Euclidean plane, the points "wind around" O and toward O, as a snail pulls back into its hut (Fig. 7.3).

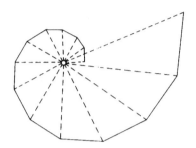

Figure 7.3

A ***spiral symmetry*** in an ordinary Euclidean plane is defined as the product of a rotation and a dilatation with the same centers. In matrix form, it can be expressed as

$$\begin{pmatrix} d\cos\theta & d\sin\theta \\ -d\sin\theta & d\cos\theta \end{pmatrix}, \tag{1}$$

where d is a constant number $\neq 0$ and θ is an angle between O and π. This spiral symmetry is the product of a dilatation with dilatation factor d and a rotation about O with angle of rotation θ.

Theorem 7.1.1 *In an ordinary Euclidean plane, every snail map is a spiral symmetry. However, not every spiral symmetry is a snail map.*

Proof We choose O as the origin of a Cartesian coordinate system. Any line w through O is mapped onto a line w^* by the rotation

$$\begin{pmatrix} \cos\theta & \sin\theta \\ -\sin\theta & \cos\theta \end{pmatrix},$$

which transforms u into v, where $0 < \theta < \pi$. As is readily seen from Figure 7.4, the image of an arbitrary point F_a under the given snail map is

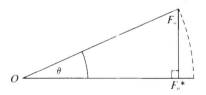

Figure 7.4

obtained by first rotating F_a about θ and then applying a dilatation with dilatation factor $d = \cos\theta$. So

$$\cos\theta\begin{pmatrix} \cos\theta & \sin\theta \\ -\sin\theta & \cos\theta \end{pmatrix} = \begin{pmatrix} \cos^2\theta & \cos\theta\sin\theta \\ -\cos\theta\sin\theta & \cos^2\theta \end{pmatrix}$$

represents the snail map, and it is clearly a spiral symmetry.
For $d \neq \cos\theta$, the spiral symmetry (1) is not a snail map. ◆

Exercises

1. (a) Is the inverse map of a snail map also a snail map?
   (b) Is the square of a snail map also a snail map?

2. (a) Show that

$$\begin{pmatrix} \tfrac{1}{2} & \tfrac{1}{2} \\ -\tfrac{1}{2} & \tfrac{1}{2} \end{pmatrix}$$

represents a snail map of the ordinary Euclidean plane.
(b) How often must this map be applied in order to shift the point $(20, 0)$ into the interior of the unit circle $x_1^2 + x_2^2 = 1$?
Find the image of $(20, 0)$.

3. In the snail map of Exercise 2 find the image and the inverse image of the set of all points of the interior of the unit circle $x_1^2 + x_2^2 = 1$.

4. Find all snail maps with center O of the ordinary Euclidean plane that move $(2, 2)$ into the interior of the unit circle $x_1^2 + x_2^2 = 1$.

5. Do all spiral symmetries with the same center form a group?

6. Show that if S is a snail map and if M is a motion of the metric plane then $M^{-1}SM$ is also a snail map.

7.2   Properties of Snail Maps

In this section we make a number of observations on snail maps in metric planes that will be needed later on.   We are given a snail map (*) as defined in Section 7.1.   Note that p_a denotes the perpendicular of a through O.

Theorem 7.2.1   *If a and b pass through O, then*

$$ab = a^*b^*.$$

Proof   By definition of a snail map, we have $aa^* = uv$ and also $bb^* = uv$; hence, $aa^* = bb^*$ and thus (multiplying by a from the left and by b^* from the right)

$$ab = a^*b^*. \qquad ◆$$

Theorem 7.2.2 *For any line a that is not a polar line of O,*

$$p_a^* = p_{a^{\bullet}}.$$

Proof If a passes through O, we find, by Theorem 7.2.1,

$$a p_a = a^* p_a^* = O.$$

Therefore, p_a^* equals the perpendicular of a^* through O, that is, $p_a^* = p_{a^{\bullet}}$.
If a does not pass through O, Theorem 7.2.2 is true, by definition (see Fig. 7.2). ◆

Theorem 7.2.3 *If $a \neq b$ then also $a^* \neq b^*$.*

Proof If O lies on a and b, this follows from the definition of a rotation. If O lies on a but not on b, then O lies on a^* but not on b^*, from which $a^* \neq b^*$ follows. A similar conclusion is made for a or b being polar of O. So let neither a nor b pass through O and let none of them be a polar line of O.

Suppose $a^* = b^*$. Then $p_a^* = p_{a^{\bullet}} = p_{b^{\bullet}} = p_b^*$ (see Theorem 7.2.2 and thus $p_a = p_b$ (a rotation being a one-to-one map for lines). Since a and b both pass through the intersection of $p_a = p_b$ and $a^* = b^*$ (by definition of a snail map), a and b would both be perpendicular to p_a at the same point and, hence, would be equal (see Section 1.2, Axiom P.2), a contradiction. ◆

Theorem 7.2.1 readily implies the following theorem.

Theorem 7.2.4 *Let a pass through O. Then b is perpendicular to a if and only if b^* is perpendicular to a^*.*

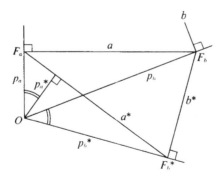

Figure 7.5

Theorem 7.2.5 *a passes through F_b if and only if a^* passes through F_b^*.*

Proof The theorem clearly is true when b is the polar line of O. If b is not the polar line of O, we have, by definition of *, $p_a p_a^* = p_b p_b^*$ and thus, by Theorem 7.2.2,

$$p_{a^{\bullet}} = p_a p_b p_{b^{\bullet}}.$$

Since F_b lies on a, we conclude from the first theorem on three reflections that ap_bb^* is a line and, hence, that

$$F_ap_a^*F_{b^\cdot} = F_a(p_ap_bp_{b^\cdot})F_{b^\cdot} = ap_bb^*$$

is a line. By Theorem 1.17.1 or by Hjelmslev's theorem (see Section 1.17), this implies that $F_{b^\cdot}$ lies on a^* (see Fig. 7.5).

All these conclusions apply in the reverse order. So if a^* passes through $F_{b^\cdot}$ then a passes through F_b. ◆

Theorem 7.2.6 *Let b pass through O. Then abc is a line if and only if $a^*b^*c^*$ is a line.*

Proof First we treat the case that b and c also pass through O. By Theorem 7.2.1, we find: $ba = (bac)c$ implies $b^*a^* = (bac)^*c^*$, and conversely. Therefore, $(bac)^* = b^*a^*c^*$, from which Theorem 7.2.6 readily follows.

Now let b and c be arbitrary lines. If one of them coincides with a, the theorem is trivially true. If $a = c$, then $abc = aba$ is always a line and so is $a^*b^*a^*$ (see Section 1.11). Therefore, let $a \neq c$ and a, c both be unequal to b.

We now test the assertion that if F_adF_c is a line for some d passing through O, then $F_{a^\cdot}d^*F_{c^\cdot}$ is also a line, where d^* passes through O, and conversely. In fact, by Section 1.17, $F_a\,dF_c$ is a line if and only if the line joining F_a and F_c is perpendicular to d. By Theorem 7.2.4, this is also equivalent to saying that d^* and the line determined by $F_{a^\cdot}$, $F_{c^\cdot}$ are perpendicular. So, by applying Section 1.17 once more, we have proved the assertion.

Now the proof of Theorem 7.2.6 is accomplished quickly:

$$abc = F_a(p_abp_c)F_c$$

is a line if and only if

$$F_{a^\cdot}(p_{a^\cdot}b^*p_{c^\cdot})F_{c^\cdot} = a^*b^*c^*$$

is a line. ◆

Theorem 7.2.7 **Transitivity theorem** *If $a \neq b$ and if abc, abd are lines, then acd is also a line.*

Proof If a and b possess a common point or a common perpendicular, the theorem follows from Theorem 1.16.1 or 1.16.2, respectively. Otherwise we choose a point O on a. If v is the perpendicular of b through O, then a, v determine a snail map $*$ with $a^* = v$. As is readily seen from Figure 7.6, a^* and b^* meet in a point F_b. By Theorem 7.2.6, abc, abd are lines if

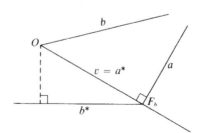

Figure 7.6

and only if $a*b*c*$, $a*b*d*$ are lines, respectively. If $a*b*c*$ and $a*b*d*$ are lines, then $a*$, $b*$, $c*$, $d*$ all pass through F_b and thus $a*c*d*$ is a line (by the first theorem on three reflections). So, by Theorem 7.2.6, acd is a line. ◆

From the preceding proof we find, in addition:

Theorem 7.2.8 *If $a \neq b$ and if abc is a line, then acb and bac are also lines.*

The proof of Theorem 7.2.7 also substantiates this one. Now we are able to prove Theorem 7.2.6 without the restriction that O lies on b:

Theorem 7.2.9 Preservation theorem abc is a line if and only if $a*b*c*$ is a line.

Proof By Section 1.17, there exists a line d through O such that abd is a line. So, if abc is a line, we conclude from Theorem 7.2.7 that acd is a line. Applying Theorem 7.2.6, we find that $a*b*d*$ and $a*c*d*$ are lines and hence, by Theorem 7.2.8, $a*d*b*$ and $a*d*c*$ are lines, also. So by the transitivity theorem, 7.2.7, $a*b*c*$ is a line. All conclusions may be applied in the reverse order, thus showing that if $a*b*c*$ is a line, abc is also a line. ◆

309

[7.3]

Foundation of Euclidean and Non-Euclidean Geometry

Exercises

1. Show that $F_{a^*} = F_a^*$ unless O is the pole of a.
2. Is perpendicularity preserved under a snail map?
3. Show that under a snail map no line is mapped onto a perpendicular of itself.
4. Carry out explicitly the proof of Theorem 7.2.8.
5. Show that the image of a metric plane under a snail map can also be considered a metric plane. (Find appropriate definitions for perpendicularity and reflections.)
6. Show that if under a snail map not every point of the metric plane has an inverse image the metric plane is neither elliptic nor Euclidean.

7.3   Pencils of Lines

Let $\alpha = ab$ be an arbitrary product of two different lines (reflections). We denote the set of all lines x such that abx is also a line by

$$L(\alpha)$$

and call this set a *pencil of lines*. If α is a point P (that is, if $a \perp b$), the pencil $L(\alpha)$ consists of all lines through P. It is a pencil of the first kind (Fig. 7.7), also called a *proper pencil*. Otherwise it is a pencil of the second or third kind (Section 1.16).

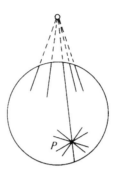

Figure 7.7

The set of all lines perpendicular to a line s is called a *pencil of perpendiculars* and is denoted by $L(s)$. So whenever α is involutoric, $L(\alpha)$ is the pencil of all x such that αx is involutoric.

Theorem 7.3.1 *If a, b, c belong to $L(\alpha)$, then abc is a line.*

Proof Let $\alpha = uv$. We may assume $a \neq u, v$. Now uva, uvb, uvc are lines. By the transitivity theorem (7.2.7) uab, uac are lines, and hence, by Theorem 7.2.8, aub, auc are lines. Applying Theorem 7.2.7 once more, we find that abc is a line. ◆

Theorem 7.3.2 *If a, b lie in $L(\alpha)$, $a \neq b$, and if abc is a line, then c lies in $L(\alpha)$.*

Proof Again let $\alpha = uv$. Consider the pencil $L(ab)$. By assumption, abc and $bvu = uvb = \alpha b$ are lines. Since $abb = a$ and since $(ab)(bvu) = avu = uva = \alpha a$ are also lines, c, b and $b\alpha^{-1}$ $(= bvu)$ belong to $L(ab)$. By Theorem 7.3.1, $cb(b\alpha^{-1}) = c\alpha^{-1}$ is a line d, and hence $d = d^{-1} = (c\alpha^{-1})^{-1} = \alpha c$ is also a line. Therefore c lies in $L(\alpha)$. ◆

Theorem 7.3.3 *If a, b both belong to $L(\alpha)$ and $L(\beta)$, then $a = b$ or $L(\alpha) = L(\beta)$; that is, any pencil $L(\alpha)$ is uniquely determined by two of its lines.*

Proof Assume $a \neq b$ and let x be an arbitrary line of the pencil $L(\alpha)$. By Theorem 7.3.1, abx is a line; hence x belongs to $L(ab)$. Conversely, let y belong to $L(ab)$. By Theorem 7.3.2, y lies also in $L(\alpha)$. Hence $L(\alpha) = L(ab)$. In the same way $L(\beta) = L(ab)$ is shown. Therefore, $L(\alpha) = L(\beta)$. ◆

Theorem 7.3.4 *If a, b, c belong to L(α), then abc also belongs to L(α).*

Proof $a \neq b$ can be assumed. In the proof of Theorem 7.3.3, a and b may be interchanged so that $L(ab) = L(\alpha) = L(ba)$. Hence x lies in $L(\alpha)$ if and only if *bax* is a line. Now Theorem 7.3.4 follows from $ba(abc) = c$. ◆

Now we shall apply snail maps to pencils of lines. Let * be defined as in the preceding sections. Since, by Theorem 7.2.9, the relation "*abc* is a line" remains unchanged if a snail map is applied, every pencil is mapped into another pencil of lines.

First we look at those pencils $L(\alpha)$ whose lines have a common perpendicular s passing through O. We may set $L(\alpha) = L(s)$. We call such a pencil **conjugate** to O (or to the pencil of all lines through O).

Theorem 7.3.5 *The snail map * interchanges pencils conjugate to O; that is, it leaves the set of all pencils conjugate to O invariant.*

Proof By Theorem 7.2.4, all images of lines perpendicular to s are perpendicular to $s*$ (O on s and $s*$). This implies the theorem. ◆

Theorem 7.3.6 *Given an arbitrary pencil L(α) that is not conjugate to O, there exists a snail map * (with O* = O) such that the image of L(α) is contained in a proper pencil (all lines passing through a point).*

Proof By Section 1.17, there exists a line a in $L(\alpha)$ that passes through O. Let b be another line of $L(\alpha)$. We denote the perpendicular of b through O

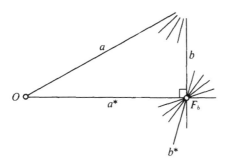

Figure 7.8

by $a*$. Clearly a is not perpendicular to $a*$, since otherwise $L(\alpha)$ would be conjugate to O. So $aa*$ defines a snail map transforming $L(\alpha) = L(ab)$ into $L(a*b*) = L(F_b)$ (compare also the proof of Theorem 7.2.7). ◆

Theorem 7.3.7 *Different pencils are mapped into different pencils under any snail map.*

Proof By Theorem 7.2.3, no two lines are mapped onto the same line. If, therefore, two pencils are mapped onto the same pencil, it is readily seen that any pencil is mapped onto this pencil, contradicting the definition of a snail map. ◆

Theorem 7.3.8   *Under a snail map every proper pencil is mapped onto a proper pencil.*

Proof   From the definition of a snail map it is clear that any proper pencil is mapped onto a proper pencil (compare Fig. 7.2).

It is left to show that every line b' through the image F_a. of a point F_a occurs as the image of a line b through F_a. We obtain b by dropping first the perpendicular $p_{b'}$ from O onto b', and then dropping the perpendicular b

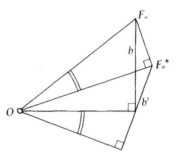

Figure 7.9

from F_a onto the inverse image of $p_{b'}$ under * (see Fig. 7.9, compare Fig. 7.5). By Hjelmslev's theorem (Section 1.17), $b* = b'$. ◆

Theorem 7.3.9   *Under a snail map *, every pencil of lines contains the image of a pencil of lines.*

Proof   Let a' join the given pencil $L(\alpha)$ to O, and let P be a point whose image $P*$ under the snail map does not lie on a'. We join $P*$ and $L(\alpha)$ by a line b'. Then $L(\alpha) = L(a'b')$. We may assume $L(\alpha) \neq L(O)$ so that b' does not pass through O. Let a be the inverse image of a' under *. (The existence of the inverse image of a' is readily seen with the aid of Theorem 7.3.8.)   By Theorem 7.3.8, b' also possesses an inverse image b under *.   Clearly, $a \neq b$ and so $L(ab)$ is a pencil that is mapped into $L(\alpha)$. ◆

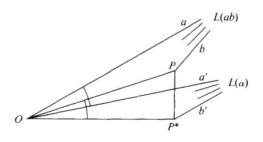

Figure 7.10

Theorem 7.3.10 *If * and ° are two snail maps with center O, then $a^{*\circ} = a^{\circ*}$ for any line a.*

Proof This is another conclusion from Hjelmslev's theorem. ◆

Exercises

1. Consider the set of all inverse images of points and lines of a metric plane under a given snail map *. Show that this set is also a metric plane.

2. Show that a hyperbolic metric plane possesses infinitely many points and lines. (*Hint*: Find a snail map that maps an improper pencil onto a proper one and apply this map over and over again.)

3. Show that in a Euclidean metric plane the product of a translation and a snail map is also a snail map.

4. Is the statement of Exercise 3 also true for non-Euclidean planes?

5. Carry out in detail the proof of Theorem 7.3.10.

6. Can a product of two different snail maps with a common center also be a snail map?

7.4   *Embedding in a Projective Plane*

Any pencil of lines is called an ***ideal point***. If, in particular, the pencil consists of lines passing through a point P of the metric plane, we say the ideal point (pencil) $L(P)$ is a ***proper point***. So the set of points of the metric plane can be considered a subset of the set of all ideal points. In the case of an elliptic plane, the two sets coincide.

Let a be an arbitrary line of the metric plane. We call the totality of all ideal points to which a belongs an ***ideal line*** or, more specifically, a ***proper ideal line*** $l(a)$.

We choose a point O of the metric plane. By the ***inverse image*** of a proper ideal line under a snail map with center O, we mean the set of all pencils which are mapped into (not necessarily onto) an ideal point of the given proper ideal line.

The inverse image of an arbitrary proper ideal line is also called an ***ideal line***. Furthermore, define the set of all pencils $L(s)$ where s passes through O as an ***ideal line*** to be the ***polar line*** $l(O)$ of O. Since the image of any pencil under a snail map * is contained in another pencil, we may consider the latter pencil as a whole the image of the former one under *. So any snail map transforms the set of ideal points into itself. Since the inverse image of a pencil is also a pencil, this map is one-to-one.

Theorem 7.4.1 *Under a snail map * any proper ideal line is mapped onto a proper ideal line, that is, $l*(a) = l(a*)$.*

Proof This follows directly from the definition of ideal points and Theorem 7.3.8. ◆

Theorem 7.4.2 *Under a snail map * the image of any ideal line is an ideal line.*

Proof If the ideal line is the polar line of O, it remains, by Theorem 7.3.5, invariant under *. Any other ideal line l is the inverse image of a proper ideal line $l'(a)$ under a snail map $°$ with center O. By Theorem 7.3.10, $l*° = l°* = (l'(a))*$, which is a proper ideal line. Therefore, $l°*$, being the inverse image of a proper ideal line, is an ideal line. ◆

Theorem 7.4.3 *If l_1, l_2 are any two different ideal lines, there exists one and only one ideal point which belongs to both, l_1 and l_2.*

Proof If $l_1 = l(O)$, then l_2 is not equal to $l(O)$ and thus it is the inverse image of a proper ideal line $l(a)$ under a snail map $°$. There is a unique perpendicular b of a through O. The inverse image under $°$ of the pole $L(b)$ of b is common to l_1 and l_2 and evidently is the only common point of l_1, l_2.

Let l_1 and l_2 both be not equal to $l(O)$. There exists a snail map * (center O) mapping l_1 onto a proper ideal line $l(a)$. Furthermore, there is a snail map $°$ (center O) mapping l_2^* onto a proper ideal line $l(b) = l_2^{*°}$. By Theorem 7.4.1, $l_1^{*°} = l°(a) = l(a°)$ is also a proper ideal line. $a° \neq b$, since, otherwise, $l_1 = l_2$ would follow. The pencil $L(a°b)$ determined by $a°$ and b is common to $l(a°)$ and $l(b)$. By the definition of ideal points, $L(a°b)$ is the only ideal point common to $l(a°)$ and $l(b)$. The pencil L for which $L^{*°} = L(a°b)$ belongs to l_1 and l_2. By Theorems 7.3.7 and 7.3.9, L is uniquely determined. ◆

Theorem 7.4.4 *Any two different ideal points are joined by one and only one ideal line.*

Proof We need only show the existence of a line joining two ideal points L, L'; the uniqueness follows from Theorem 7.4.3. If both, L and L' belong to the polar $l(O)$ of O, there is nothing to prove. So let L not lie on $l(O)$. Then there exists a snail map * (center O) and perpendicular lines a, b such that $L^* = L(ab)$.

By Theorem 1.17.3 L^* and L'^* have a line c in common and thus are joined by a proper ideal line l. Hence L and L' are joined by the inverse image of l under *. ◆

By Theorems 7.4.3 and 7.4.4 and the existence properties of a metric plane we have (see Section 5.1) the following theorem.

Theorem 7.4.5 *Ideal points and ideal lines satisfy the axioms of a projective plane.*

This theorem provides the "embedding" of a metric plane into a projective one which was our goal. The projective plane thus obtained is minimal in the following sense.

Theorem 7.4.6 *Any ideal line passing through a proper ideal point is a proper ideal line.*

Exercises

1. Show that any ideal line not containing $L(O)$ may be characterized as the set of all inverse images of an ideal point L under all snail maps with center O, together with L itself and a point on $l(O)$, provided it is not equal to $l(O)$.

2. Given a snail map * find a second snail map ° such that the composed map *° is a central collineation of the ideal projective plane (both snail maps having center O).

3. Show that the snail maps with center O and their inverses generate a commutative group Sn_0 of collineations of the ideal projective plane.

4. Let a, u, b be lines through O, let Sn be the snail map defined by au (mapping a onto u), and let Sn' be the snail map defined by bu (mapping b onto u). We let

$$\overset{a}{\underset{b}{Su}} = SnSn'^{-1}.$$

This product is an element of Sn_0.   (a) Describe how the image under $\overset{a}{\underset{b}{Su}}$ of any point of a is found metrically.   (b) Show that any line through O can be mapped onto any other line through O by such a map.

5. Show that any element S of Sn_0 (see Exercise 3) that possesses a fixed line a through O is a central collineation.   (*Hint*: Given an arbitrary line x through O, apply $\overset{x}{\underset{a}{Su}} \, S \, \overset{x}{\underset{a}{Su}}^{-1}$ for some u and use the commutativity of Sn_0.)

6. Show that any element of Sn_0 that has a fixed point different from O and not on $l(O)$, is the identity map.

7.5   *A Theorem of Hessenberg*

Our next goal is to prove Pappus' theorem in the ideal projective plane introduced in Section 7.4.   As a preparation, we prove a theorem on metric planes which (for Euclidean planes) is due to Hessenberg.   We begin with a lemma.

Lemma 7.5.1 (*Lemma on* 9 *lines*) *Let the group of all motions of a metric plane be given (generated by reflections). If* α_1, α_2, α_3, β_1, β_2, β_3 *are group elements,* $\alpha_1 \neq \alpha_2$; $\beta_1 \neq \beta_2$, *and if the eight products* $\alpha_i\beta_h$; $i, h = 1, 2, 3$ *except* $\alpha_3\beta_3$ *are lines* (= *reflections*) *then* $\alpha_3\beta_3$ *is also a line. Symbolically we may express this in the following table:*

	β_1	β_2	β_3
α_1	$\times$	$\times$	$\times$
α_2	$\times$	$\times$	$\times$
α_3	$\times$	$\times$	0

If all products designated by $\times$ *are lines, then the product designated by* 0 *is also a line.*

Proof We may assume $\beta_3 \neq \beta_1$, since, otherwise, $\alpha_3\beta_3 = \alpha_3\beta_1$ is already a line. The following identities hold in any group:

$$\beta_1^{-1}\beta_2 = \beta_1^{-1}\alpha_1^{-1}\alpha_1\beta_2 = (\alpha_1\beta_1)^{-1}(\alpha_1\beta_2)$$
$$= \beta_1^{-1}\alpha_2^{-1}\alpha_2\beta_2 = (\alpha_2\beta_1)^{-1}(\alpha_2\beta_2) \qquad (1)$$
$$= \beta_1^{-1}\alpha_3^{-1}\alpha_3\beta_2 = (\alpha_3\beta_1)^{-1}(\alpha_3\beta_2);$$

and also

$$\beta_1^{-1}\beta_3 = (\alpha_1\beta_1)^{-1}(\alpha_1\beta_3) = (\alpha_2\beta_1)^{-1}(\alpha_2\beta_3)$$
$$= (\alpha_3\beta_1)^{-1}(\alpha_3\beta_3) \qquad (2)$$

Since $\beta_1 \neq \beta_2$, β_3 and since $\alpha_i\beta_k$ is assumed to be a line a_{ik}, except for $i = k = 3$, (1) and (2) imply that $\beta_1^{-1}\beta_2$ and $\beta_1^{-1}\beta_3$ define pencils

$$L(a_{11}a_{12}) = L(a_{21}a_{22}) = L(a_{31}a_{32})$$

and

$$L'(a_{11}a_{13}) = L'(a_{21}a_{23}),$$

respectively. Clearly $a_{11} \neq a_{21}$ and thus a_{11}, a_{21} belong to both pencils. By Theorem 7.3.3, $L(a_{11}a_{21}) = L'(a_{11}a_{13})$ and thus, by Theorem 7.3.4, $\alpha_3\beta_3 = a_{31}a_{21}a_{23}$ is a line (by (2)). ◆

We say: x and y are in **symmetric position with respect to the lines** a, b if (Fig. 7.11)

$$ax = yb.$$

Figure 7.11

Theorem 7.5.2   Theorem of Hessenberg   *Let a_i, b_i, c_i, l be lines, $i = 1, 2, 3$, such that*

1. *$a_1 a_2 a_3$ is a line;*
2. *$a_1 a_2 = b_2 b_1$ and $a_1 a_3 = b_3 b_1$;*
3. *$a_k c_k c_1$ is a line for any permutation i, k, l of $1, 2, 3$;*
4. *$c_1 b_1 \neq c_2 b_2$.*

If $c_1 b_1 l$ and $c_2 b_2 l$ are lines, $c_3 b_3 l$ is also a line (Fig. 7.12).

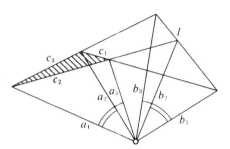

Figure 7.12

Proof   We consider the following table of products:

	$b_1 a_3 c_2$	$b_2 a_3 c_1$	l
$c_1 b_1$	$c_1 a_3 c_2$	$c_1 (b_1 b_2 a_3) c_1$	$c_1 b_1 l$
$c_2 b_2$	$c_2 (b_2 b_1 a_3) c_2$	$c_2 a_3 c_1$	$c_2 b_2 l$
$c_3 b_3$	$c_3 (b_3 b_1 a_3) c_2$	$c_3 (b_3 b_2 a_3) c_1$	$c_3 b_3 l$

By assumption 4, $c_1 b_1 \neq c_2 b_2$.   From $(a_1 a_2 a_3)^2 = 1$, we conclude $a_1 a_2 a_3 = a_3 a_2 a_1$, and using $a_1 a_2 = b_2 b_1$, $a_1 a_3 = b_3 b_1$, we find

$$a_2 a_3 = a_1 (a_1 a_2 a_3) = a_1 (a_3 a_2 a_1) = (b_3 b_1)(b_1 b_2) = b_3 b_2.$$

From $a_1 a_3 = b_3 b_1$ we obtain $b_1 a_3 = b_3 a_1$, and from $a_2 a_3 = b_3 b_2$ we have $b_2 a_3 = b_3 a_2$.

Suppose $b_1 a_3 c_2 = b_2 a_3 c_1$; then $b_3 a_1 c_2 = b_3 a_2 c_1$ and hence $a_1 c_2 = a_2 c_1$; also $a_1 a_2 = c_2 c_1 = b_2 b_1$, contrary to $c_1 b_1 \neq c_2 b_2$.   Therefore, $b_1 a_3 c_2$ is not equal to $b_2 a_3 c_1$.   By Section 1.16, all lines a_1, a_2, a_3, b_1, b_2, b_3 belong to one pencil.   Using this fact together with assumption 3, it is now seen that Lemma 7.5.1 may be applied to the above table.   So $c_3 b_3 l$ is a line.   ◆

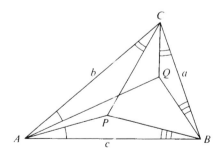

Figure 7.13

Exercises

1. In an ordinary hyperbolic or Euclidean metric plane, let P, Q be two different points in the interior of a triangle $\triangle ABC$, and suppose $\angle CAQ = \angle PAB$ and $\angle ACP = \angle QCB$. Prove that $\angle ABP = \angle QBC$ (see Fig. 7.13).

2. State the theorem of Exercise 1 for arbitrary metric planes and show that it is true in this generality.

3. Prove anew the theorem of angular bisectors (compare Exercise 1).

4. Is the theorem of Exercise 1 also true for P, Q not in the interior of $\triangle ABC$?

5. Prove and explain the geometrical meaning of the following theorem: Let $da' = a''a$; $ab' = b''b$; $bc' = c''c$; and $cd' = d''d$. If $a'b' = d'c'$, then $a''b'' = d''c''$.

6. We replace, in Hessenberg's theorem, the lines $c_1c_2c_3$, l by points C_1, C_2, C_3, L. The theorem thus obtained is proved in strict analogy to Hessenberg's theorem.
   (a) Explain the geometrical meaning of this new theorem (Fig. 7.14).
   (b) Carry out the proof.

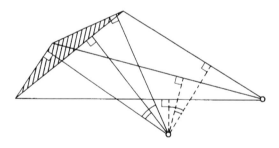

Figure 7.14

7.6   *Proof of Pappus' Theorem*

In order to prove Pappus' theorem it is sufficient to prove Brianchon's theorem (see Section 5.18). Before we carry out the proof for the ideal projective plane, we show Brianchon's theorem in a version that is valid for metric planes.

Theorem 7.6.1 *Let the pencil A contain the lines a_1, a_2, a_3; let the pencil B contain the lines b_1, b_2, b_3 but not a_1, a_2, a_3; and let c_{ik} ($i, k = 1, 2, 3; i \neq k$) be six lines of a pencil C such that no a_i is in C and such that $a_i b_k c_{ik}$ is a line. We assume B and C to be proper pencils. Then any two of the following equations imply the third one:*

$$c_{12} = c_{21}; \qquad c_{13} = c_{31}; \qquad c_{23} = c_{32}$$

[7.6]

(see Fig. 7.15).

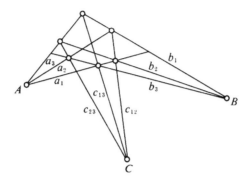

Figure 7.15

Proof C and A, B can be joined by lines a, b, respectively (Section 1.17). Let $d_{ik} = a c_{ik} b$; that is, let c_{ik}, d_{ik} be in a symmetric position to a, b.

We may assume $c_{23} = c_{32}$; $c_{13} = c_{31}$, and assert: $c_{12} = c_{21}$. From $c_{23} = c_{32}$ it follows that $d_{23} = d_{32}$ and from $c_{13} = c_{31}$ we have $d_{13} = d_{31}$ (Fig. 7.16). Since B does not contain a_1, a_2, a_3, the pencils $L(a_1 d_{23})$, $L(a_2 d_{13})$, $L(a_3 d_{12})$, and $L(a_3 d_{21})$ are different from B.

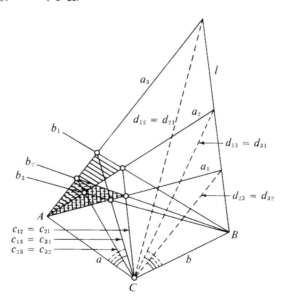

Figure 7.16

Applying Hessenberg's theorem four times (see Figures 7.12 and 7.16), we see that B, $L(a_1 d_{23})$, $L(a_2 d_{13})$ are on a line; B, $L(a_1 d_{23})$, $L(a_3 d_{12})$ are on a line; B, $L(a_2 d_{13})$, $L(a_3 d_{12})$ are on a line; and B, $L(a_2 d_{13})$, $L(a_3 d_{21})$ are on a line. This readily implies that there exists a common line l of B, $L(a_1 d_{23})$, $L(a_2 d_{13})$, $L(a_3 d_{12})$, and $L(a_3 d_{21})$. Clearly $l \neq a_3$ (since a_3 is not in B). Since $L(a_3 d_{12})$ and $L(a_3 d_{21})$ both contain a_3 and l, we conclude $L(a_3 d_{12}) = L(a_3 d_{21})$ and thus $d_{12} = d_{21}$ (otherwise $C = L(a_3 d_{12})$ and, therefore, a_3 in C would follow). Now

$$c_{12} = ad_{12}b = ad_{21}b = c_{21}. \qquad \blacklozenge$$

Theorem 7.6.2   *In the ideal projective plane, Pappus' theorem is valid.*

Proof   By Sections 3.13 and 5.1, we need only show the affine version of Brianchon's theorem—that is, show it for one configuration line chosen as line at infinity. We assume the polar $l(O)$ of O to be this line.

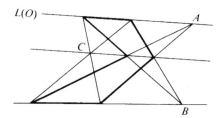

Figure 7.17

We are given a hexagon the sides of which meet alternatively in A, B (see Fig. 7.17). It is to be shown that the diagonals of this hexagon meet in a point C. One side of the hexagon which passes through A may be chosen as the "line at infinity" $l(O)$. Now B and C are not on $l(O)$ and may thus, by a product of two snail maps (center O), be mapped onto proper points. Applying Theorem 7.6.1 and mapping back, we see that C lies on the third diagonal line. $\blacklozenge$

So far, the construction of ideal lines has been dependent on the special choice of O. With the aid of Pappus' theorem, we now are able to show that we obtain the same ideal lines no matter which O we choose.

Theorem 7.6.3   *The definition of ideal lines and, thus, of the ideal projective plane is independent of the choice of O.*

Proof   Let A_1, A_2, A_3 be different and lie on an ideal line. We choose a proper ideal point B_2 not on this line. The lines joining B_2 and A_1, A_3, respectively, are proper ones (by Theorem 7.4.6); let C_3, C_1 be proper ideal points on them. On the line joining C_3 and C_1 we choose a point C_2. Joining C_2 and A_1, A_3, we obtain proper ideal lines which intersect the lines joining A_2, C_1 and A_2, C_3 in ideal points B_1, B_3, respectively (Fig. 7.18).

By Pappus' theorem, B_1, B_2, B_3 are on an ideal line which (by Theorem 7.4.6) is proper. The eight proper lines thus constructed provide a figure in the original metric plane independent of the point O. So if O is replaced by another point, A_1, A_2, A_3 are, by Pappus' theorem, still on an ideal line. ◆

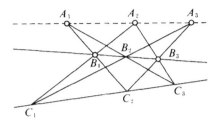

Figure 7.18

Exercises

1. Prove the following special case of Pappus' theorem without using Hessenberg's theorem: Three configuration points lie on $l(O)$; the two configuration lines not passing through these three configuration points meet in O. (*Hint*: Use the equality

$$Sc_1\genfrac{}{}{0pt}{}{a}{b}\genfrac{}{}{0pt}{}{b}{a}\genfrac{}{}{0pt}{}{a}{b}Sc_2Sc_3 = Sc_3\genfrac{}{}{0pt}{}{a}{b}\genfrac{}{}{0pt}{}{b}{a}\genfrac{}{}{0pt}{}{a}{b}Sc_2Sc_1$$

of products introduced in Section 7.4, where c_1, c_2, c_3 are chosen as indicated in Fig. 7.19.)

2. Following the ideas of Exercise 1, prove Pappus' theorem (affine version) for Euclidean metric planes.

3. Prove directly the affine Desargues' theorem with center O. (*Hint*: Use the product $Sc_1\genfrac{}{}{0pt}{}{b}{c}Sc_2\genfrac{}{}{0pt}{}{c}{a}Sc_3\genfrac{}{}{0pt}{}{a}{b}$ with a notation as indicated in Figure 7.20; apply Exercise 6 of Section 7.4.)

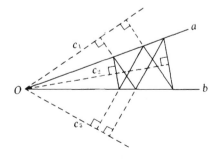

Figure 7.19

4. Prove the general theorem of Desargues.

5. Prove Theorem 7.6.3 with the aid of Desargues' theorem.

6. Let an ordered metric plane (Euclidean or hyperbolic) be given (Section 2.2), and let $\triangle$ be a triangle. Show: A closure theorem on points and lines is true in general if it is true for the configuration points lying in the interior of $\triangle$.

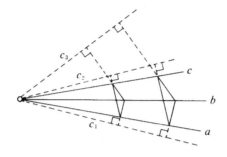

Figure 7.20

7.7 *Extended Motions*

Having embedded the metric plane into a projective plane, we may describe the motions of the metric plane in projective terms. First we show a lemma.

Lemma 7.7.1 *In a metric plane let Sn be a snail map (center O) defined by a product uv. Let c be a line through O. Then $R_c Sn R_c$ is the snail map $\widetilde{Sn}$ defined by vu (setting again R_c for the reflection c).*

Proof We denote the reflection at c by a prime and Sn by a star.

Let p be a line through O. We find ($cuv = vuc$, since cuv is a line through O):

$$(p)R_c Sn R_c = p'^{*'} = c(p'uv)c = c(cpcuv)c = pvucc = pvu = (p)\widetilde{Sn}.$$

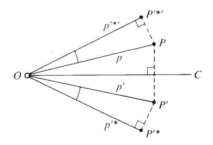

Figure 7.21

Let P be a point on p not on the pole of O (if it exists in the metric plane). P'^* is the foot of the perpendicular of p'^* through P' on p'^* and thus $P'^{*'}$ is the foot of $P'' = P$ onto $p'^{*'} = (p)\widetilde{Sn}$. Therefore,

$$P'^{*'} = (P)\widetilde{Sn}.$$

If P lies on the polar of O, then also $P'^{*'} = (P)\widetilde{Sn}$. ◆

Theorem 7.7.2 *The reflection at a line c of a metric plane induces an involutoric central collineation with center $L(c)$ and axis $l(c)$ in the corresponding ideal projective plane.*

Proof Any pencil $L(ab)$ is mapped under the given reflection R_c onto $L(a'b') = L(caccbc) = L(cabc)$. This does not depend on the choice of the pencil lines a, b, since $abx = dey$ implies

$$a'b'x' = caccbccxc = cabxc = cdeyc = d'e'y'.$$

Any proper ideal line $l(a)$ is clearly mapped onto the proper ideal line $l(cac)$. If a general ideal line $l(a)$ is given, choose a point O on c, not the pole of $l(a)$, and a snail map Sn (center O) that transforms $l(a)$ onto a proper ideal line $l(b)$. Now, by Lemma 7.7.1,

$$(l(a))R_c = (l(a))SnR_c\widetilde{Sn}^{-1} = (l(b))R_c\widetilde{Sn}^{-1} = (l(b'))\widetilde{Sn}^{-1},$$

which, by Theorems 7.3.8 and 7.4.2 is again an ideal line. Clearly, $l(c)$ remains pointwise fixed under R_c and also $L(c)$ remains invariant under R_c. ◆

Theorem 7.7.2 implies (see Section 5.6 for projective collineations) the following theorem.

Theorem 7.7.3 *Any motion of a metric plane induces a projective collineation in the corresponding ideal projective plane.*

Later on we shall need the following theorem:

Theorem 7.7.4 *In the ideal projective plane, the "Fano-Axiom" holds: The diagonal points of a complete quadrilateral are not collinear.*

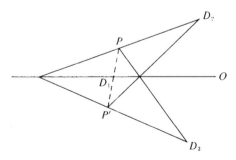

Figure 7.22

Proof We have seen in Section 5.10 that there exist involutoric harmonic homologies. As is readily seen from Figure 7.22, this implies the existence of complete quadrilaterals in which the diagonal points D_1, D_2, D_3 are not collinear. ◆

Exercises

1. Classify all collineations of the ideal projective plane which are induced by motions of the metric plane (use Section 1.18).

2. Describe, in an ordinary Euclidean plane, the group of all collineations generated by all snail maps with the same center O and all reflections at lines through O.

3. Show that if, in a metric plane, $(ab)^2 \neq 1$, then $L((ab)^2) = L(ab)$.

4. Show that if $(abc)^2 \neq 1$ in a metric plane, then $w(abc)w = bac$ is equivalent to w being perpendicular to c and lying in $L(ab)$.

5. Interpret the result of Exercise 4 geometrically.

6. Carry out explicitly the proof of Theorem 7.7.4.

7.8   *Perpendiculars and Polars*

In an elliptic plane, perpendicularity induces a polarity in the sense of Section 5.12. This follows from the results of Section 1.12. We shall show also that, in the case of a hyperbolic plane, perpendicularity induces a polarity in the corresponding ideal projective plane. However, for Euclidean planes, such a result cannot be expected; the "poles" $L(a)$ of all lines lie on one line (line at infinity). We obtain only a "degenerate" kind of polarity. So the Euclidean planes require a separate treatment when polarity is involved. In fact, it is useful to subdivide the set of all metric planes into two classes, those in which there exist rectangles (representing the Euclidean case) and those in which there are no rectangles (representing the non-Euclidean case):

We call a metric plane *metric-Euclidean* if it satisfies the following additional axiom.

Axiom **R**   *There exist lines a, b, c, d with $a \neq b$; $c \neq d$ such that $a \perp c$; $a \perp d$; $b \perp c$; $b \perp d$* (Fig. 7.23).

Figure 7.23

We say the metric plane is ***metric-non-Euclidean*** if the logical contrary of Axiom R is true:

Axiom $\overline{\text{R}}$ *There exist no four lines a, b, c, d with $a \neq b$; $c \neq d$ such that $a \perp c$; $a \perp d$; $b \perp c$; $b \perp d$.*
So every metric plane is either metric-Euclidean or metric-non-Euclidean.

Theorem 7.8.1 *Let $A \neq B$. The product ABC is a point D if and only if there exists a line through C which has the same perpendiculars as the line through A, B has* (Fig. 7.24).

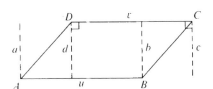

Figure 7.24

Proof I. Let ABC be a point D so that $AB = DC$. From $A \neq B$ we conclude $D \neq C$. Let u join A, B and let v join C, D. We let $A = au$; $B = bu$; $C = cv$; $D = dv$. By Theorem 1.17.1, a line x is perpendicular to u if and only if AxB is a reflection. This is equivalent to postulating that axb is a line or (by the transitivity theorem, see Section 1.16, Exercise 1) that abx is a line. But

$$abx = auubx = ABx = DCx = dvvcx = dcx,$$

and so x is perpendicular to u if and only if it is perpendicular to v.
    II. Let $x \perp u$ be equivalent to $x \perp v$. We find $ABC = auubcv = abcv = dv$ (by the second theorem on three reflections, $abc = d$). Since $d \perp u$, we have, by assumption, $d \perp v$ and hence dv is a point D. ◆

Theorem 7.8.2 *If two lines possess two different perpendiculars in common, they have all perpendiculars in common.*

Proof This follows directly from Theorem 1.16.2 ◆

Theorem 7.8.3 *If Axiom R is satisfied, $a \perp c$, $b \perp c$, and $a \perp d$ imply $b \perp d$.*

Proof Let a_0, b_0, c_0, d_0 be the sides of the rectangle which exists according to Axiom R (Fig. 7.25). Let U, V be different points on a, and let u, v be the perpendiculars of a_0 through U, V, respectively. If $a = u$ ($= v$), we interchange the roles of a and c (the assumptions of our theorem are symmetric in a, c). So we may suppose $a \neq u, v$. By Theorem 7.8.2, $u \perp b_0$; $v \perp b_0$, and so, by Theorems 7.8.1 and 7.8.2 for a point $W \neq U$ on u we conclude that WUV is a point. Again applying Theorems 7.8.1

and 7.8.2, we see that there is a line w through W which has the same perpendiculars as a has. Clearly $w \neq a$ and thus, by Theorem 7.8.2, $b \perp d$. ◆

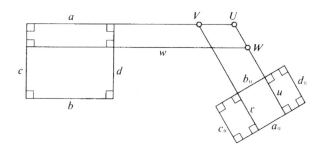

Figure 7.25

Theorem 7.8.4 *If Axiom R is true, any two pencils of the second kind (consisting of perpendiculars of a line) that have a line in common are identical.*

Proof This follows directly from Theorem 7.8.3. ◆

Theorem 7.8.5 *If Axiom R is true, ABC is a point for any three points A, B, C.*

Proof For $A = B$ this is trivial; for $A \neq B$ it is seen with the aid of Theorems 7.8.1 and 7.8.3. ◆

We turn again to general metric planes. If $l(a)$ is a proper ideal line, we call $L(a)$ the **pole** of $l(a)$ and $l(a)$ the **polar** (**line**) of $L(a)$. Similarly, $l(P)$ is said to be the **polar** (**line**) of the proper ideal point $L(P)$, and $L(P)$ is the **pole** of $l(P)$. (See Fig. 7.26.)

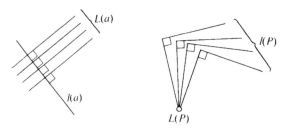

Figure 7.26

Theorem 7.8.6 *If $l(a)$, $l(b)$, $l(c)$ have a point in common, their poles $L(a)$, $L(b)$, $L(c)$ are collinear.*

Proof Let P be the common point of $l(a)$, $l(b)$, $l(c)$. If P is a proper ideal point, we may, by Theorem 7.6.3, choose it as a point O in the construction of the ideal projective plane. Then, by definition, $L(a)$, $L(b)$, $L(c)$ are on a line.

If P is not a proper ideal point, we choose a point O such that P does not lie on $l(O)$. By a snail map *, P may be shifted to a proper ideal point, and so $L(a^*)$, $L(b^*)$, $L(c^*)$ are on a line. Since snail maps induce collineations in the ideal projective plane (see Exercise 2), this implies that $L(a)$, $L(b)$, $L(c)$ are also on a line. ◆

The converse of Theorem 7.8.6 is, in general, not true: $L(a)$, $L(b)$, $L(c)$ may be collinear without $l(a)$, $l(b)$, $l(c)$ having a point in common. The following two theorems will clarify this in detail.

Theorem 7.8.7 *If Axiom R is true, all proper ideal lines have their poles on the same line (line at infinity).*

Proof Let O be a point of the metric plane. The line $l(O)$ consists of all poles of lines $l(a)$ with O on a. Let b be an arbitrary line of the metric plane. Letting $O = p_b b'$, we conclude from Theorem 7.8.3 that any perpendicular of b' is also a perpendicular of b, and conversely. Therefore $L(b) = L(b')$. ◆

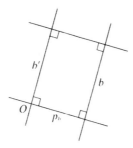

Figure 7.27

Theorem 7.8.8 *If Axiom R̄ is true and if $L(a)$, $L(b)$, $L(c)$ are collinear, then $l(a)$, $l(b)$, $l(c)$ have a point in common.*

Proof Suppose a, b, c have no point in common. We may assume that a, b intersect in a proper point P, possibly after applying a snail map (Fig. 7.28). Let p be a perpendicular of c through P and let $P = pc'$. Clearly,

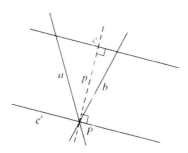

Figure 7.28

$L(a)$, $L(b)$, $L(c')$ are different points on the same ideal line $l(P)$. Now the intersection of $l(p)$ and $l(P)$ equals both $L(c)$ and $L(c')$. So c and c' have all perpendiculars in common. Since $c \neq c'$, this implies the existence of a rectangle, contrary to Axiom $\bar{\mathrm{R}}$. ◆

In case Axiom R holds, we may assign a pole to every ideal line except the line at infinity, which is the only nonproper ideal line. In case Axiom $\bar{\mathrm{R}}$ holds, Theorem 7.8.8 guarantees a one-to-one correspondence between the set of ideal points and the set of ideal lines. In both cases, we can show the following theorem.

Theorem 7.8.9 *Given a proper ideal point $L(P)$, the mapping*

$$l(a) \longrightarrow L(a)$$

defined for all lines a passing through P is a one-to-one map of the set of all ideal lines through P onto the set of all points on the polar $l(P)$ of $L(P)$.

Proof We may, by Theorem 7.6.3, choose P as point O. Theorem 7.8.9 then follows by the definition of the line $l(O)$. ◆

Exercises

1. Show that Axiom R is equivalent to the following: Let A, B, C be the vertices of a triangle with right angle at C; let B' be the midpoint of A, B and let C' be the midpoint of A, C. Then the line through A, C is perpendicular to the line through C', B'.

2. Show explicitly that snail maps induce collineations in the "ideal" projective plane (compare Section 7.4).

3. Show that Axiom **R** and Axiom **Euc** (Section 1.5) are equivalent.

4. Show that, for any Euclidean or hyperbolic metric plane, proper ideal lines have improper poles and proper ideal points have improper polars.

5. Show that a metric plane is Euclidean if and only if all poles of proper ideal lines lie on one line.

7.9   *Bilinear Forms in Pencils*

We introduce homogeneous coordinates in the ideal projective plane. Since Pappus' theorem holds, this is possible (see Sections 5.4 and 3.13). So there exists a three-dimensional vector space $\mathscr{V} = \mathscr{V}_3(\mathscr{K})$ over a field $\mathscr{K}$ such that the one-dimensional and two-dimensional subspace of $\mathscr{V}$ correspond to the points and lines of the projective plane, respectively. Any two-dimensional subspace is given by an equation

$$\mathbf{ux} = 0 \tag{1}$$

with a constant vector $\mathbf{u} = \mathbf{0}$. Any multiple $a\mathbf{u}$ with $a \neq O$ represents the same plane. As in Section 5.4, we denote the set of all multiples of a vector $\mathbf{x}$ by $\mathscr{K}\mathbf{x}$, and we have the following one-to-one correspondence ($\mathbf{x} \neq \mathbf{0}$; $\mathbf{u} \neq \mathbf{0}$):

$$\mathscr{K}\mathbf{x} \longrightarrow \text{point of the projective plane,}$$
$$\mathscr{K}\mathbf{u} \longrightarrow \text{line of the projective plane.}$$

A point is on a line if and only if (1) holds. We consider now the set $\mathscr{P}$ of all those vectors $\mathbf{x}$ for which $\mathscr{K}\mathbf{x}$ is the pole of a proper ideal line, together with $\mathbf{0}$. Also let $\mathscr{L}$ be the set of all vectors $\mathbf{u}$ such that $\mathscr{K}\mathbf{u}$ represents a proper ideal line, together with $\mathbf{0}$. (Concerning Fig. 7.29 see Exercise 1

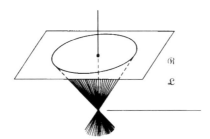

Figure 7.29

below.) By $\mathscr{L}_0$ and $\mathscr{P}_0$ we denote the sets of one-dimensional subspaces of $\mathscr{V}$ lying in $\mathscr{L}$ and $\mathscr{P}$, respectively (that is, the homogeneous coordinates of proper ideal lines and their poles). Assigning to every proper ideal line $l(a)$ its pole $L(a)$, we obtain a mapping

$$\psi: \mathscr{L}_0 \longrightarrow \mathscr{P}_0$$

of $\mathscr{L}_0$ onto $\mathscr{P}_0$. If Axiom $\bar{\mathrm{R}}$ holds, this mapping is one-to-one; if Axiom R holds, it is not one-to-one.

We wish to extend ψ to a mapping of $\mathscr{L}$ onto $\mathscr{P}$. This can be done in many ways. For example, one may pick out at random a vector $\mathbf{u}_1 \neq \mathbf{0}$ of any one-dimensional subspace $\mathscr{K}\mathbf{u}$ of $\mathscr{L}_0$ and assign to it an arbitrarily chosen vector $\mathbf{x}_1 \neq \mathbf{0}$ of the subspace $\mathscr{K}\mathbf{x}$ of $\mathscr{V}$ corresponding to $\mathscr{K}\mathbf{u}$ under ψ. After having done this for all $\mathscr{K}\mathbf{u}$, one can map

$$a\mathbf{u}_1 \longrightarrow a\mathbf{x}_1$$

for all $a \in \mathscr{K}$ and thus extend ψ to a mapping φ of $\mathscr{L}$ onto $\mathscr{P}$. We call such a φ a **refinement** of ψ. Clearly, φ is linear in any one-dimensional subspace $\mathscr{K}\mathbf{u}$ of $\mathscr{L}$, but, on the whole, it is not linear. If one of the vectors $\mathbf{u}_1$ is replaced by another vector $\mathbf{u}_2 \neq \mathbf{0}$ of $\mathscr{K}\mathbf{u}$, we obtain again a refinement φ' of ψ. We say: φ' is obtained from φ by a **linear change in $\mathscr{K}\mathbf{u}$**.

Another refinement is deduced from φ by letting

$$(\mathbf{u})(a\varphi) = a \cdot (\mathbf{u})\varphi \quad \text{for all } \mathbf{u},$$

where $a \neq 0$ is a constant. So $a\varphi$ is a "multiple" of φ.

It is also possible to choose φ in such a way that it has a linear extension from $\mathscr{L}$ onto all of $\mathscr{V}$. In fact, we are aiming at such a map (see next section). We let

$$B_\varphi(\mathbf{x}, \mathbf{y}) = \mathbf{x}(\mathbf{y}\varphi),$$

where $\mathbf{x}$ is an arbitrary vector of $\mathscr{V}$ and $\mathbf{y}$ a vector of $\mathscr{L}$. This is a generalized bilinear form (see Section 4.7). By linear changes of φ, we shall slowly arrange B_φ to be a bilinear form. Linearity in the first argument is already clear, by definition. We find:

$$B_\varphi(\mathbf{x}_1 + \mathbf{x}_2, \mathbf{y}) = B_\varphi(\mathbf{x}_1, \mathbf{y}) + B_\varphi(\mathbf{x}_2, \mathbf{y}), \tag{2}$$

$$B_\varphi(a\mathbf{x}, \mathbf{y}) = B_\varphi(\mathbf{x}, a\mathbf{y}) = aB_\varphi(\mathbf{x}, \mathbf{y}). \tag{3}$$

For $\mathbf{x}, \mathbf{y}$ in $\mathscr{L}$,

$$B_\varphi(\mathbf{x}, \mathbf{y}) = 0 \tag{4}$$

if and only if

$$B_\varphi(\mathbf{y}, \mathbf{x}) = 0.$$

Equation (4) is seen as follows:

$$\mathbf{x} \cdot (\mathbf{y}\varphi) = 0$$

means geometrically that the proper ideal line $l(x) = \mathscr{K}\,\mathbf{x}$ passes through the pole $L(y)$ of the proper ideal line $l(y) = \mathscr{K}\mathbf{y}$ (that is, $\mathbf{x}$ and $\mathbf{y}$ are perpendicular in the metric plane). (See Fig. 7.30.)

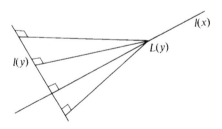

Figure 7.30

This relationship is, however, symmetric—that is, $l(y)$ also passes through the pole $L(x)$ of $l(x)$. This implies (4). By Theorem 7.7.2, a reflection at a line a of the given metric plane may be extended to an involutoric homology with center $L(a)$ and axis $l(a)$. Such a harmonic homology may, by Sections 5.11 and 5.12, be represented in $\mathscr{V}$ as

$$\mathbf{x}' = -\mathbf{x} + 2\frac{\mathbf{x} \cdot \mathbf{a}}{\mathbf{u} \cdot \mathbf{a}} \cdot \mathbf{u}, \tag{5}$$

where $l(a) = \mathcal{K}\mathbf{u}$ and $L(a) = \mathcal{K}\mathbf{a}$. If a refinement φ of ψ is given, we may let (possibly after a linear change in $\mathcal{K}\mathbf{u}$):

$$\mathbf{a} = (\mathbf{u})\varphi,$$

and so (5) obtains the form

$$\mathbf{x}' = -\mathbf{x} + 2\frac{B_\varphi(\mathbf{x}, \mathbf{u})}{B_\varphi(\mathbf{u}, \mathbf{u})}\cdot\mathbf{u}. \tag{5'}$$

Now let $\mathbf{u}, \mathbf{v}$ represent linearly independent vectors of $\mathcal{L}$ and let $\mathbf{w} = \mathbf{u} + \mathbf{v}$. We denote again by R_u, R_v, R_w the involutoric homologies (5') corresponding to the lines $\mathcal{K}\mathbf{u}, \mathcal{K}\mathbf{v}, \mathcal{K}\mathbf{w}$, respectively. Since these lines lie in a pencil, we obtain (using the transitivity theorem, 7.2.7)

$$(\mathbf{x})R_u R_v R_w = (\mathbf{x})R_w R_v R_u$$

for all $\mathbf{x}$ in $\mathcal{V}$. $\tag{6}$

This implies, by direct calculation, the following identity:

$$B_\varphi(\mathbf{u}, \mathbf{v})\cdot B_\varphi(\mathbf{v}, \mathbf{w})\cdot B_\varphi(\mathbf{w}, \mathbf{u}) = B_\varphi(\mathbf{v}, \mathbf{u})B_\varphi(\mathbf{w}, \mathbf{v})B_\varphi(\mathbf{u}, \mathbf{w}). \tag{7}$$

Note that this identity holds for all refinements φ of ψ.

Theorem 7.9.1 *If $\mathcal{U}$ is a two-dimensional subspace of $\mathcal{V}$ contained in $\mathcal{L}$ (and hence represents a proper pencil of lines), there exists a refinement φ of ψ which induces on $\mathcal{U}$ an inner product, that is, a symmetric bilinear form B_φ with the following property:*

$$B_\varphi(\mathbf{x}, \mathbf{y}) = 0 \quad \text{for all } \mathbf{y} \text{ if and only if } \quad \mathbf{x} = \mathbf{0}.$$

Proof Every vector $\mathbf{u} \neq \mathbf{0}$ of $\mathcal{U}$ represents a line of a pencil. This pencil

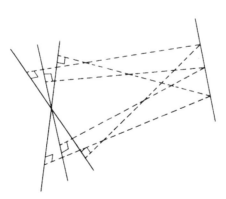

Figure 7.31

is a proper one, since it contains only proper ideal lines. Let $\mathbf{u}, \mathbf{v}$ be two vectors of $\mathcal{U}$ such that $\mathcal{K}\mathbf{u}, \mathcal{K}\mathbf{v}$ represent lines that are not perpendicular. So $B_\varphi(\mathbf{u}, \mathbf{v}) \neq 0$. By a linear change in $\mathcal{K}\mathbf{v}$ we can arrange that

$$B_\varphi(\mathbf{u}, \mathbf{v}) = B_\varphi(\mathbf{v}, \mathbf{u}). \tag{8}$$

Let $\mathbf{w} = a\mathbf{u} + b\mathbf{v}$ be any linear combination of $\mathbf{u}, \mathbf{v}$ with $a \neq 0$, $b \neq 0$. The application of Theorem 7.8.6 now brings linearity into the picture: Since $\mathcal{K}\mathbf{u}$, $\mathcal{K}\mathbf{v}$, $\mathcal{K}\mathbf{w}$ represent lines through a point, the points $\mathcal{K}(\mathbf{u})\varphi$, $\mathcal{K}(\mathbf{v})\varphi$, $\mathcal{K}(\mathbf{w})\varphi$ are collinear:

$$(\mathbf{w})\varphi = a'(\mathbf{u})\varphi + b'(\mathbf{v})\varphi. \tag{9}$$

Multiplying (9) for $a' \neq 0$ by aa'^{-1} yields

$$(\mathbf{w})(aa'^{-1}\varphi) = a(\mathbf{u})\varphi + b''(\mathbf{v})\varphi,$$

where $b'' = aa'^{-1}b'$. For every $\mathcal{K}\mathbf{w} \neq \mathcal{K}\mathbf{u}$, $\mathcal{K}\mathbf{v}$, we change φ linearly: We replace φ by $aa'^{-1}\varphi$ and again use φ for the changed refinement. So we have $(\mathbf{w})\varphi = a(\mathbf{u})\varphi + b''(\mathbf{v})\varphi$. We are going to show $b'' = b$. In fact, since $B_\varphi(\mathbf{u}, \mathbf{v}) = B_\varphi(\mathbf{v}, \mathbf{u}) \neq 0$, we obtain from the identity (7):

$$B_\varphi(\mathbf{v}, \mathbf{w})B_\varphi(\mathbf{w}, \mathbf{u}) = B_\varphi(\mathbf{w}, \mathbf{v})B_\varphi(\mathbf{u}, \mathbf{w}).$$

We now evaluate this identity (using Equations (2) and (3)):

$$\begin{aligned}
B_\varphi(\mathbf{v}, \mathbf{w})B_\varphi(\mathbf{w}, \mathbf{u}) &= (\mathbf{v} \cdot (\mathbf{w})\varphi)(\mathbf{w} \cdot (\mathbf{u})\varphi) \\
&= (\mathbf{v} \cdot (a \cdot (\mathbf{u})\varphi + b''(\mathbf{v})\varphi))((a\mathbf{u} + b\mathbf{v}) \cdot (\mathbf{u})\varphi) \\
&= a^2 B_\varphi(\mathbf{v}, \mathbf{u})B_\varphi(\mathbf{u}, \mathbf{u}) + ab B_\varphi(\mathbf{v}, \mathbf{u})^2 + ab'' B_\varphi(\mathbf{v}, \mathbf{v})B_\varphi(\mathbf{u}, \mathbf{u}) \\
&\quad + bb'' B_\varphi(\mathbf{v}, \mathbf{v})B_\varphi(\mathbf{v}, \mathbf{u}) \\
&= B_\varphi(\mathbf{w}, \mathbf{v})B_\varphi(\mathbf{u}, \mathbf{w}) \\
&= a^2 B_\varphi(\mathbf{u}, \mathbf{v})B_\varphi(\mathbf{u}, \mathbf{u}) + ab'' B_\varphi(\mathbf{u}, \mathbf{v})^2 + ab B_\varphi(\mathbf{v}, \mathbf{v})B_\varphi(\mathbf{u}, \mathbf{u}) \\
&\quad + bb'' B_\varphi(\mathbf{v}\mathbf{v})B_\varphi(\mathbf{u}, \mathbf{v}).
\end{aligned}$$

This implies

$$(b - b'')(B_\varphi(\mathbf{u}, \mathbf{v})^2 - B_\varphi(\mathbf{u}, \mathbf{u})B_\varphi(\mathbf{v}, \mathbf{v})) = 0.$$

We may conclude $b = b''$ if the second term in the product has been shown to be not equal to 0. Suppose it is equal to 0. We consider the vector

$$\mathbf{z} = B_\varphi(\mathbf{u}, \mathbf{v})\mathbf{u} - B_\varphi(\mathbf{u}, \mathbf{u})\mathbf{v}.$$

$K\mathbf{z}$ represents another line of the pencil defined by $\mathcal{K}\mathbf{u}$, $\mathcal{K}\mathbf{v}$. Since

$$\mathbf{z}(\mathbf{u}\varphi) = B_\varphi(\mathbf{u}, \mathbf{v})B_\varphi(\mathbf{u}, \mathbf{u}) - B_\varphi(\mathbf{u}, \mathbf{u})B_\varphi(\mathbf{v}, \mathbf{u}) = 0$$

(using $B_\varphi(\mathbf{u}, \mathbf{v}) = B_\varphi(\mathbf{v}, \mathbf{u})$) and since also

$$\mathbf{z}(\mathbf{v}\varphi) = B_\varphi(\mathbf{u}, \mathbf{v})^2 - B_\varphi(\mathbf{u}, \mathbf{u})B_\varphi(\mathbf{v}, \mathbf{v}) = 0$$

(from our assumption), we see that $\mathcal{K}\mathbf{z}$ is perpendicular to the lines $\mathcal{K}\mathbf{u}$, $\mathcal{K}\mathbf{v}$ which pass through a proper ideal point; hence $\mathcal{K}\mathbf{u} = \mathcal{K}\mathbf{v}$, contrary to the choice of $\mathbf{u}, \mathbf{v}$, to be linearly independent. This completes the proof that $b = b''$ and hence that φ is linear on $\mathcal{U}$. As a consequence, $B_\varphi(\mathbf{x}, \mathbf{y})$ is a bilinear form on $\mathcal{U}$. From Equation (8), it is readily deduced that

$$B_\varphi(\mathbf{x}, \mathbf{y}) = B_\varphi(\mathbf{y}, \mathbf{x})$$

for all $\mathbf{x}, \mathbf{y}$ in $\mathcal{U}$.

It is left to show that if $B_\varphi(\mathbf{x}, \mathbf{y}) = 0$ for all $\mathbf{x}$, then $\mathbf{y} = \mathbf{0}$. Suppose, on the contrary, $\mathbf{x} \cdot (\mathbf{y})\varphi = 0$ for all $\mathbf{x}$ in $\mathcal{U}$ and $a\mathbf{y} \neq \mathbf{0}$. Then, in particular, $\mathbf{y} \cdot (\mathbf{y})\varphi = 0$ and the proper ideal line $\mathcal{K}\mathbf{y}$ would be perpendicular to itself, contradicting the axioms of a metric plane. ◆

It should be noted that the refinement φ by which $B_\varphi(\mathbf{x}, \mathbf{y})$ has been defined need not be a one-to-one mapping of $\mathcal{L}$ onto $\mathcal{P}$ but is one-to-one on the subspace $\mathcal{U}$ of $\mathcal{L}$.

Exercises

1. Let an ordinary Euclidean plane be the plane $x_3 = 1$ in a Euclidean 3-space (compare Section 5.14). If O is joined by lines to all points of the unit circle $x_1^2 + x_2^2 = 1$ in the plane $x_3 = 1$, we obtain a circular cone. Consider all points (vectors) outside this cone, that is, points (x_1, x_2, x_3) for which $x_1^2 + x_2^2 > x_3^2$. Show that this set serves as the set $\mathcal{L}$ and at the same time as the set $\mathcal{P}$ for an ordinary hyperbolic plane.

2. Find linear mappings of $\mathcal{V}$ onto itself which induce on $\mathcal{L}$ refinements of ψ, in the following way: Let α, β be planes of the affine space coordinatized by $\mathcal{V}$ where α, β do not contain O. Consider the maps induced on α by ψ and extend it linearly.

3. Replace, in Exercise 2, β by a line not containing O, and define an analogous refinement as in Exercise 1. Discuss its relationship to Axiom R.

4. Carry out the calculation that proves the identity (7).

7.10   Bilinear Form in the Plane

With the aid of Theorem 7.9.1, we are now in the position to construct a bilinear form on $\mathcal{V}$ which characterizes the perpendicularity of lines in the metric plane. The subdivision of metric planes into metric-Euclidean ones (Axiom R) and metric-non-Euclidean ones is relevant. We explore first the metric-Euclidean case.

As in Section 7.9, we consider a two-dimensional subspace $\mathcal{U}$ of $\mathcal{L}$ representing all proper ideal lines in a proper pencil $L(P)$. By Theorem 7.9.1, there exists a symmetric bilinear form B_φ on $\mathcal{U}$ which is also an inner

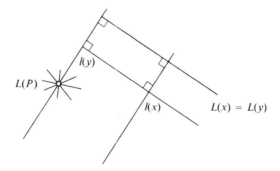

Figure 7.32

product on $\mathcal{U}$. Given any proper ideal line $l(x) = \mathcal{K}\mathbf{x}$, we can find a proper ideal line $l(y) = \mathcal{K}(\mathbf{y})$ through $L(P)$ such that x and y have a perpendicular in common. Assuming that Axiom R holds, this implies $L(x) = L(y)$. The refinement φ by which B_φ on $\mathcal{U}$ has been defined may be changed linearly on every $\mathcal{K}\mathbf{x}$ for $\mathbf{x}$ not in $\mathcal{U}$ without affecting B_φ (on $\mathcal{U}$). We may, therefore, assume, that $(\mathbf{x})\varphi = (\mathbf{y})\varphi$.

Let $\mathcal{K}\mathbf{z}$ represent the ideal line z that carries all poles of proper ideal lines. This ideal line is not a proper one. We define

$$(\mathbf{z})\varphi = 0.$$

Let $\mathbf{x} \neq \mathbf{0}$ belong neither to $\mathcal{K}\mathbf{z}$ nor to $\mathcal{U}$. The ideal line $\mathcal{K}\mathbf{x}$ intersects z in a point $L(a)$. We join $L(a)$ and $L(P)$ by a line $l(y) = \mathcal{K}\mathbf{y}$, and let

$$(\mathbf{x})\varphi = (\mathbf{y})\varphi.$$

ψ is now a linear map of $\mathcal{V}$ into itself. On $\mathcal{L}$ it induces a refinement of the map φ as defined in the preceding section.

Theorem 7.10.1 *Let Axiom* R *be true. The form B defined by*

$$B(\mathbf{x}, \mathbf{y}) = \mathbf{x}(\mathbf{y}\varphi)$$

is a symmetric bilinear form on $\mathcal{V}$ and induces an inner product on every two-dimensional subspace that represents a pencil of proper ideal lines.

Proof The linearity of B in both arguments follows from the linearity of φ and the definition of B. The symmetry is seen as follows:

Since $\mathbf{z} \neq \mathbf{0}$ is not in $\mathcal{U}$ and since $\mathcal{U}$ is two-dimensional, any vector of $\mathcal{V}$ can be expressed as a linear combination of $\mathbf{z}$ and a vector of $\mathcal{U}$. Letting

$$\mathbf{x} = \mathbf{u}_1 + \mathbf{z}_1; \quad \mathbf{u}_1 \in \mathcal{U}; \quad \mathbf{z}_1 \in K\mathbf{z},$$
$$\mathbf{y} = \mathbf{u}_2 + \mathbf{z}_2; \quad \mathbf{u}_2 \in \mathcal{U}; \quad \mathbf{z}_2 \in K\mathbf{z},$$

we have

$$B(\mathbf{x}, \mathbf{y}) = B(\mathbf{u}_1, \mathbf{u}_2) + B(\mathbf{u}_1, \mathbf{z}_2) + B(\mathbf{z}_1, \mathbf{u}_2) + B(\mathbf{z}_1, \mathbf{z}_2).$$

$(\mathbf{z})\varphi = 0$ implies $(\mathbf{z}_2)\varphi = 0$, and thus $B(\mathbf{u}_1, \mathbf{z}_2) = 0$. Since $(\mathbf{u}_2)\varphi$ represents a point on z, we have

$$\mathbf{z}_1 \cdot (\mathbf{u}_2)\varphi = B(\mathbf{z}_1, \mathbf{u}_2) = 0.$$

Finally, by definition, $B(\mathbf{z}_1, \mathbf{z}_2) = 0$. Therefore,

$$B(\mathbf{x}, \mathbf{y}) = B(\mathbf{u}_1, \mathbf{u}_2).$$

Similarly it is seen that

$$B(\mathbf{y}, \mathbf{x}) = B(\mathbf{u}_2, \mathbf{u}_1).$$

By Theorem 7.9.1, $B(\mathbf{u}_1, \mathbf{u}_2) = B(\mathbf{u}_2, \mathbf{u}_1)$ and so $B(\mathbf{x}, \mathbf{y}) = B(\mathbf{y}, \mathbf{x})$. Also, by Theorem 7.9.1, B induces an inner product on any two-dimensional subspace $\mathcal{U}$ that represents lines of a proper pencil. ◆

B is called a **metric fundamental form** of the metric-Euclidean plane. Now we turn to the construction of a symmetric bilinear form in the metric-non-Euclidean case.

Theorem 7.10.2 *Let Axiom $\bar{\mathbf{R}}$ be true. There exists a refinement $\bar{\varphi}$ of ψ such that the form is an inner product $B(\mathbf{x}, \mathbf{y}) = \mathbf{x}(\mathbf{y}\bar{\varphi})$ in $\mathscr{V}$.*

Proof Let three proper pencils $L(P)$, $L(P')$, $L(P'')$ be given which are (as ideal points) not collinear (that is, they have no line in common). To each of these pencils there corresponds a two-dimensional subspace $\mathscr{U}$, $\mathscr{U}'$, $\mathscr{U}''$ of $\mathscr{V}$ respectively which represents its lines. Let u join $L(P)$ and $L(P')$, let v join $L(P)$ and $L(P'')$, and let w be common to $L(P'')$ and

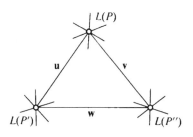

Figure 7.33

$L(P')$. $\mathscr{K}\mathbf{u}$, $\mathscr{K}\mathbf{v}$, $\mathscr{K}\mathbf{w}$ are to represent u, v, w, respectively. If B_φ is an inner product on $\mathscr{U}$, so is $B_{a\varphi}$, for $a \neq 0$. So we may assume that

$$(\mathbf{u})\varphi = (\mathbf{u})\varphi', \qquad (\mathbf{v})\varphi = (\mathbf{v})\varphi''.$$

In other words, we assume that φ coincides with φ', φ'' on the intersections $\mathscr{U} \cap \mathscr{U}'$, $\mathscr{U} \cap \mathscr{U}''$, respectively. We define a further refinement:

$$(\mathbf{x})\bar{\varphi} = \begin{cases} (\mathbf{x})\varphi, & \text{for } \mathbf{x} \text{ in } \mathscr{U} \\ (\mathbf{x})\varphi' & \text{for } \mathbf{x} \text{ in } \mathscr{U}' \\ (\mathbf{x})\varphi'' & \text{for } \mathbf{x} \text{ not in } \mathscr{U} \text{ or } \mathscr{U}' \end{cases}$$

By some bilinear changes we shall make $B_{\bar{\varphi}}$ a symmetric linear form: $\mathbf{u}$, $\mathbf{v}$, $\mathbf{w}$ are linearly independent and may thus be chosen as a basis of $\mathscr{V}$. For an arbitrary $\mathbf{x}$,

$$\mathbf{x} = a_1\mathbf{u} + a_2\mathbf{v} + a_3\mathbf{w}.$$

For $\mathbf{x}$ in $\mathscr{U}$ or $\mathscr{U}'$ we have, since $\bar{\varphi}$ is linear on $\mathscr{U}$, $\mathscr{U}'$,

$$(\mathbf{x})\bar{\varphi} = a_1(\mathbf{u})\bar{\varphi} + a_2(\mathbf{v})\bar{\varphi} + a_3(\mathbf{w})\bar{\varphi}.$$

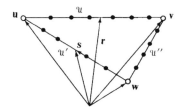

Figure 7.34

Now let **x** be neither in $\mathscr{U}$ nor in $\mathscr{U}'$.

Let for a moment (see Fig. 7.34)

$$\mathbf{r} = a_1\mathbf{u} + a_2\mathbf{v} \quad \text{and} \quad \mathbf{s} = a_1\mathbf{u} + a_3\mathbf{w},$$

so that $\mathbf{x} = \mathbf{r} + a_3\mathbf{w} = a_2\mathbf{v} + \mathbf{s}$. Since **x**, **r**, **w** are linearly dependent, $(\mathbf{x})\bar{\varphi}$, $(\mathbf{r})\bar{\varphi}$, $(\mathbf{w})\bar{\varphi}$ are, by Theorem 7.8.6, also linearly dependent; up to a linear change of $\bar{\varphi}$ in $\mathscr{K}\mathbf{x}$, we obtain

$$(\mathbf{x})\bar{\varphi} = (\mathbf{r})\bar{\varphi} + a_3'(\mathbf{w})\bar{\varphi} \tag{1}$$

for some a_3' of $\mathscr{K}$.

From the linearity of $\bar{\varphi}$ on $\mathscr{U}$, we have

$$(\mathbf{r})\bar{\varphi} = a_1(\mathbf{u})\bar{\varphi} + a_2(\mathbf{v})\bar{\varphi}$$

and thus, substituting in (1), we obtain

$$(\mathbf{x})\bar{\varphi} = a_1(\mathbf{u})\bar{\varphi} + a_2(\mathbf{v})\bar{\varphi} + a_3'(\mathbf{w})\bar{\varphi}. \tag{2}$$

Similarly, we deduce from the linear dependence of **x**, **v**, **s** that there exist elements a_2', b of $\mathscr{K}$ for which

$$(\mathbf{x})\bar{\varphi} = a_2'(\mathbf{v})\bar{\varphi} + b(\mathbf{s})\bar{\varphi}.$$

Using the linearity of $\bar{\varphi}$ on $\mathscr{U}'$, we find

$$\begin{aligned}
(\mathbf{x})\bar{\varphi} &= a_2'(\mathbf{v})\bar{\varphi} + b(a_1\mathbf{u} + a_3\mathbf{w})\bar{\varphi} \\
&= a_2'(\mathbf{v})\bar{\varphi} + ba_1(\mathbf{u})\bar{\varphi} + ba_3(\mathbf{w})\bar{\varphi} \\
&= ba_1(\mathbf{u})\bar{\varphi} + a_2'(\mathbf{v})\bar{\varphi} + ba_3(\mathbf{w})\bar{\varphi}.
\end{aligned} \tag{3}$$

Thus (2) and (3) yield

$$(a_1 - ba_1)(\mathbf{u})\bar{\varphi} + (a_2 - a_2')(\mathbf{v})\bar{\varphi} + (a_3' - ba_3)(\mathbf{w})\bar{\varphi} = 0. \tag{4}$$

Since **u**, **v**, **w** are linearly independent, so are, by Theorem 7.8.8, $(\mathbf{u})\bar{\varphi}$, $(\mathbf{v})\bar{\varphi}$, $(\mathbf{w})\bar{\varphi}$. Therefore, all coefficients in (4) vanish and we have

$$a_1 = ba_1, \qquad a_2 = a_2', \qquad a_3' = ba_3.$$

We distinguish two cases:

(a) $a_1 \neq 0$. In this case, $b = 1$, and hence $a_3' = a_3$, so that

$$(\mathbf{x})\bar{\varphi} = a_1(\mathbf{u})\bar{\varphi} + a_2(\mathbf{v})\bar{\varphi} + a_3(\mathbf{w})\bar{\varphi}.$$

(b) $a_1 = 0$. We let $\mathbf{v}' = \mathbf{u} + \mathbf{v}$ and write **x** in the form

$$\mathbf{x} = -a_2\mathbf{u} + a_2\mathbf{v}' + a_3\mathbf{w}.$$

If also $a_2 = 0$, there is nothing to prove; so let $a_2 \neq 0$. Using the same conclusions as above, we find

$$(\mathbf{x})\bar{\varphi} = -a_2(\mathbf{u})\bar{\varphi} + a_2(\mathbf{v}')\bar{\varphi} + a_3(\mathbf{w})\bar{\varphi}.$$

Applying once again the linearity of $\bar{\varphi}$ on $\mathscr{U}$, we see

$$(\mathbf{v}')\bar{\varphi} = (\mathbf{u})\bar{\varphi} + (\mathbf{v})\bar{\varphi}$$

and so

$$(\mathbf{x})\bar{\varphi} = a_2(\mathbf{v})\bar{\varphi} + a_3(\mathbf{w})\bar{\varphi}.$$

Letting also

$$(\mathbf{x})\bar{\varphi} = a_1(\mathbf{u})\bar{\varphi} + a_2(\mathbf{v})\bar{\varphi} + a_3(\mathbf{w})\bar{\varphi},$$

for $\mathbf{x}$ not in $\mathscr{L}$, we have thus proved that $\bar{\varphi}$ is a linear mapping of $\mathscr{V}$ onto itself.

It is readily seen that $\bar{\varphi}$ coincides now on $\mathscr{U}''$ with φ''. So $B(\mathbf{v}, \mathbf{w}) = B(\mathbf{w}, \mathbf{v})$, from the symmetry of $B_{\varphi''}$ on $\mathscr{U}''$. Since also, by assumption, $B(\mathbf{u}, \mathbf{v}) = B(\mathbf{v}, \mathbf{u}); B(\mathbf{u}, \mathbf{w}) = B(\mathbf{w}, \mathbf{u})$, we obtain by direct verification:

$$B(\mathbf{x}, \mathbf{y}) = B(\mathbf{y}, \mathbf{x}) \quad \text{for all } \mathbf{x}, \mathbf{y} \text{ in } \mathscr{V}.$$

Since $\bar{\varphi}$ is one-to-one, $B(\mathbf{x}, \mathbf{y}) = \mathbf{x}(\mathbf{y})\varphi = 0$ for all $\mathbf{x}$ implies $\mathbf{y} = \mathbf{0}$. Therefore, B is an inner product on $\mathscr{V}$. ◆

B is also called *metric fundamental form* of the metric-non-Euclidean plane. So we have arrived at what is probably the deepest theorem proved in this book. Its implications will be studied in the next section.

Exercises

1. Find the metric fundamental form of the ordinary hyperbolic plane as given by the circle model (see Exercise 1, Section 7.9). (*Hint*: Express φ in matrix form and write $B(\mathbf{x}, \mathbf{y})$ in the form $\mathbf{x}A\mathbf{y}$.)

2. Suppose

$$B(\mathbf{x}, \mathbf{y}) = x_1^2 + 2x_1x_2 + 5x_2^2 + 4x_2x_3 + x_3^2.$$

Show that this form belongs to a metric-Euclidean plane.

3. Show that $x_1^2 - x_2^2$ cannot be a metric fundamental form.

7.11 Fundamental Theorem

Now we need only a few more steps to reach our main goal. So far we have embedded the given metric plane into a Pappusian projective plane coordinatized by a three-dimensional vector space $\mathscr{V}$ over a field $\mathscr{K}$ (homogeneous coordinates). In addition, we have deduced a bilinear form B in $\mathscr{V}$ from the "metric" properties of the metric plane (perpendicularity, facts about reflections). A vector space $\mathscr{V}$ together with a bilinear form is called a *metric vector space*.

We consider all linear mappings of $\mathscr{V}$ onto itself that leave B invariant. This set clearly is a group under successive application. It is called the *orthogonal group* $O_3(\mathscr{K}, B)$ of the three-dimensional metric vector space over $\mathscr{K}$ with bilinear form B. This name stems from the fact that B is a generalized inner product and from the definition of an orthogonal matrix (see Section 4.11). We say a mapping $\mathbf{x} \to \mathbf{x}'$ of $O_3(\mathscr{K}, B)$ has determinant $+1$ if, for any $\mathbf{a}_1, \mathbf{a}_2, \mathbf{a}_3$ the determinant equation

$$[\mathbf{a}_1', \mathbf{a}_2', \mathbf{a}_3'] = [\mathbf{a}_1, \mathbf{a}_2, \mathbf{a}_3]$$

holds. Clearly the subset of $O_3(\mathscr{K}, B)$ consisting of those elements that have determinant $+1$ is a subgroup; we denote it by $O_3^+(\mathscr{K}, B)$ and call it the *proper orthogonal group*.

Theorem 7.11.1 *Every reflection of the metric plane induces in the corresponding ideal projective plane a harmonic homology which in $\mathscr{V}$ is represented by an element of $O_3^+(\mathscr{K}, B)$.*

Proof The first part of the theorem is contained in Theorem 7.7.2. If $\mathscr{K}\mathbf{u}$ is the fixed line of a harmonic homology, this homology can, by Equation (5′) in Section 7.9, be expressed as

$$\mathbf{x}' = -\mathbf{x} + 2\frac{B(\mathbf{x}, \mathbf{u})}{B(\mathbf{u}, \mathbf{u})}\mathbf{u}, \tag{1}$$

where $\mathscr{K}\mathbf{x}$ is an arbitrary line. We find by direct computation, letting

$$c = \frac{1}{B(\mathbf{u}, \mathbf{u})} \quad \text{and} \quad \mathbf{u} = u_1\mathbf{a}_1 + u_2\mathbf{a}_2 + u_3\mathbf{a}_3$$

($\mathbf{a}_1, \mathbf{a}_2, \mathbf{a}_3$ linearly independent):

$$\begin{aligned}
[\mathbf{a}_1', \mathbf{a}_2', \mathbf{a}_3'] &= [-\mathbf{a}_1 + 2cB(\mathbf{a}_1, \mathbf{u})\mathbf{u}, -\mathbf{a}_2 + 2cB(\mathbf{a}_2, \mathbf{u})\mathbf{u}, -\mathbf{a}_3 + 2cB(\mathbf{a}_3, \mathbf{u})\mathbf{u}] \\
&= [\mathbf{a}_1, \mathbf{a}_2, \mathbf{a}_3](-1 + 2cB(\mathbf{a}_1, \mathbf{u})u_1 + 2cB(\mathbf{a}_2, \mathbf{u})u_2 + 2cB(\mathbf{a}_3, \mathbf{u})u_3) \\
&= [\mathbf{a}_1, \mathbf{a}_2, \mathbf{a}_3](-1 + 2cB(u_1\mathbf{a}_1 + u_2\mathbf{a}_2 + u_3\mathbf{a}_3, \mathbf{u})) \\
&= [\mathbf{a}_1, \mathbf{a}_2, \mathbf{a}_3].
\end{aligned}$$

For $\mathbf{a}_1, \mathbf{a}_2, \mathbf{a}_3$ linearly dependent,

$$[\mathbf{a}_1, \mathbf{a}_2, \mathbf{a}_3] = 0 = [\mathbf{a}_1', \mathbf{a}_2', \mathbf{a}_3'].$$

Furthermore, (1) leaves B fixed, as is seen by a routine calculation:

$$B(\mathbf{x}', \mathbf{y}') = B(-\mathbf{x} + 2cB(\mathbf{x}, \mathbf{u})\mathbf{u}, -\mathbf{y} + 2cB(\mathbf{y}, \mathbf{u})\mathbf{u}) = B(\mathbf{x}, \mathbf{y}).$$

So (1) belongs to $O_3^+(\mathscr{K}, B)$. ◆

The projective plane coordinatized by the metric vector space $\mathscr{V}$ is called a *projective-metric plane* Π. $O_3^+(\mathscr{K}, B)$ induces a subgroup of the group of all collineations of Π, called the *group of motions* of Π. Theorem 7.11.1 implies the following one.

Theorem 7.11.2 Fundamental theorem *Any metric plane can be embedded into a projective-metric plane Π such that the group of all motions of the metric plane is a subgroup $\mathscr{G}$ of the group of motions of Π.*

In particular, any reflection of the metric plane induces a harmonic homology in Π. So every element of $\mathscr{G}$ is a projective collineation.

Two lines x, y of the metric plane are perpendicular if and only if, for the corresponding lines $\mathscr{K}x, \mathscr{K}y$ in Π,

$$B(\mathbf{x}, \mathbf{y}) = 0.$$

If we introduce a basis in $\mathscr{V}$, the bilinear form B may be represented by a matrix $\mathbf{A}$ (see Section 4.10).

$$B(\mathbf{x}, \mathbf{y}) = \mathbf{x}A\mathbf{y}.$$

We say B has **rank** r if the matrix $\mathbf{A}^*$ has rank r. In Section 7.8 we saw that when Axiom $\bar{\mathrm{R}}$ is true the assignment pole $\rightarrow$ polar is one to one, and so $\mathbf{A}$ has rank 3. If Axiom R is satisfied, different lines may have the same pole (see Theorem 7.8.7), and so $\mathbf{A}$ has rank less than 3. Since, however, there exists a line on which a one-to-one map is induced (see Section 7.9), $\mathbf{A}$ has rank 2.

We say a projective metric plane is *ordinary* or *singular* if its metric fundamental form B has rank 3 or rank 2, respectively.

Theorem 7.11.3 *The group of motions of a metric plane is a subgroup of the group of motions*

(a) *of an ordinary projective-metric plane if Axiom $\bar{\mathrm{R}}$ holds;*

(b) *of a singular projective-metric plane if Axiom R is true.*

Now that we have described all metric planes, Euclidean and non-Euclidean, in terms of projective geometry, the converse question arises: Which subgroups of the groups of motions of projective-metric planes belong to a metric plane in the sense of the fundamental theorem? This question is still unanswered. It has been answered under a variety of additional conditions which we do not discuss here. However, it remains an unsolved problem in mathematics. In the next section we shall pick out only the essential cases of ordinary Euclidean, hyperbolic, and elliptic geometry.

Exercises

1. Carry out the calculations in the proof of Theorem 7.11.1.

2. Why do the following matrices not define metric fundamental forms:

$$\begin{pmatrix} 1 & 1 & 1 \\ 1 & 1 & 1 \\ -1 & -1 & -1 \end{pmatrix}, \begin{pmatrix} 1 & 0 & 1 \\ 0 & 1 & 0 \\ 2 & 0 & 1 \end{pmatrix}, \begin{pmatrix} 1 & 1 & 1 \\ 1 & 1 & 0 \\ 0 & 1 & 1 \end{pmatrix}, \begin{pmatrix} 0 & 0 & 0 \\ 0 & 1 & 0 \\ 0 & 0 & -1 \end{pmatrix}.$$

3. Express an arbitrary quasi-rotation (see Section 1.18) by a vector equation in $\mathscr{V}$. (*Hint*: Use Equation (1) of the preceding section.)

7.12   *Embedding of Ordinary Elliptic, Hyperbolic, and Euclidean Planes*

In order to study more specific implications of the Fundamental theorem, we assume now that the given metric plane is elliptic, hyperbolic, or Euclidean. Furthermore, we shall introduce axioms on order and continuity.

In the case of an elliptic plane, we choose a line g as the "line at infinity" and postulate for the affine plane obtained from the elliptic plane by leaving out g (see Section 5.1), the order Axioms O.1–O.4 (Section 2.2), and the Dedekind Axiom C (Section 2.9). By Section 3.16, the coordinate field coordinatizing the affine plane is the field $\mathscr{R}$ of real numbers. If we introduce homogeneous coordinates over the same field $\mathscr{R}$, we have coordinatized the given metric plane (as a projective plane). We call it, under the above assumptions, an ***ordinary elliptic plane***. If the plane is Euclidean or hyperbolic, we introduce Axioms O.1–O.4 and C as we did in Chapter 2 when introducing ordinary Euclidean and hyperbolic planes.

In the case of a Euclidean plane, order and continuity carry directly over to the field $\mathscr{K}$ of scalars of the vector space $\mathscr{V}$ that coordinatizes the corresponding ideal plane.

In the case of a hyperbolic plane, we choose a nonproper ideal line g_∞ of the ideal plane as the line at infinity. Let $a \neq g_\infty$ be a line of the metric plane, let $l(a)$ be the corresponding ideal line and let $\bar{l}(a)$ be obtained from $l(a)$ by leaving out the point A in which $l(a)$ and g_∞ intersect. Let P, Q, R be three noncollinear points of the metric plane such that Q lies on a and

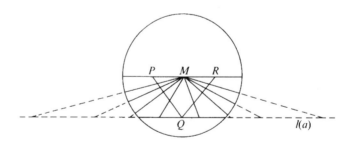

Figure 7.35

such that A, P, R are collinear. On the segment (PR) (see Section 2.2) we choose a point M and project from M all points of the line $\bar{l}(a)$ onto the union U of the segments (PQ), (QR) and the point Q (see Fig. 7.35).

This map is one-to-one and onto, since, by Pasch's Axiom O.4, every line through M different from the line joining P, R meets Q or one of the sides (PQ), (QR). It is readily seen that U satisfies, after appropriate definitions, the order Axioms O.1–O.4. So these carry over to $\bar{l}(a)$. Also the Dedekind Axiom C is extended in this way to $\bar{l}(a)$. This implies again that the field $\mathscr{K}$ is the field $\mathscr{R}$ of real numbers. If we introduce the metric fundamental form B according to the fundamental theorem and represent it by a matrix **A**, the matrix **A** is a symmetric matrix of rank 2 or 3 with real numbers in its entries. (It should be noted that here the matrix **A** represents what was called in Section 5.13 the adjoint matrix of the matrix **A**; it represents the bilinear form for lines.) By Section 5.9 we can choose the coordinate system (basis of $\mathscr{V}$) in such a way that **A** has diagonal form with $1, -1, 0$ as diagonal elements. We assume that to ordinary elliptic, hyperbolic, and Euclidean planes the following matrices can be assigned, respectively,

$$\mathbf{A}_{\text{ell}} = \begin{pmatrix} 1 & 0 & 0 \\ 0 & 1 & 0 \\ 0 & 0 & 1 \end{pmatrix},$$

$$\mathbf{A}_{\text{hyp}} = \begin{pmatrix} 1 & 0 & 0 \\ 0 & 1 & 0 \\ 0 & 0 & -1 \end{pmatrix},$$

$$\mathbf{A}_{\text{euc}} = \begin{pmatrix} 1 & 0 & 0 \\ 0 & 1 & 0 \\ 0 & 0 & 0 \end{pmatrix}.$$

In fact, since **A** and $c\mathbf{A}$ represent, for $c \neq 0$, projectively the same fundamental form, we choose either **A** or $-\mathbf{A}$. Up to an interchange of coordinates, the element $+1$ can always be assumed to be in the upper left corner. So for rank 3 we obtain two possibilities which are covered by $\mathbf{A}_{\text{ell}}$ and $\mathbf{A}_{\text{hyp}}$. In case of rank 2 the nonzero entries must have the same sign, since, otherwise, in the Euclidean plane self-perpendicular lines would exist. So $\mathbf{A}_{\text{euc}}$ can be chosen in the Euclidean case.

We shall now show that each of the three ordinary metric planes can be represented by a well-known model. These models are not just examples of metric planes but represent, up to an "isomorphy" (see below), the only ordinary metric planes. Also the proofs that each of these models does satisfy the axioms of an ordinary metric plane will be completed.

Ordinary Elliptic Plane

The bilinear form given by $\mathbf{A}_{\text{ell}}$ reads

$$B(\mathbf{x}, \mathbf{y}) = \mathbf{x}\mathbf{A}_{\text{ell}}\mathbf{y} = x_1 y_1 + x_2 y_2 + x_3 y_3 = \mathbf{x} \cdot \mathbf{y}.$$

$\mathbf{A}_{\text{ell}} = \mathbf{I}$ is the adjoint (and inverse) matrix of itself. So the dual form is identical with B, and $B(\mathbf{x}, \mathbf{y})$ may also be considered to be defined for points

$\mathscr{K}$x, $\mathscr{K}$y (instead of lines). The reflection at a line $\mathscr{K}$a can be represented by

$$\mathbf{x}' = -\mathbf{x} + 2(\mathbf{x} \cdot \mathbf{a})\mathbf{a}$$

if $\mathbf{a}^2 = 1$ is chosen (see Equation (1) in Section 7.11). Since any motion of the elliptic plane (product of reflections) leaves B fixed, it maps, in particular, the sphere

$$\mathbf{x} \cdot \mathbf{x} = x_1^2 + x_2^2 + x_3^2 = 1 \tag{1}$$

onto itself. Instead of representing the points of the elliptic plane by lines of $\mathscr{V}$ (homogeneous coordinates) we may represent them by the "antipodal" pairs of points in which these lines intersect the sphere (1). Lines of the elliptic plane are then represented by the circles in which the planes through 0 intersect (1) (Fig. 7.36).

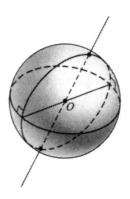

Figure 7.36

Two lines $\mathscr{K}$a, $\mathscr{K}$b are perpendicular if and only if

$$B(\mathbf{a}, \mathbf{b}) = \mathbf{a} \cdot \mathbf{b} = 0.$$

Interpreting this in $\mathscr{V}$ we see that $\mathscr{K}$a, $\mathscr{K}$b as planes are perpendicular in the sense of three-dimensional Euclidean geometry. Therefore, any "equator" of the sphere (1) has "poles" in the usual meaning of the word as "poles" in the sense of the polarity defined by the fundamental form B. So we have arrived, in a natural way, at Klein's model of an elliptic plane.

Theorem 7.12.1 *Any ordinary elliptic metric plane can be represented by Klein's model. Conversely, Klein's model satisfies the axioms of an ordinary elliptic metric plane.*

Proof Only the second part of the theorem is left to be shown. We *define* points and lines and perpendicularity as in Section 1.5. Clearly, perpendicularity is represented by the bilinear form $B(\mathbf{u}, \mathbf{v}) = \mathbf{u}A_{\text{ell}}\mathbf{v} = \mathbf{u} \cdot \mathbf{v}$. Reflections are defined by the equation at the top of the page.

The axioms of an elliptic metric plane can readily be verified: The axioms of incidence I.1 to I.3 follow from the definition of a projective plane. We denote this plane by Π. Perpendicularity is symmetric, since

A_{ell} is symmetric; so Axiom P.1 is true. Axiom P.2 is seen as follows: If a is a line, Q is the pole of a, and P is an arbitrary point, we join P and Q by a line b (in $\mathscr{V}$ we join the corresponding lines through 0 by a plane). Since b passes through the pole of a, it is perpendicular to a. If P is on a, clearly $P \neq Q$ and so b is unique. Axiom P.3 is obvious.

Axioms M.1 and M.2 follow from the definition of a reflection. The reflection in a is characterized as an involutoric homology of Π with center at the pole of a. We denote it by R_a. If a, b, c are lines through a point P, the product $R_a R_b R_c$ is a reflection R_d with axis d passing through P. It leaves the fundamental form $B(\mathbf{x}, \mathbf{y}) = \mathbf{x} \cdot \mathbf{y}$ unchanged and leaves the line $d = \mathscr{K}\mathbf{d}$ pointwise fixed. Hence any line

$$\mathbf{x} \cdot \mathbf{d} = 0$$

is mapped onto itself, and so the center of the harmonic homology $R_a R_b R_c$ is the pole of d; that is, $R_a R_b R_c = R_d$. If we define any product $R_a R_b$ as a **rotation**, we conclude that $R_a R_b = R_d R_c$, where c can be chosen arbitrarily through the intersection of a, b. Therefore, if a is mapped onto a' by a rotation about a point R on a, we may assume this rotation to be of the form $R_a R_b$. Since R_a leaves a pointwise fixed, R_b maps a onto a' in the same way as $R_a R_b$ does. This proves M.3. Since any two lines intersect, there exists no translation in the sense of Section 1.4; and so M.4 is trivially true. If Π is diminished by a line g_∞, we obtain an ordinary Euclidean plane as an affine subplane. So order and continuity of this plane (see Section 2.10) can be introduced, in particular Axioms O.1–O.5 and C. ◆

Ordinary Hyperbolic Plane

The bilinear form given by A_{hyp} reads

$$B(\mathbf{x}, \mathbf{y}) = \mathbf{x} A_{hyp} \mathbf{y} = x_1 y_1 + x_2 y_2 - x_3 y_3.$$

A_{hyp} is again the adjoint and inverse matrix of itself. As in the case of an elliptic plane, the dual form is identical with B, so that $B(\mathbf{x}, \mathbf{y})$ may also be considered to be defined for points $\mathscr{K}\mathbf{x}$, $\mathscr{K}\mathbf{y}$.

All motions leave B fixed, in particular they map

$$B(\mathbf{x}, \mathbf{x}) = x_1^2 + x_2^2 - x_3^2 = 0 \qquad (2)$$

onto itself. This time it is reasonable to introduce inhomogeneous coordinates (see Section 5.13):

$$\xi_1 = \frac{x_1}{x_3}, \qquad \xi_2 = \frac{x_2}{x_3},$$

thus obtaining an ordinary Euclidean plane. (2) becomes

$$\xi_1^2 + \xi_2^2 = 1. \qquad (2')$$

Since we lose the line $x_3 = 0$ anyway by introducing inhomogeneous co-ordinates, we assume this line as line l_∞ in the above sense. Now we assert that the lines of the metric plane are the chords of the unit circle $(2')$. We denote the circle $(2')$ by $\mathscr{C}$. In fact, by Theorem 7.4.6, any ideal line passing through a proper ideal point is a proper ideal line. If a point on or outside

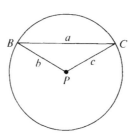

Figure 7.37

$\mathscr{C}$ were a proper ideal point, the tangent of $\mathscr{C}$ through this point would be a proper ideal line perpendicular to itself, contrary to Axiom P.3. So all points of the metric plane are inside $\mathscr{C}$. Let a be a line of the metric plane and let P not lie on a (Fig. 7.37). The ideal line $l(a)$ intersects $\mathscr{C}$ in two ideal points B, C (see Section 2.14). If $l(b)$, $l(c)$ are the ideal lines joining P and B, C, respectively, the lines a, b have no point of intersection (in the metric plane) and also no common perpendicular (the polar of B not being a proper ideal line); that is, b is an asymptote of a through P. Also c is an asymptote of a. Any ideal point on the intersection of $l(a)$ and the interior of $\mathscr{C}$ (the interior of $\mathscr{C}$ consisting of all points (ξ_1, ξ_2) satisfying $\xi_1^2 + \xi_2^2 < 1$) must be a proper ideal point, since, otherwise, the line joining P and such a point would be a third asymptote of a, contrary to the hyperbolic axiom. This proves our assertion. We are now ready to discuss the disc model of a hyperbolic plane.

Theorem 7.12.2 *Any ordinary hyperbolic plane can be represented by the disc model. Conversely, the disc model satisfies the axioms of a hyperbolic plane.*

Proof Again we need show only the second part of the theorem. We choose all points interior to the circle $(2')$ of a Euclidean plane as points of a metric plane. Lines are the (open) chords of this circle. Axioms I.1–I.3 are readily seen to be true.

We consider the projective extension Π of the given Euclidean plane. Its points and lines are called ***ideal points*** and ***ideal lines***. The line at infinity may be denoted by l_∞. We call two lines a, b ***perpendicular*** if their corresponding ideal lines $l(a)$, $l(b)$ are conjugate with respect to the circle $(2')$ (one passes through the pole of the other). Axioms P.1 and P.2 follow, as in the proof of Theorem 7.12.1. P.3 is seen as follows: Let A, B be the ideal points in which $l(a)$ meets the circle $\mathscr{C}$ and let the tangents of $\mathscr{C}$ at A, B meet in Q. If Q is not on l_∞, any line joining a point of the

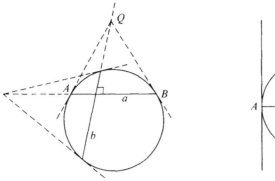
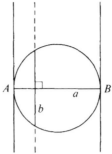

Figure 7.38   **Figure 7.39**

unit disc to Q passes through the angular space bounded by the rays $\overrightarrow{QA}$, $\overrightarrow{QB}$ and, therefore, through the interior of $\triangle QAB$ (in the Euclidean plane). So, by Theorem 2.7.1, it intersects the metric line a. In the case that Q is on l_∞, the perpendiculars of a are the Euclidean perpendiculars of a, which clearly intersect a (see Figs. 7.38 and 7.39).

Reflections are harmonic homologies with center outside $\mathscr{C}$ and axis equal to the polar of the center (with respect to $\mathscr{C}$). Axioms M.1 and M.2 readily follow from this definition. In order to prove M.4, let a, b, c be perpendiculars

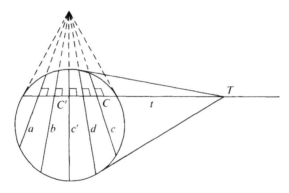

Figure 7.40

of a line t and let c intersect t at C. We consider $C' = (C)R_bR_a$, which is also a point on t. Let c' be the perpendicular of t through C'. With the aid of Figure 7.40, we find a reflection R_d which maps c' onto c such that d is perpendicular to t. The product $R_aR_bR_c$ is involutoric as is seen from Theorems 5.16.2 and 5.10.1 (multiply respective cross-ratios). By Theorem 5.7.3, it is a homology. Then

$$(C')R_aR_bR_c = (C)R_bR_aR_aR_bR_c = C$$

also implies $(c')R_aR_bR_c = c$. M.3 is proved analogously; t being the (ideal) polar line of R.

By Theorem 2.5.3, $R_a R_b R_c$ does not change the sides of t; so if we join corresponding ends of c and c' by ideal lines, these lines meet in a nonproper ideal point T' of $l(t)$, the center of $R_a R_b R_c$. We see from Figure 7.40 that $T' = T$, so that $R_a R_b R_c = R_d$ follows. The proof of Axiom M.4 can now be completed by the same argument used for Axiom M.3 in the proof of Theorem 7.12.1.

The hyperbolic axiom, the order Axioms O.1–O.5 and Axiom C are readily verified. ◆

Ordinary Euclidean Plane

Finally, we consider the bilinear form given by A_{euc}:

$$B(\mathbf{u}, \mathbf{v}) = \mathbf{u}A_{euc}\mathbf{v} = u_1 v_1 + u_2 v_2.$$

We again introduce inhomogeneous coordinates,

$$\xi_1 = \frac{x_1}{x_3}, \qquad \xi_2 = \frac{x_2}{x_3}.$$

Any two lines

$$u_1 \xi_1 + u_2 \xi_2 + u_3 = 0,$$
$$v_1 \xi_1 + v_2 \xi_2 + v_3 = 0$$

are perpendicular if and only if $u_1 v_1 + u_2 v_2 = 0$ (see Section 7.10). So perpendicularity is the ordinary perpendicularity as defined in a Euclidean coordinate plane over the field of real numbers. We obtain, in fact, an ordinary Euclidean plane.

Assigning to every line of a pencil its perpendicular in the pencil provides an involution with no real fixed points. It can also be considered an involution of the line at infinity (see Section 5.10). Conversely, every Euclidean coordinate plane over the reals satisfies Axioms I.1–I.3, P.1–P.3, M.1–M.4, O.1–O.5 and C (see Appendix Theorem 3.III.5):

Theorem 7.12.3 *An ordinary Euclidean plane can be represented as a Euclidean coordinate plane over the reals. Conversely, any Euclidean coordinate plane over the reals satisfies the axioms of an ordinary Euclidean plane.*

We are now able to prove a theorem mentioned on page 77.

Theorem 7.12.4 Categoricity theorem *The systems of axioms of ordinary Euclidean, hyperbolic, and elliptic planes are all categoric.*

Proof Given one of these systems, we have shown above (Theorems 7.12.1–7.12.3) that it can be represented by either Klein's model or the disc model, or the Euclidean coordinate plane over the reals. So any other model is isomorphic to one of these models. ◆

Whereas we considered only one model for an ordinary elliptic plane and one model for an ordinary Euclidean plane, we had altogether five models for a hyperbolic Euclidean plane. The diagram of Figure 7.41 indicates how the isomorphies between these models can be worked out explicitly.

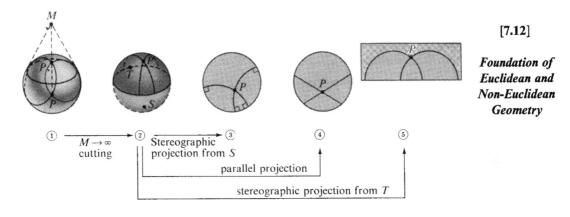

Figure 7.41

Exercises

1. Give an alternative proof of the second part of Theorem 7.12.2 by using Theorem 6.10.4 and a stereographic projection as described in Section 6.10.

2. Establish all isomorphisms indicated in Figure 7.41. (*Hint*: In some cases a stereographic projection as defined in Section 6.10 can be used.)

3. Given the vector space $\mathscr{V}_3(\mathscr{R})$ over the reals and a symmetric 3×3 matrix $\mathbf{A}$ with real entries. Show that $\mathbf{A}$ defines an elliptic or a hyperbolic plane (via a projective-metric plane) if and only if $\mathbf{A}$ has rank 3.

4. Given, as in Exercise 3, $\mathscr{V}_3(\mathscr{R})$ and a symmetric 3×3 matrix $\mathbf{A}$ with real entries, set up necessary and sufficient conditions for $\mathbf{A}$ to define a Euclidean plane.

Appendix

Chapter 1

1.1 Kant and Einstein

According to Aristotle there are two sources for obtaining knowledge: (1) logic, (2) experience. The first one is that of obtaining statements from other statements by logical deduction. The classical example reads: All men are mortal, Gaius is a man, hence Gaius is mortal. In other words, if we take as known facts that all men are mortal and that Gaius is a man, then via logic we obtain the knowledge that Gaius will have to die. That all men are mortal, however, we know by experience. Also we realize by experience that objects get warm if the sun shines on them and that blowing into a trumpet makes a sound; experience is a universal source of knowledge.

Following Aristotle, it was assumed for a long time that any knowledge is either logical or experiential. The philosopher Kant eventually challenged this belief by naming another source of knowledge: (3) pure intuition. One of Kant's main issues concerns the axioms of geometry. These are not obtained by logical conclusions because of the very nature of axioms. Are geometrical axioms obtained by experience? Kant says that any experience preassumes space and time as "conditions of its possibility." If there is no space, there is no experience, in particular no geometrical experience. Geometrical truths are based on an intuition that underlies all spatial considerations. Hence, the fundamentals of geometry which we call geometrical axioms are obtained neither by logic nor by experience. They are obtained by pure intuition, Kant says. Space and time are considered to be forms of pure intuition. Geometry is built mainly in the frame of the first one, space.

This reasoning implies that there can be but one geometry. Since men are familiar with Euclidean geometry, Kant concludes, there can be only Euclidean geometry. Kant, however, refers merely to the physical space we live in. It has been pointed out more recently that his reasoning does not exclude the construction of logical systems involving "points" and "lines" based upon non-Euclidean axioms. Kantian philosophy does

---

Paris, Louvre. Persian Bowmen. Woermann, *Geschichte der Kunst*, Vol. I, Bibliographisches Institut, Leipzig und Wien, 1915.

not deny the possibility of those systems.   It simply does not accept them as "geometries" in a real sense referring to physical space.

In order to prove that Kant is right, a deduction would have to be shown from space as a "form of pure intuition" to geometrical axioms, in particular to the axiom of parallels.   Neither Kant nor any of his followers has been able to show such a deduction.   Therefore Kant's ideas have remained unverified assertions.

When Einstein developed his Theory of Relativity, he challenged the idea of pure intuition in geometry.   If we apply geometry to physics, we apply it to rays of light or edges of crystals, rather than to "abstract" lines.   We cannot eliminate the physical idea of a line from things in space, from matter in our universe.   If several geometries are all logically possible, we may use them in physics whenever they lead to reasonable physical theories.   So, if there are large triangles defined by rays of light in which the sum of angles is not 180 degrees, we still may consider these rays to be "straight" and build up our geometry by means of this assumption.

Einstein's "non-Euclidean" geometry was much more complicated than hyperbolic or elliptic geometry.   It was the so-called Riemannian geometry. Though Einstein did not discover non-Euclidean geometry, he was the first to give Riemannian geometry a definite meaning in physics.   As did Gauss, Bolyai, and Lobachevsky before him, he made it clear that there is a deeper reason for Euclidean geometry not having been deduced from forms of pure intuition than that nobody was able to do the job.   Such a deduction turns out to be impossible.   We are free to create geometrical structures that are logically consistent and to apply them to physics whenever it makes sense.

Does this mean that we have to cancel number 3 of the above "sources of knowledge"?   No! Einstein's ideas can be misleading in this matter.   Space still may be considered a form of pure intuition, applied to geometry in the following way: The ability of our minds to apply abstract thoughts on points and lines from the physical world is based upon spatial intuition in general. This intuition, however, does not prescribe any definite method of procedure. That man's mind is a creative mind is becoming beautifully apparent in **geometry.**

Appendix

Chapter 2

2.1 Finite Groups of Motions

We wish to encounter all possible finite groups of motions that can occur in an ordered Euclidean or hyperbolic plane.

Theorem 2.1.1 *If a group of motions contains a translation $\neq I$, it is infinite.*

Proof Consider a point P and its images P, $P_1 = (P)T$, $P_2 = (P)T^2, \ldots$, $P_n = (P)T^n, \ldots$ under successive application of the given translation T (Fig. A.2.1). Any ray P_n, r, where P_{n+1} is on r, does not

$$P \quad P_1 \quad P_2 \qquad\qquad\qquad P_{n-1} \, P_n \quad P_{n+1}$$

Figure A.2.1

contain the points $P, P_1, \ldots, P_{n-1}$. This implies the next theorem. ◆

Theorem 2.1.2 *Every finite group of proper motions consists only of rotations about one and the same point.*

Proof Suppose the group contains rotations S_1, S_2 about different centers O_1, O_2, both $\neq I$. If $S_1 S_2$ were a translation, by Theorem 2.1.1 the group would be infinite. So $S_1 S_2$ is a rotation, as is $S_1^{-1} S_2^{-1}$. We assert: $S_1^{-1} S_2^{-1} S_1 S_2$ is a translation.

First we note that under a rotation all pairs composed of a line and its image line have the same angle (Fig. A.2.2). So it follows that under $S_1^{-1} S_2^{-1} S_1 S_2$ every line is mapped onto a parallel of itself. Even rays are mapped onto parallel rays.

If $S_1^{-1} S_2^{-1} S_1 S_2$ were a rotation, it would follow that

$$S_1^{-1} S_2^{-1} S_1 S_2 = I;$$

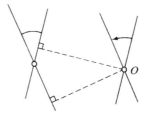

Figure A.2.2

hence $S_1 = S_2^{-1}S_1S_2$, which would mean that

$$(O_2)S_1 = (O_2)S_2^{-1}S_1S_2 = (O_2)S_1S_2;$$

and so $(O_2)S_1 = O_2$ (S_2 possessing only O_2 as a fixed point). Since $O_1 \neq O_2$ has been assumed, $S_1 = I$ would follow, a contradiction.

Therefore, $S_1^{-1}S_2^{-1}S_1S_2$ is a translation and thus, because of Theorem 2.I.1, is equal to I. This yields $S_2^{-1}S_1S_2 = S_1$, and hence

$$(O_2)S_2^{-1}S_1S_2 = (O_2)S_1S_2 = (O_2)S_1.$$

By Theorem 2.4.5, a rotation having two fixed points is the identity map. Therefore, we conclude from $S_2 \neq I$:

$$(O_2)S_1 = O_2.$$

By the same argument, this implies $O_2 = O_1$. ◆

Theorem 2.I.3 *Every finite group of motions containing improper motions consists of rotations about one point O and reflections in axes through O.*

Proof Suppose the group contains a glide reflection TR_a. Then $(TR_a)^2$ is a translation. By Theorem 2.I.1, $(TR_a)^2 = I$; hence for a point X on a, $(X)T^2 = X$, which implies $T^2 = I$ and $T = I$. So $TR_a = R_a$.

If S is a rotation of the given group, R_aSR_a is a rotation with the same center O that S has (by Theorem 2.I.2). Therefore,

$$(O)R_aSR_a = O \quad \text{and} \quad (O)R_aS = (O)R_a,$$

which implies $(O)R_a = O$; hence O is on a. ◆

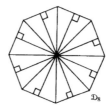

Figure A.2.3

Let $\mathscr{G}_n$ be the group in Theorem 2.I.2 and let the ray r emanate from the
common center of all rotations of $\mathscr{G}_n$, where $\mathscr{G}_n$ contains n elements $S_1 = I$,
$S_2, \ldots, S_n$.   Set $(r)S_i = r_i$.   Among the angles $\angle(r, r_i)$ there is a smallest one,
say $\angle(r, r_2)$.   Any of the others can be covered by placing $\angle(r, r_2)$ finitely
often into it; otherwise a remainder smaller than $\angle(r, r_2)$ would exist.   Any
angle $\angle(r, r_i)$ belongs to a rotation S_i and its inverse S_i^{-1}, and, therefore,
every S_i can be considered a (positive or negative) power of S_2.   However,
S_i^{-1} is itself a positive power of S_i: There exists a k such that $S_i^k = I$ (otherwise the group would be infinite, see Section 1.10), so $S_i^{-1} = S_i^{k-1}$.

We obtain thus a cyclic group $\mathscr{S}_n$.   It can be considered the group of all
proper motions leaving a "regular" n-gon fixed (Fig. A.2.3).   If the group
contains an improper motion M as in Theorem 2.I.3, consider an arbitrary
improper motion of the group—that is, a reflection R.   The product of two
improper motions is proper, and we can set

$$M = (MR^{-1})R,$$

thus we see that all improper motions are obtained as products of rotations
and one of the reflections.   We obtain as many reflections as there are
rotations $\neq I$.   This group is called the

dihedral group $\mathscr{D}_n$

if the subgroup of rotations has order n, that is, if the total number of
elements is $2n$. $\mathscr{D}_n$ can be considered the group of all motions that leaves a
"regular n-gon" fixed (compare Fig. A.2.3).

Figure A.2.4

Generally, the group of all motions leaving a given figure fixed, is called
the **symmetry group** of this figure.   In particular, $\mathscr{S}_n$ and $\mathscr{D}_n$ are called
rosette groups, for obvious reasons (see Fig. A.2.4).   Figure A.2.5(a) has the
group consisting of the identity element alone as symmetry group; we call

such a figure unsymmetric. In (b), $\mathscr{D}_1$ is the symmetry group consisting of I and a reflection in a "vertical" line. The symmetry group of (c) is $\mathscr{D}_4$.

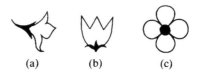

(a)   (b)   (c)

Figure A.2.5

Exercises

1. Find the multiplication table (see Exercise 4, Section 1.10) of $\mathscr{D}_4$ and the diagram of all subgroups of $\mathscr{D}_4$.

2. Find the symmetry group of a rectangle that is not a square. Is it a dihedral group?

3. Find the symmetry groups (possibly consisting only of I) for all capital letters written in a most symmetric form.

2.II   Finitely Generated Groups

A group is said to be ***finitely generated*** if there exist finitely many elements, called ***generators*** of the group, such that every group element is a product of those elements. Trivially, every finite group is finitely generated. The simplest example of an infinite but finitely generated group is the infinite cyclic group $\mathscr{Z}$. It possesses only one generator. We define this group as a group ℓ of translations in any ordered plane by repeating a translation over and over again.

We may consider ℓ as a symmetry group $\mathscr{G}_1$ of a stripe pattern as in Figure A.2.6 (infinitely extended "to the right" and "to the left"). If we consider the stripe pattern of Figure A.2.8(a), we see that it contains a group ℓ as

Figure A.2.6

subgroup of its symmetry group $\mathscr{G}_2$. However, there are also reflections in $\mathscr{G}_2$: There are those in lines of the figure and those in parallel bisectors of two adjacent lines of the figure. Let l_1 be a line of the figure and l_2 a parallel bisector of l_1 and an adjacent line of the figure. It is readily seen that

Figure A.2.7

$T = R_{l_1}R_{l_2}$ is a generating element of ℓ (Fig. A.2.7). Furthermore, $R_{l_1}T$ and $R_{l_2}T$ represent, for arbitrary $T \in \ell$, all possible reflections that belong to $\mathscr{G}_2$. So $\mathscr{G}_2$ possesses two generators, R_{l_1} and R_{l_2}.

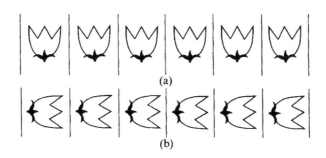

Figure A.2.8

In Figure A.2.8(b) there is one line l perpendicular to the stripe lines such that R_l belongs to the symmetry group $\mathscr{G}_3$ of that figure. $\mathscr{G}_3$ clearly possesses two generators, T, R_l.

Finally, the symmetry group $\mathscr{G}_4$ of Figure A.2.9 consists of all elements of $\mathscr{G}_2$ and $\mathscr{G}_3$; that is, it possesses generators T, R_{l_1}, R_l. For obvious reasons, $\mathscr{G}_1$–$\mathscr{G}_4$ are called *frieze groups*.

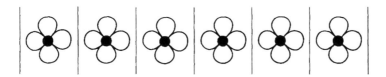

Figure A.2.9

Now we proceed to the case in which there are two translations T, T' in different directions that belong to the symmetry group of a pattern, as in

Figure A.2.10. We choose T and T′ in such a way that each flower is mapped onto the next flower "to the right" or "above," respectively (the meaning of

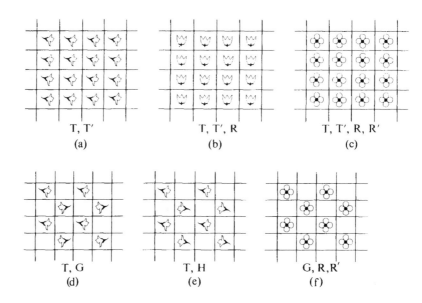

Figure A.2.10

"to the right" and "above" can easily be made precise, see Exercise 1). Then the group $\mathscr{T}$ generated by T, T′ is a subgroup of the symmetry group consisting of all translations in that group.

In pattern (a), $\mathscr{T}$ is obviously the full symmetry group. As a further example, we investigate the full symmetry group $\mathscr{G}$ of (d), leaving the others as exercises. In pattern (d), there exists a glide reflection G mapping each little square onto one touching it in a vertex (the "upper left" or "lower right"). Clearly $G^2 = T$. Any element of $\mathscr{G}$ maps each flower onto another one; this mapping can be achieved by a translation from $\mathscr{T}$ and G or G^{-1}. Therefore, $\mathscr{G}$ is the group generated by T and G.

Exercises

1. (a) Give a precise definition of the generators T, T′ of the group $\mathscr{T}$ used above.
   (b) Replace T, T′ by two other generators of $\mathscr{T}$.

2. Show that the symmetry groups of patterns (a)–(f) in Figure A.2.10 possess generators written "below," T, T′ being translations, G, a glide reflection, and H being a half-turn.

3. In pattern (f), replace all squares by rectangles (of equal size) and find the symmetry group.

2.III   Tessellations

An alternative way of studying symmetry groups is that of discussing tessellations of the plane. If $\mathscr{F}$ is a point set and $\mathscr{G}$ some group of motions, it may happen that the following is true:

(1) Every point of the plane is covered by at least one image of $\mathscr{F}$ under a motion $\mathscr{G}$.

(2) No two such images have the interior of a triangle in common.

Then we say that the totality of all images of $\mathscr{F}$ under $\mathscr{G}$-motions is a ***tessellation*** of the plane. $\mathscr{F}$ is said to be a ***fundamental region*** of this tessellation (or of $\mathscr{G}$). If $\mathscr{G}$ is finite, $\mathscr{F}$ obviously must be "infinitely extended" (which may, for example, be defined as "not contained in the interior of a triangle"). The simplest fundamental regions for finite $\mathscr{G}$ are angular regions, as is illustrated in Figure A.2.11(a) (for $\mathscr{S}_6$) and Figure A.2.12(a) (for $\mathscr{D}_6$). How-

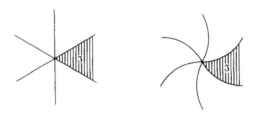

Figure A.2.11

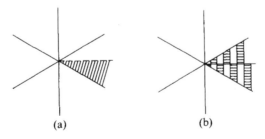

(a)   (b)

Figure A.2.12

ever, as is seen from Figures A.2.11(b) and A.2.12(b), the fundamental regions need not be angular regions. Of greater interest are tessellations belonging to infinite groups, in particular those possessing t (see Section 2.II) as a subgroup. (For examples see Figure A.2.13.) Tessellations of the largest symmetry groups containing t as a subgroup are the ***regular tessellations***. These can be characterized by possessing ***regular p-gons*** as fundamental regions (Fig. A.2.14), regular p-gons being defined in the usual

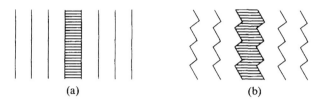

(a)   (b)

Figure A.2.13

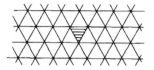 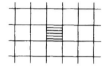 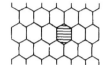

Figure A.2.14

manner (hatched parts of Fig. A.2.14). If the number of p-gons possessing a common vertex is q, we denote the tessellation by the so-called *Schläfli symbol*

$$\{p, q\}.$$

We assert: *The only regular tessellations of the plane are*

$$\{3, 6\}, \quad \{4, 4\}, \quad and \quad \{6, 3\}.$$

In fact, the angle of a p-gon is

$$\frac{p - 2}{p}\, \pi,$$

so we find

$$\frac{p - 2}{p}\, \pi = \frac{2\pi}{q} \quad \text{or} \quad (p - 2)(q - 2) = 4.$$

The only positive integers p, q solving this equation provide the tessellations $\{3, 6\}$, $\{4, 4\}$, $\{6, 3\}$.

Tessellations of groups containing ℓ as subgroups can be applied in many ways as, for example, in brick patterns (Fig. A.2.15), and floor patterns (Fig. A.2.16). They are of particular importance in crystallography. As in the

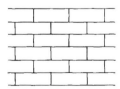   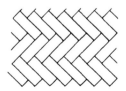

Figure A.2.15   **Figure A.2.16**

case of finite groups, the tessellations of infinite groups can have rather general fundamental regions, as is illustrated in Figure A.2.17.

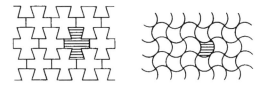

[2.III]

Appendix

Figure A.2.17

Exercises

1. Find tessellations for the groups $\mathscr{G}_1$–$\mathscr{G}_4$ of Section 2.II.
2. Find the tessellations for all symmetry groups of Figures A.2.10(a)–(f).
3. Call a point set ***strongly connected*** if any two points P, Q can be joined by a finite chain of triangles $T_1, \ldots, T_r$ such that P lies in T; Q lies in T_r; and T_i, T_{i+1} have an interior point in common $(i = 1, \ldots, r - 1)$. Show that the only strongly connected fundamental regions of $\mathscr{D}_n$ are angular regions (compare Fig. A.2.11).

Appendix

Chapter 3

3.1 Coordinatization of Euclidean Planes

Euclidean planes were defined in Chapter 1 as metric planes satisfying the axiom of parallels. So Euclidean planes are special affine planes. Whereas in a general affine plane Desargues' theorem cannot be proved (see Section 3.12), the additional assumptions in the definition of a Euclidean plane permit, as we shall see in this section, the proof of Pappus' theorem. Therefore, every Euclidean plane can be coordinatized by a field. An alternative proof of this fact is contained in the results of Chapter 7, where an analogous statement is proved for any metric plane.

We can restrict ourselves to proving the special case of Pappus' theorem (P) (see Section 3.13), in which PQ and $\bar{P}\bar{Q}$ meet in a point, since only this case has been used in the proof of Theorem 3.13.1:

(P_0) Let P_1, P_2, P_3 lie on p and let Q_1, Q_2, Q_3 lie on q such that p, q intersect in a point O different from P_1, P_2, P_3, Q_1, Q_2, Q_3. If P_1Q_2 is properly parallel to Q_1P_2 and if P_1Q_3 is properly parallel to Q_1P_3, then P_2Q_3 is properly parallel to Q_2P_3.

In order to prove (P_0), we need the following lemmas:
(a) The altitudes of a triangle meet in a point, the orthocenter of the triangle.
(b) A quadrangle $ABCD$ is a parallelogram if and only if $H_A H_B = H_D H_C$.

Proofs of (a) and (b) are outlined in Exercises 6 of Section 2.13 and 5 of Section 2.12.
(c) If A, B, C, D, E, F are different points such that $H_A H_B = H_D H_C$ and furthermore AE is parallel to BF and CF is parallel to DE, then $H_A H_B = H_E H_F$ unless $AE = ED$.

To prove (c), let $H_A H_B H_F = H_{E'}$ (by (b)). Applying (b) again shows that E' is on the parallel AE of BF. Substituting $H_D H_C$ for $H_A H_B$ yields

$$H_D H_C H_F = H_{E'}$$

so that E' is on DE, too. Now, if $AE \neq ED$, we find $E = E'$ (Fig. A.3.1).

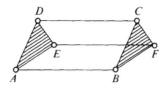

Figure A.3.1

Theorem 3.I.1 *In a Euclidean plane, that is, in a metric plane satisfying the Euclidean axiom of parallels, (P_0) is satisfied.*

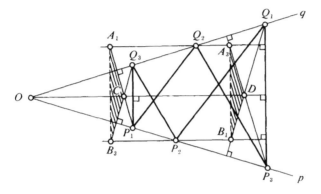

Figure A.3.2

Proof We introduce the following points: O, P_1, P_2, P_3 on p and O, Q_1, Q_2, Q_3 on $q \neq p$, all points being different; furthermore:

C is the orthocenter of $\triangle OP_1Q_3$,
D is the orthocenter of $\triangle OP_3Q_1$,
A_1 is the orthocenter of $\triangle P_1Q_2Q_3$,
A_3 is the orthocenter of $\triangle P_3Q_1Q_2$,
B_1 is the orthocenter of $\triangle Q_1P_2P_3$,
B_3 is the orthocenter of $\triangle Q_3P_1P_2$.

First we prove $H_{A_1}H_{A_3} = H_{B_3}H_{B_1}$ if and only if P_1Q_3 is parallel to P_3Q_1. In fact, let P_1Q_3 be parallel to Q_1P_3. Then clearly

$$A_1Q_2 = A_3Q_2, \qquad B_1P_2 = B_3P_2, \qquad CO = DO,$$

and all these lines are perpendicular to P_1Q_3 as well as to Q_1P_3. Since, furthermore,

$$\begin{cases} A_1C \text{ is parallel to } A_3D & \text{(both perpendicular to } q) \\ B_3C \text{ is parallel to } B_1D & \text{(both perpendicular to } p), \end{cases} \qquad (*)$$

we conclude from (b):

$$H_{A_1}H_{A_3} = H_CH_D \quad \text{and} \quad H_{B_3}H_{B_1} = H_CH_D;$$

hence $H_{A_1}H_{A_3} = H_{B_3}H_{B_1}$.

Conversely, let $H_{A_1}H_{A_3} = H_{B_3}H_{B_1}$. From (∗) and $A_1C \neq B_3C$, we find, by using (c),

$$H_{A_1}H_{A_3} = H_C H_D.$$

By (b), therefore, A_1A_3 is parallel to CD. Since CD is perpendicular to P_1Q_3, so is A_1A_3; and thus P_1Q_3 is parallel to Q_1P_3.

If A_2, B_2 are the orthocenters of $\triangle P_2Q_1Q_3$ and $\triangle Q_2P_1P_3$, respectively, we find correspondingly (by permuting 1, 2, 3):

$$H_{A_1}H_{A_2} = H_{B_2}H_{B_1} \quad \text{if and only if } P_1Q_2 \text{ is parallel to } Q_1P_2,$$
$$H_{A_2}H_{A_3} = H_{B_3}H_{B_2} \quad \text{if and only if } P_2Q_3 \text{ is parallel to } Q_2P_3.$$

The assumptions of (P_0) imply now that

$$H_{A_2}H_{A_1} = H_{B_1}H_{B_2} \quad \text{and} \quad H_{A_1}H_{A_3} = H_{B_3}H_{B_1}.$$

Since $H_{B_1}H_{B_2}H_{B_3}$ is, by (b), involutoric, we can write

$$H_{B_1}H_{B_2}H_{B_3} = H_{B_3}H_{B_2}H_{B_1};$$

hence it follows that

$$\begin{aligned} H_{B_1}H_{B_2}H_{B_3} \cdot H_{B_1} &= H_{B_3}H_{B_2}H_{B_1} \cdot H_{B_1} = H_{B_3}H_{B_2} \\ &= H_{A_2}H_{A_1}H_{A_1}H_{A_3} = H_{A_2}H_{A_3}. \end{aligned}$$

Therefore, P_2Q_3 is parallel to Q_2P_3. ◆

3.II   Formulas for Perpendicularity and Reflections

Throughout this section we consider a Euclidean plane with free mobility, that is, a metric plane satisfying Axioms Euc and FM (see Section 2.1). As we have seen in the preceding section, it can be coordinatized by a field $\mathscr{K}$. We assume the coordinate axes c_1, c_2 to be perpendicular.

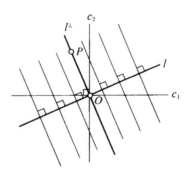

Figure A.3.3

Theorem 3.II.1 *If l is an arbitrary line through O, the reflection* R_l *can be represented by a linear mapping*

$$(x_1, x_2) \longrightarrow (ax_1 + bx_2, dx_1 + ex_2)$$

with $ae - bd \neq 0$.

Proof Let $l^\perp$ be the perpendicular of l through O and let $P \neq O$ be a point on $l^\perp$ (Fig. A.3.3). Let $P' = (P)R_l$. By axiom FM, there is a reflection mapping l onto c_1. Combining this with a stretch that possesses c_1 as axis and c_2 as a trace we find an affinity A mapping l onto c_1 and P onto $(0, 1)$. Set $(P')A = (0, y)$. The mapping

$$\check{S}: (x_1, x_2) \longrightarrow (x_1, yx_2)$$

is a stretch with c_1 as axis and c_2 as a trace. It maps $(P)A$ onto $(P')A$. By Theorem 3.3.5, $A\check{S}A^{-1}$ is a stretch that maps P onto P'. It has axis $l = (c_1)A^{-1}$ and a trace $l^\perp = (c_2)A^{-1}$. Since, by Theorem 3.5.2, there is only one stretch with axis l that maps P onto P', we conclude:

$$A\check{S}A^{-1} = R_l.$$

Since A, $\check{S}$ are linear mappings, so is R_l. ◆

Let $P = (x_1, x_2)$ lie on l. By the *slope* of l we mean the ratio

$$s = \frac{x_2}{x_1},$$

which is the same for all $P \neq (0, 0)$ on l. If $l = c_2$, we set formally

$$s = \infty,$$

defining $x/\infty = 0$ and $x/0 = \infty$ for $x \neq 0$. Let S be a rotation that maps

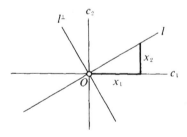

Figure A.3.4

every line l through 0 onto $l^\perp$. Since S is a product of two reflections, it can be represented by a linear mapping

$$(x_1, x_2) \longrightarrow (fx_1 + gx_2, hx_1 + kx_2) = (x_1', x_2'),$$

where $fk - gh \neq 0$.

For the slopes, S induces a map

$$S^*: s \longrightarrow s' = \frac{x_2'}{x_1'} = \frac{hx_1 + kx_2}{fx_1 + gx_2} = \frac{h + ks}{f + gs}$$

for $s \neq 0, \infty$. For $s = 0$, that is, $l = c_1$, and for $s = \infty$, that is, $l = c_2$, we obtain

$$0 \longrightarrow \infty,$$
$$\infty \longrightarrow 0.$$

This yields $f = 0$ and $k = 0$. Since $fk - gh \neq 0$, we have $g, h \neq 0$. Now

$$s' = \frac{h}{g \cdot s} = \frac{q}{s} \quad \left(q = \frac{h}{g} \right),$$

or (if $s, s' \neq 0$),

$$ss' = q.$$

q is a constant that is not a square in the coordinate field $\mathscr{K}$ (otherwise $q = r^2$; hence $s = r$; $s' = r$ or $s = -r$; $s' = -r$ which would imply $l = l^\perp$). q is called a **perpendicularity constant**.

If, conversely, $ss' = q$, we assert that the lines l, l' through O with slopes s, s' are perpendicular. In fact, the perpendicular of l through O has slope $q/s = s'$, from which $l' = l^\perp$ follows.

Theorem 3.II.2 *If, in a Euclidean plane with free mobility, the coordinate axes are chosen to be perpendicular, the slopes s, s' of perpendicular lines satisfy (for $s, s' \neq 0$)*

$$ss' = q,$$

where q is a constant that is not a square in $\mathscr{F}$.

Reflections can now be represented as follows.

Theorem 3.II.3 *If l is a line passing through O and if $A = (a_1, a_2)$ is a point $\neq 0$ on l, then the reflection R_l is represented by*

$$(x_1, x_2) \longrightarrow (-c(qa_1^2 + a_2^2)x_1 + 2ca_1a_2x_2, \, -2cqa_1a_2x_1 + c(qa_1^2 + a_2^2)x_2),$$

where

$$c = (-qa_1^2 + a_2^2)^{-1}.$$

Proof One verifies this formula by first substituting for (x_1, x_2) any point on l, that is, a point (a_1t, a_2t), to see that l remains pointwise fixed, and then checking the image of $l^\perp$,

$$x_2 = q\frac{a_1}{a_2}x_1 \quad \text{(if } a_2 \neq 0, \text{ otherwise } x_1 = d),$$

to see that $l^\perp$ is a fixed line and, finally, showing that perpendicularity is preserved. We omit the calculations (see Exercise 1). ◆

In ordinary Euclidean planes, the perpendicularity constant is usually taken to be

$$q = -1.$$

Then $-qa_1^2 + a_2^2 = a_1^2 + a_2^2$ is the "square of the distance between 0 and A."

Exercises

1. Carry out explicitly the calculations in the proof of Theorem 3.II.3.
2. Show that, in Theorem 3.II.3, $-qa_1^2 + a_2^2 \neq 0$.
3. Extend Theorem 3.II.3 to any line l (not necessarily containing O). *Hint*: Combine Theorem 3.II.3 with a translation.
4. Can the field of complex numbers be the coordinate field of a Euclidean plane?

3.III   Ordered Fields and Ordered Affine Planes

By an ***ordered affine plane*** we mean an affine plane in which an order relation is defined that satisfies order Axioms O.1–O.4 (see Section 2.2). We shall now discuss the coordinatization of ordered affine planes, in particular that of ordinary Euclidean planes. The definition of line segments and rays can be taken directly from Sections 2.2 and 2.4. Also Theorems 2.2.2, 2.2.3, and 2.4.1 are valid in ordered affine planes.

A ***parallel projection*** of a line a onto a line a' is defined by intersecting a, a' with all lines of a parallel pencil that does not contain a or a'. To every point A on a there corresponds a uniquely determined point A' on a', where A, A' are on the same line of the defining parallel pencil (Fig. A.3.5).

Figure A.3.5

Theorem 3.III.1   *The relation "between" is preserved under parallel projections.*

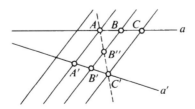

Figure A.3.6

Proof Let A, B, C lie on a and $[ABC]$ and let A', B', C' be the images of A, B, C under the given parallel projection (Fig. A.3.6). We may assume $C \neq C'$. First we project a onto AC', denoting the image of B by B''. From Axiom O.4 applied to $\triangle ACC'$ it follows that $[AB''C']$. If $a' \neq AC'$, we apply Axiom O.4 once again to $\triangle AA'C'$, obtaining $[A'B'C']$. ◆

An ***ordered field*** $\mathcal{K}$ is defined as a field whose elements satisfy a relation "$<$" with the following properties:

O.I For any two elements x, y of $\mathcal{K}$ one and only one of the relations

$$x < y \qquad x = y \qquad y < x$$

is true.

O.II If $x < y$ and $y < z$, then $x < z$.

O.III If $0 < x$ and $0 < y$, then $0 < xy$.

O.IV If $x < y$, then $x + z < y + z$ for any z.

We now state a few properties of ordered fields: We can assume $0 < 1$, since otherwise $1 < 0$, and introduce a new symbol "$\prec$" by saying

$$x \prec y \quad \text{if} \quad y < x$$

we obtain a new order relation satisfying again Properties O.I–O.IV, and, in addition, $0 \prec 1$.

Theorem 3.III.2 *In an ordered field $\mathcal{K}$ for which $0 < 1$ we have*:

(a) $0 < 1 < 2 < 3 < \cdots$,
(b) $x < x + y$ *and* $x - y < x$ *if* $y > 0$,
(c) $0 < x$ *implies* $-x < 0$,
(d) $0 < x$ *implies* $0 < x^{-1}$.

Proof (a) follows from $0 < 1$ and a successive application of Property O.IV. To verify (c), let $0 < x$ and suppose $0 < -x$. Then, by Property O.IV, $x < -x + x = 0$, contradicting Property O.I. To prove (d), let $0 < x$ and assume $x^{-1} < 0$. Then, by (c), $0 < -x^{-1}$ and, by Property O.III, $0 < x(-x^{-1}) = -1$. This would, by (c), imply $1 < 0$, contrary to $0 < 1$. Finally, (b) follows from Property O.IV and (c). ◆

The following two theorems show how the order of an affine plane and the order of the coordinate field of the plane are linked.

Theorem 3.III.3 *If an affine plane is coordinatized by a field $\mathcal{K}$ and if Axioms O.1–O.4 are satisfied, then the order of the plane induces, in a natural way, an order of the field.*

Proof Let c_1 be a coordinate axis with points O and E as 0 and 1 of $\mathcal{K}$, respectively. We call the ray emanating from O and containing E ***positive***, its elements x we call ***positive***, too:

$$0 < x.$$

The elements y of the opposite ray are said to be **negative**:

$$y < 0.$$

If x, y are two elements of $\mathcal{K}$, we let

$$x < y$$

if and only if one of the following cases applies:

(a) y is positive and x is between O and y;
(b) y is positive and x is negative;
(c) y is negative and y is between O and x.

O.I and O.II follow from Axiom O.3 and Theorem 2.2.3. To show Property O.III, we construct xy geometrically as indicated in Figure A.3.7. 1, y can be mapped by a parallel projection onto points U, V on a

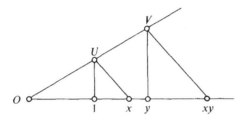

Figure A.3.7

line through O.   Again U, V can be mapped, by a parallel projection, onto x, xy, respectively.   By Theorem 3.III.1, Property O.III follows.

To prove Property O.IV, we proceed in an analogous fashion: Construct $x + z$ and $y + z$ as indicated in Figure A.3.8.   Then a two-fold application of Theorem 3.III.1 proves Property O.IV.   ◆

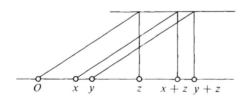

Figure A.3.8

Theorem 3.III.4   *Let an affine plane be coordinatized by an ordered field. Then an order relation can be introduced in the plane such that Axioms O.1–O.4 are satisfied.*

Proof   Let $P = (p_1, p_2)$, $Q = (q_1, q_2)$, $R = (r_1, r_2)$ be three different points on a line l.   If l is parallel to the x_2 axis, that is, $p_1 = q_1 = r_1$, then we say Q is between P and R if

$$p_2 < q_2 < r_2 \quad \text{or} \quad r_2 < q_2 < p_2.$$

If l is not parallel to the x_2 axis, we say Q is between P and R if

$$p_1 < q_1 < r_1 \quad \text{or} \quad r_1 < q_1 < p_1.$$

For "Q between P and R," we write

$$[P\,Q\,R].$$

Axiom O.1 (that is, "$[PQR]$ implies $[RQP]$") follows by definition. Axiom O.2 is readily reduced to showing: If $a, b \in \mathcal{K}$ and $a < b$, then there exist elements $c, d \in \mathcal{K}$ such that

$$a < c < b < d.$$

In fact, we let

$$c = \frac{a+b}{2} \quad \text{and} \quad d = b + 1.$$

By Theorem 3.III.2(d), $0 < \frac{1}{2}$ and thus, by Property O.III, $0 < (b-a)/2$. Hence, by Theorem 3.III.2(b),

$$a < a + \frac{b-a}{2} = \frac{a+b}{2} = c = b - \frac{b-a}{2} < b.$$

Axiom O.3 follows from Property O.I.

To prove Pasch's Axiom O.4, we proceed as follows: Our definition of "betweenness" for points on a line is defined in such a way that, by Property O.IV, it remains invariant under translations

$$(x_1, x_2) \longrightarrow (x_1 + c_1, x_2 + c_2)$$

and under mappings

$$(x_1, x_2) \longrightarrow (x_1 + c_1 x_2, x_2)$$

or

$$(x_1, x_2) \longrightarrow (x_1, x_2 + d_2 x_1),$$

which are seen to be shears with axes $x_2 = 0$ or $x_1 = 0$, respectively.

By these mappings we can shift an arbitrary triangle so that one of its vertices become O, one vertex lies on the x_1 axis, the third one lies on the x_2 axis. Let $A = (0, 0)$, $B = (b_1, 0)$, $C = (0, c_2)$ be these points. We can assume $b_1 > 0$ and $c_2 > 0$ (by shifting the vertices in an appropriate order). The line BC is represented by

$$\frac{x_1}{b_1} + \frac{x_2}{c_2} = 1.$$

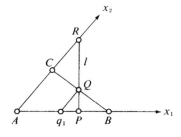

Figure A.3.9

Let P be a point on (AB), and let l pass through P, but not through A, B, or C (Fig. A.3.9). If l is parallel to AC, it intersects CB in a point Q for which, by definition, $[BQC]$. If l is not parallel to AC, let

$$\frac{x_1}{p} + \frac{x_2}{r} = 1$$

be its equation, $R = (0, r)$ being its point of intersection with the x_2-axis. In case $0 < r < c_2$, we have $[ARC]$. So we can assume either $r > c_2$ or $r < 0$. The intersection of l and BC is a point $Q = (q_1, q_2)$, with

$$q_1 = \frac{pr - c_2p}{b_1r - c_2p}\, b_1.$$

From $p < b_1$ and $r > c_2 > 0$ we find (with the aid of Theorem 3.III.2)

$$pr - c_2p < b_1r - c_2p;$$

hence

$$q_1 = \frac{pr - c_2p}{b_1r - c_2p}\, b_1 < b_1.$$

Furthermore, we conclude, in case $r > c_2$, that $b_1r > b_1c_2$. Since $b_1 > p$, we also have $b_1c_2 > pc_2$ and thus

$$b_1r > pc_2 \quad \text{and} \quad 0 < \frac{r - c_2}{b_1r - c_2p}\, pb_1 = q_1.$$

This inequality clearly holds for $r < 0$, too. Therefore,

$$0 < q_1 < b_1;$$

hence $[BQC]$. This proves Axiom O.4. ◆

Axiom C of Section 2.9 can be restated for the coordinate field as follows:

Axiom C′ *Let $\mathcal{K}$ be subdivided into two nonempty subsets $\mathcal{S}_1$, $\mathcal{S}_2$ such that $x < y$ for any $x \in \mathcal{S}_1$, $y \in \mathcal{S}_2$. Then there exists a number $z \in \mathcal{K}$ such that either $z \in \mathcal{S}_1$ and $x < z$ for all $x \in \mathcal{S}_1$, $x \neq z$, or $z \in \mathcal{S}_2$ and $z < y$ for all $y \in \mathcal{S}_2$, $y \neq z$.*

An ordered field satisfying Axiom C′ is called a *field of real numbers*. We denote it by $\mathcal{R}$. The ordinary field of real numbers is usually defined by starting with the "natural numbers," $1, 2, \ldots$, and extending this set first to include all integers, then to include all rationals; completing it finally by assuming the Dedekind Axiom C′. Since all natural numbers belong to any ordered field in their "natural order" (see Theorem 3.III.2), we can consider the field of ordinary real numbers to be constructed in a given field $\mathcal{K}$ that satisfies Axiom C′. It even covers, by Axiom C′, the whole field $\mathcal{K}$. *So we can identify $\mathcal{K}$ with the field of ordinary real numbers.*

So Theorems 3.III.3 and 3.III.4 imply the following one.

Theorem 3.III.5 *Any ordered Pappian Euclidean plane satisfying Axiom* C *can be coordinatized by the field of ordinary real numbers.   Conversely, the field of ordinary real numbers provides an ordered Pappian Euclidean plane satisfying Axiom* C.

Now the question of categoricity raised in Section 2.10 can be answered.

Theorem 3.III.6 *The system of axioms that defines the ordinary Euclidean (metric) plane (see Section* 2.10) *is categoric.*

Proof By Theorems 3.I.1 and 3.III.5, an ordinary Euclidean plane can be coordinatized by the ordinary field of real numbers; that is, any model of such a plane is isomorphic to the plane defined by the ordinary field of real numbers. This, in turn, implies that any two models of ordinary Euclidean planes are isomorphic. ◆

Exercises

1. Does the field of complex numbers allow an order relation " $<$ " that makes it an ordered field?
2. Can every ordered field be considered a subfield of the field of real numbers? (Compare Section 3.15.)

Appendix

Chapter 4

4.1 Bilinear Forms and Inner Products

We present here a few remarks on inner products. The definition given in Sections 4.6 and 4.11,

$$\mathbf{x} \cdot \mathbf{x} = x_1 x_2 + \cdots + x_n x_n,$$

refers to a coordinate system. Our new definition is to be coordinate free. Furthermore, we shall consider inner products as special cases of so-called bilinear forms. An interesting example of a bilinear form arises from physics: In a 4-dimensional vector space over the field $\mathscr{R}$ of real numbers, let the number

$$\mathbf{x} \cdot \mathbf{y} = x_1 y_1 + x_2 y_2 + x_3 y_3 - cts$$

be assigned (c = velocity of light) to any vectors

$$\mathbf{x} = (x_1, x_2, x_3, t), \quad \mathbf{y} = (y_1, y_2, y_3, s).$$
$$\mathbf{x} \cdot \mathbf{x} = x_1^2 + x_2^2 + x_3^2 - ct^2$$

is called the *Lorentz form*. Here $\mathbf{x} \cdot \mathbf{x} = 0$ and $\mathbf{x} \cdot \mathbf{x} < 0$ can occur for $\mathbf{x} \neq \mathbf{0}$, which is impossible in the case of an ordinary inner product. (Figure A.4.1 provides a picture for $x_2 = x_3 = 0$, the shaded area is, in the theory of relativity, called the *cone of light*.)

Let a vector space over an arbitrary field $\mathscr{K}$ be given. A *bilinear form* is generally defined as a mapping which assigns to every pair of vectors $\mathbf{x}, \mathbf{y}$ an element of $\mathscr{K}$ denoted by $\langle \mathbf{x}, \mathbf{y} \rangle$, such that the following rules are satisfied:

B.1 $\langle \mathbf{x} + \mathbf{y}, \mathbf{z} \rangle = \langle \mathbf{x}, \mathbf{z} \rangle + \langle \mathbf{y}, \mathbf{z} \rangle$,

B.2 $\langle \mathbf{x}, \mathbf{y} + \mathbf{z} \rangle = \langle \mathbf{x}, \mathbf{y} \rangle + \langle \mathbf{x}, \mathbf{z} \rangle$,

B.3 $\langle a\mathbf{x}, \mathbf{y} \rangle = \langle \mathbf{x}, a\mathbf{y} \rangle = a\langle \mathbf{x}, \mathbf{y} \rangle$ for any $a \in \mathscr{K}$.

This axiomatic definition is coordinate free. However, if a coordinate system is introduced, that is, a basis of the vector space, the bilinear

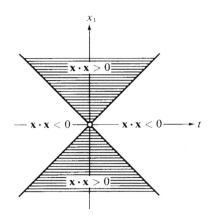

Figure A.4.1

form can also be written in coordinates: Let $\mathbf{a}_1, \ldots, \mathbf{a}_n$ be such a basis. We obtain from Rules B.1, B.2, and B.3, for

$$\mathbf{x} = x_1\mathbf{a}_1 + \cdots + x_n\mathbf{a}_n,$$
$$\mathbf{y} = y_1\mathbf{a}_1 + \cdots + y_n\mathbf{a}_n:$$

$$\langle \mathbf{x}, \mathbf{y} \rangle = x_1\langle \mathbf{a}_1, \mathbf{a}_1 \rangle y_1 + \cdots + x_1\langle \mathbf{a}_1, \mathbf{a}_n \rangle y_n$$
$$+ x_2\langle \mathbf{a}_2, \mathbf{a}_1 \rangle y_1 + \cdots + x_2\langle \mathbf{a}_2, \mathbf{a}_n \rangle y_n$$
$$\cdots\cdots\cdots$$
$$+ x_n\langle \mathbf{a}_n, \mathbf{a}_1 \rangle y_1 + \cdots + x_n\langle \mathbf{a}_n, \mathbf{a}_n \rangle y_n.$$

Using the summation symbol and letting

$$a_{ik} = \langle \mathbf{a}_i, \mathbf{a}_k \rangle,$$

we find

$$\langle \mathbf{x}, \mathbf{y} \rangle = \sum_{i,k=1}^{n} x_i a_{ik} y_k.$$

With respect to any given bilinear form in a vector space, we call two vectors $\mathbf{x}, \mathbf{y}$ *perpendicular* if $\langle \mathbf{x}, \mathbf{y} \rangle = 0$. If $\langle \mathbf{x}, \mathbf{x} \rangle = 1$, we say $\mathbf{x}$ is a *unit vector*. A basis $\mathbf{a}_1, \ldots, \mathbf{a}_n$ of the space is called an *orthonormal basis* if all $\mathbf{a}_i$ are unit vectors and are two-by-two perpendicular. In this case, $\langle \mathbf{x}, \mathbf{y} \rangle$ reduces to

$$\langle \mathbf{x}, \mathbf{y} \rangle = x_1 y_1 + \cdots + x_n y_n = \mathbf{x} \cdot \mathbf{y}.$$

It may be, of course, that a vector not equal to $\mathbf{0}$ is perpendicular to all vectors of the space, for example, if all $a_{ik} = 0$. If there is no such vector, we call the bilinear form an *inner product*.

Exercises

1. In the case of a Lorentz form, find all vectors that are perpendicular to themselves.

2. Let $\mathcal{K}$ be the field of complex numbers. If now the Lorentz form is defined as above ($c > 0$, c real), show that there exists an orthonormal basis with respect to this form. Why is this not so when $\mathcal{K}$ is the field of real numbers?

3. Find a bilinear form, not identically 0, with respect to which every vector is self-perpendicular.

4.II   *Matrix Calculus*

A matrix has been defined as a square array of elements of a skew field $\mathcal{F}$. It has *n* **rows** and *n* **columns**. However, for many purposes, it is useful to consider more generally an arbitrary rectangular array of numbers and call it a *matrix*:

$$\mathbf{A} = \begin{pmatrix} a_{11} & \cdots & a_{1m} \\ \vdots & & \vdots \\ a_{n1} & \cdots & a_{nm} \end{pmatrix} = (a_{ik}).$$

For example, the representation of a vector $\mathbf{x}$:

$$(x_1, \ldots, x_n) \quad \text{or} \quad \begin{pmatrix} x_1 \\ \vdots \\ x_n \end{pmatrix}$$

is a representation by a matrix of one row or one column. We often identify vectors with such matrices.

Let $\mathbf{A}$ and $\mathbf{B}$ be $n \times n$ matrices which stand for linear mappings of a vector space onto itself:

$$\mathbf{x}' = \mathbf{x}\mathbf{A}, \qquad \mathbf{x}'' = \mathbf{x}'\mathbf{B}.$$

If we apply these two mappings in succession, we obtain, by substitution:

$$\mathbf{x}'' = (\mathbf{x}\mathbf{A})\mathbf{B}.$$

Written in coordinates, this means we have to substitute $x_1', \ldots, x_n'$ of the first of the following systems of equations into the second:

$$\begin{aligned} x_1' &= x_1 a_{11} + \cdots + x_n a_{n1}, & x_1'' &= x_1' b_{11} + \cdots + x_n' b_{n1}, \\ &\;\;\vdots & &\;\;\vdots \\ x_n' &= x_1 a_{1n} + \cdots + x_n a_{nn}, & x_n'' &= x_1' b_{1n} + \cdots + x_n' b_{nn}. \end{aligned}$$

We obtain the resulting mapping:

$$\begin{aligned} x_1'' &= x_1(a_{11}b_{11} + \cdots + a_{1n}b_{n1}) + \cdots + x_n(a_{n1}b_{11} + \cdots + a_{nn}b_{n1}), \\ &\;\;\vdots \\ x_n'' &= x_1(a_{11}b_{1n} + \cdots + a_{1n}b_{nn}) + \cdots + x_n(a_{n1}b_{1n} + \cdots + a_{1n}b_{nn}). \end{aligned}$$

This suggests the general definition of a *matrix product*:

$$
\mathbf{A} \cdot \mathbf{B} = \begin{pmatrix} a_{11} & \cdots & a_{1m} \\ \vdots & & \vdots \\ a_{n1} & \cdots & a_{nm} \end{pmatrix} \begin{pmatrix} b_{11} & \cdots & b_{1l} \\ \vdots & & \vdots \\ b_{m1} & \cdots & b_{ml} \end{pmatrix} = \begin{pmatrix} \sum_{i=1}^{m} a_{1i} \cdot b_{i1} & \cdots & \sum_{i=1}^{m} a_{1i} b_{il} \\ \vdots & & \vdots \\ \sum_{i=1}^{m} a_{ni} \cdot b_{i1} & \cdots & \sum_{i=1}^{m} a_{ni} b_{il} \end{pmatrix}.
$$

Two matrices can be multiplied only if the first one has as many columns as the second one has rows. In case of two $n \times n$ matrices, as mentioned above, we obtain as the resulting mapping

$$
\mathbf{x}'' = (\mathbf{x}\mathbf{A})\mathbf{B} = \mathbf{x}(\mathbf{A}\mathbf{B}).
$$

Example Let **A**, **B** be the matrices of Examples (a), (b) in Section 4.10. As a result of applying first the reflection (a), then the dilatation (b), we obtain the transformation matrix

$$
\mathbf{A} \cdot \mathbf{B} = \begin{pmatrix} 0 & 1 & 0 \\ 1 & 0 & 0 \\ 0 & 0 & 1 \end{pmatrix} \begin{pmatrix} d & 0 & 0 \\ 0 & d & 0 \\ 0 & 0 & d \end{pmatrix} = \begin{pmatrix} 0 & d & 0 \\ d & 0 & 0 \\ 0 & 0 & d \end{pmatrix}.
$$

The right side of $\mathbf{x}' = \mathbf{x}\mathbf{A}$ can also be looked upon as a matrix multiplication:

$$
\mathbf{x}\mathbf{A} = (x_1, \ldots, x_n) \begin{pmatrix} a_{11} & \cdots & a_{1n} \\ \vdots & & \vdots \\ a_{n1} & \cdots & a_{nn} \end{pmatrix} = \left(\sum_{i=1}^{n} x_i a_{i1}, \ldots, \sum_{i=1}^{n} x_i a_{in} \right).
$$

Returning to the general product of an $n \times m$ matrix and an $m \times l$ matrix, we introduce still another way of expressing it: Let the rows of **A** be considered as vectors $\mathbf{a}^1, \ldots, \mathbf{a}^n$ and let the columns of **B** be considered as vectors $\mathbf{b}^1, \ldots, \mathbf{b}^l$. Using the inner product notation

$$
\mathbf{a}^i \cdot \mathbf{b}^i = a_{i1} b_{1i} + \cdots + a_{im} b_{mi},
$$

we obtain

$$
\mathbf{A} \cdot \mathbf{B} = \begin{pmatrix} \mathbf{a}^1\mathbf{b}^1 & \cdots & \mathbf{a}^1\mathbf{b}^l \\ \vdots & & \vdots \\ \mathbf{a}^n\mathbf{b}^1 & \cdots & \mathbf{a}^n\mathbf{b}^l \end{pmatrix}.
$$

This hints at a formal rule for obtaining the matrix product:

$$
\begin{pmatrix} - & i & - \end{pmatrix} \begin{pmatrix} | \\ k \\ | \end{pmatrix} = \begin{pmatrix} & \vdots & \\ \cdots & \blacksquare & \cdots \\ & \vdots & \\ & k & \end{pmatrix} i.
$$

The "ith row times the kth column provides the entity in the ith row and the kth column of the product matrix."

The inner product of vectors **x** and **y**,

$$\mathbf{x} \cdot \mathbf{y} = x_1 y_1 + \cdots + x_n y_n,$$

can also be considered as a matrix product by letting

$$\mathbf{x} = (x_1, \ldots, x_n), \qquad \mathbf{y} = \begin{pmatrix} y_1 \\ \vdots \\ y_n \end{pmatrix}.$$

[4.II]

Appendix

The resulting matrix has one row and one column; such a matrix we identify with the one entity it contains.

Since we leave open the question of whether or not a given vector is represented by a row or a column, we introduce the following agreement: If a vector **x** occurs as the left element in a product, it is to be considered a row vector. If it occurs as right element, it is to be considered a column vector. For general matrices, however, we must distinguish between a matrix **A** and the matrix $^T\mathbf{A}$ obtained from **A** by interchanging rows and columns:

$$\mathbf{A} = \begin{pmatrix} a_{11} & \cdots & a_{1m} \\ \vdots & & \vdots \\ a_{n1} & \cdots & a_{nm} \end{pmatrix}, \qquad ^T\mathbf{A} = \begin{pmatrix} a_{11} & \cdots & a_{n1} \\ \vdots & & \vdots \\ a_{1m} & \cdots & a_{nm} \end{pmatrix}.$$

We call $^T\mathbf{A}$ the *transposed* matrix of **A**. Clearly,

$$^T(^T\mathbf{A}) = \mathbf{A}.$$

We shall now prove some rules on matrix multiplication.

Theorem 4.II.1 $\mathbf{A} \cdot (\mathbf{B} \cdot \mathbf{C}) = (\mathbf{A} \cdot \mathbf{B}) \cdot \mathbf{C}$ *if all products are defined.*

Proof If **A**, **B**, **C** are $n \times n$ matrices, we can interpret $\mathbf{C} \cdot (\mathbf{B} \cdot \mathbf{A})$ as a succession of mappings. This is clearly associative (compare Section 1.8). Since, however, the matrices **A**, **B**, **C** need not be $n \times n$ matrices, we need a formal proof of the theorem. We calculate both sides explicitly:

$$\mathbf{A} \cdot (\mathbf{B} \cdot \mathbf{C}) = (a_{ik}) \Big(\sum_\rho b_{j\rho} c_{\rho l} \Big) = \Big(\sum_{\rho,\sigma} a_{i\sigma} b_{\sigma\rho} c_{\rho l} \Big),$$

$$(\mathbf{A} \cdot \mathbf{B}) \cdot \mathbf{C} = \Big(\sum_\sigma a_{i\sigma} b_{\sigma m} \Big) (c_{rl}) = \Big(\sum_{\sigma,\rho} a_{i\sigma} b_{\sigma\rho} c_{\rho l} \Big).$$

Carrying out the summations properly, we see that the right sides are equal. ◆

Though matrix multiplication is associative it is *not commutative*, even if the skew field $\mathscr{F}$ is commutative.

Example

$$\begin{pmatrix} 1 & 0 \\ 1 & 1 \end{pmatrix} \begin{pmatrix} 1 & 0 \\ 0 & -1 \end{pmatrix} = \begin{pmatrix} 1 & 0 \\ 1 & -1 \end{pmatrix} \neq \begin{pmatrix} 1 & 0 \\ -1 & -1 \end{pmatrix} = \begin{pmatrix} 1 & 0 \\ 0 & -1 \end{pmatrix} \begin{pmatrix} 1 & 0 \\ 1 & 1 \end{pmatrix}.$$

Figure A.4.2 illustrates this geometrically, the first matrix represents a reflection, the second a shear.

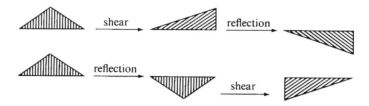

Figure A.4.2

Theorem 4.II.2 *Let the skew field $\mathscr{F}$ be commutative, that is, let it be a field $\mathscr{K}$. Then*

$$^{\mathrm{T}}(\mathbf{B} \cdot \mathbf{A}) = {}^{\mathrm{T}}\mathbf{A} \cdot {}^{\mathrm{T}}\mathbf{B}.$$

Proof We use the inner product notation:

$$^{\mathrm{T}}(\mathbf{B} \cdot \mathbf{A}) = \begin{pmatrix} \mathbf{b}^1\mathbf{a}^1 & \cdots & \mathbf{b}^n\mathbf{a}^1 \\ \vdots & & \vdots \\ \mathbf{b}^1\mathbf{a}^l & \cdots & \mathbf{b}^n\mathbf{a}^l \end{pmatrix}.$$

On the other hand,

$$^{\mathrm{T}}\mathbf{A} = \begin{pmatrix} \mathbf{a}^1 \\ \vdots \\ \mathbf{a}^l \end{pmatrix} \qquad {}^{\mathrm{T}}\mathbf{B} = (\mathbf{b}^1 \cdots \mathbf{b}^n),$$

$$^{\mathrm{T}}\mathbf{A} \cdot {}^{\mathrm{T}}\mathbf{B} = \begin{pmatrix} \mathbf{a}^1\mathbf{b}^1 & \cdots & \mathbf{a}^1\mathbf{b}^n \\ \vdots & & \vdots \\ \mathbf{a}^l\mathbf{b}^1 & \cdots & \mathbf{a}^l\mathbf{b}^n \end{pmatrix}.$$

Since $\mathscr{K}$ is commutative, $\mathbf{a}^i\mathbf{b}^h = \mathbf{b}^h \cdot \mathbf{a}^i$, and so Theorem 4.II.2 follows. ◆

There exists an $n \times n$ matrix $\mathbf{I}$ which, for any $n \times n$ matrix $\mathbf{A}$, satisfies $\mathbf{AI} = \mathbf{IA} = \mathbf{A}$, namely,

$$\mathbf{I} = \begin{pmatrix} 1 & 0 & \cdots & 0 \\ 0 & 1 & & \vdots \\ \vdots & & \ddots & 0 \\ 0 & \cdots & 0 & 1 \end{pmatrix},$$

called the **unit matrix** or **identity matrix**.

Matrices that have the same number of rows and columns can also be **added** "componentwise," as we did earlier in the special case of vectors:

$$\mathbf{A} + \mathbf{B} = \begin{pmatrix} a_{11} & \cdots & a_{1m} \\ \vdots & & \vdots \\ a_{n1} & \cdots & a_{nm} \end{pmatrix} + \begin{pmatrix} b_{11} & \cdots & b_{1m} \\ \vdots & & \vdots \\ b_{n1} & \cdots & b_{nm} \end{pmatrix}$$

$$= \begin{pmatrix} a_{11} + b_{11} & \cdots & a_{1m} + b_{1m} \\ \vdots & & \vdots \\ a_{n1} + b_{n1} & \cdots & a_{nm} + b_{nm} \end{pmatrix}.$$

If, in particular, $\mathbf{B}$ is the $n \times m$ matrix $\mathbf{0}$ consisting of zeros only, we have

$$\mathbf{A} + \mathbf{0} = \mathbf{0} + \mathbf{A} = \mathbf{A}.$$

$\mathbf{0}$ is called a *zero matrix*.

The following rules link matrix addition and matrix multiplication:

Theorem 4.II.3   *The distributive laws*

$$\mathbf{C} \cdot (\mathbf{A} + \mathbf{B}) = \mathbf{C} \cdot \mathbf{A} + \mathbf{C} \cdot \mathbf{B},$$

and

$$(\mathbf{A} + \mathbf{B}) \cdot \mathbf{C} = \mathbf{A} \cdot \mathbf{C} + \mathbf{B} \cdot \mathbf{C}$$

hold, provided all additions and multiplications are well defined.

Proof

$$\mathbf{C} \cdot (\mathbf{A} + \mathbf{B}) = (c_{ih})(a_{lj} + b_{lj}) = \left(\sum_{\rho} c_{i\rho}(a_{\rho j} + b_{\rho j}) \right)$$
$$= \left(\sum_{\rho} c_{i\rho}a_{\rho j} + \sum_{\rho} c_{i\rho}b_{\rho j} \right) = \mathbf{C} \cdot \mathbf{A} + \mathbf{C} \cdot \mathbf{B}.$$

The second law is proved correspondingly.   ◆

Finally, a multiplication of matrices by scalars can be defined:

$$a\mathbf{A} = a \begin{pmatrix} a_{11} & \cdots & a_{1m} \\ \vdots & & \vdots \\ a_{n1} & \cdots & a_{nm} \end{pmatrix} = \begin{pmatrix} aa_{11} & \cdots & aa_{1m} \\ \vdots & & \vdots \\ aa_{n1} & \cdots & aa_{nm} \end{pmatrix};$$

$$\mathbf{A}a = \begin{pmatrix} a_{11}a & \cdots & a_{1m}a \\ \vdots & & \vdots \\ a_{n1}a & \cdots & a_{nm}a \end{pmatrix}.$$

Theorem 4.II.4   *The set of all $n \times m$ matrices is a vector space under matrix addition and multiplication of matrices (from the left) by scalars.*

Exercises

1. Show that for the matrix $\mathbf{A}$ in Section 4.10, example (a), $\mathbf{A}^2 = \mathbf{I}$ holds. What is the geometrical meaning of this result?

2. Let

$$\mathbf{A} = \begin{pmatrix} 0 & 1 & 1 \\ 0 & 0 & 1 \\ 0 & 0 & 0 \end{pmatrix}.$$

Show that $\mathbf{A} \cdot \mathbf{A} \cdot \mathbf{A} = 0$.

3. Prove: The system of all $n \times n$ matrices with elements from a skew field $\mathscr{F}$ is a group under matrix addition. Matrix multiplication is also defined; it is associative and there exists a unit element **I**. Furthermore, the distributive laws hold (such a system is called a *ring* with a unit element). Which further laws of a skew field are satisfied by this system of matrices, and which are not?

4. Carry out all calculations which prove Theorem 4.II.4.

5. Show that the system of all $n \times n$ matrices of the form

$$\begin{pmatrix} d & 0 & \cdots & 0 \\ 0 & d & & \vdots \\ \vdots & & \ddots & 0 \\ 0 & \cdots & 0 & d \end{pmatrix},$$

with d an arbitrary element of a skew field $\mathscr{F}$, provide a skew field $\mathscr{F}'$ under matrix addition and matrix multiplication.

6. Two skew fields $\mathscr{F}_1$, $\mathscr{F}_2$ are called isomorphic if their additive groups and their multiplicative groups are isomorphic in the sense of Section 4.2. Show that in Exercise 5 the skew fields $\mathscr{F}$ and $\mathscr{F}'$ are isomorphic.

4.III *Inverse Mapping and the Inverse Matrix*

If a linear mapping

$$\mathbf{x}' = \mathbf{x}\mathbf{A} \tag{1}$$

is one-to-one, the inverse map exists, and it assigns to every $\mathbf{x}'$ a vector $\mathbf{x}$. We will expect that this mapping is again linear. In order to show this, we look for a matrix $\mathbf{A}^*$ such that

$$\mathbf{x} = \mathbf{x}'\mathbf{A}^*. \tag{2}$$

Substituting (1) in (2) we see that $\mathbf{A}^*$ must satisfy

$$\mathbf{A}\mathbf{A}^* = \mathbf{I}. \tag{3}$$

We call the matrix $\mathbf{A}^*$ the *inverse matrix* $\mathbf{A}^{-1}$ of $\mathbf{A}$ and write

$$\mathbf{A} \cdot \mathbf{A}^{-1} = \mathbf{I}. \tag{3'}$$

Before calculating $\mathbf{A}^*$, we express $\mathbf{A}$ as

$$\mathbf{A} = \begin{pmatrix} \mathbf{a}'_1 \\ \vdots \\ \mathbf{a}'_n \end{pmatrix} \qquad (\mathbf{a}_1, \ldots, \mathbf{a}_n \text{ rows of } \mathbf{I}),$$

where $\mathbf{a}'_1, \ldots, \mathbf{a}'_n$ are the images of $\mathbf{a}_1, \ldots, \mathbf{a}_n$ under (1). Now we can express (1) as

$$\mathbf{x}' = x_1\mathbf{a}'_1 + \cdots + x_n\mathbf{a}'_n. \tag{1'}$$

Example 379

$$
\mathbf{A} = \begin{pmatrix} 1 & 1 & -1 \\ 0 & 2 & 0 \\ 1 & 0 & 2 \end{pmatrix}, \qquad \mathbf{x}' = x_1(1, 1, -1) + x_2(0, 2, 0) + x_3(1, 0, 2).
$$

We now consider the spar products

$$
[\mathbf{a}'_1, \ldots, \mathbf{a}'_{i-1}, \mathbf{x}', \mathbf{a}'_{i+1}, \ldots, \mathbf{a}'_n] = x_i[\mathbf{a}'_1, \ldots, \mathbf{a}'_n] \qquad (i = 1, \ldots, n).
$$

Considering that (1) is one-to-one if and only if $|\mathbf{A}| = [\mathbf{a}'_1, \ldots, \mathbf{a}'_n] \neq 0$, we find *Cramer's rule*:

$$
x_i = \frac{1}{|\mathbf{A}|} [\mathbf{a}'_1, \ldots, \mathbf{a}'_{i-1}, \mathbf{x}', \mathbf{a}'_{i+1}, \ldots, \mathbf{a}'_n].
$$

Using minors in the sense of Section 4.12, we can write this as

$$
x_i = \frac{1}{|\mathbf{A}|} (x'_1 \mathbf{A}_{i1} + \cdots + x'_n \mathbf{A}_{in}).
$$

Thus, we obtain for $\mathbf{A}^* = \mathbf{A}^{-1}$:

$$
\mathbf{A}^* = \mathbf{A}^{-1} = \frac{1}{|\mathbf{A}|} \begin{pmatrix} \mathbf{A}_{11} & \cdots & \mathbf{A}_{1n} \\ \vdots & & \vdots \\ \mathbf{A}_{n1} & \cdots & \mathbf{A}_{nn} \end{pmatrix}. \tag{4}
$$

This completes the proof of the following theorem.

Theorem 4.III.1 *If a linear mapping*

$$
\mathbf{x}' = \mathbf{x}\mathbf{A}
$$

is one-to-one, the inverse mapping is also linear and can be written in the form (2), *where* $\mathbf{A}^*$ *is given by* (4).

It should be noted, however, that (4) does not, in general, provide the most convenient way of calculating $\mathbf{A}^{-1}$. Rather, (4) is of great theoretical interest. For the above example we find:

$$
\mathbf{A}^{-1} = \tfrac{1}{6} \begin{pmatrix} 4 & -2 & 2 \\ 0 & 3 & 0 \\ -2 & 1 & 2 \end{pmatrix}.
$$

Exercises

1. Verify by direct calculation $\mathbf{A}\mathbf{A}^{-1} = \mathbf{I}$ for the matrix $\mathbf{A}$ in the text.
2. Show that $\mathbf{A}^{-1}\mathbf{A} = \mathbf{I}$.

3. Find the inverse matrix of

$$A = \begin{pmatrix} 1 & 0 & 0 & 1 \\ 0 & -1 & 2 & 0 \\ 2 & 0 & 0 & 1 \\ -1 & 0 & 1 & 2 \end{pmatrix}.$$

4. Solve directly by using Cramer's rule the following system of linear equations:

$$x_1 + x_2 - x_4 = 0$$
$$2x_2 + 3x_3 = 1$$
$$x_1 - x_2 + x_3 = 0$$
$$x_3 + 2x_4 = -1.$$

Appendix

Chapter 6

6.1   Laguerre's Geometry

In this section we outline the geometry of "spears" and "cycles" and see that it shows an interesting analogy to inversive geometry.   There is a duality, however incomplete, between both geometries.

Let the ordinary Euclidean plane be given.   We assign to every line it's two orientations (see Section 2.4) and call any oriented line a ***spear***.   So every line "carries" two spears (Figs. A.6.1–A.6.3).   The totality of all

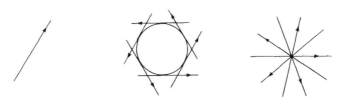

Figure A.6.1   **Figure A.6.2**   **Figure A.6.3**

spears that can be obtained from one spear by rotating it about a point R is called a ***cycle***.   Clearly the nearest points to R on the spears form a circle or a point.   Every circle thus "carries" two cycles.   However, every point "carries" only one cycle.

Two spears are said to be ***parallel*** if they can be translated into each other (with their orientations).   Parallel spears that are different do not belong simultaneously to a cycle.   This exception limits the analogy (in the sense of a duality) to inversive geometry (points corresponding to spears and circles corresponding to cycles).

We say two cycles are ***tangent*** to each other if they have one and only one spear in common.

Theorem 6.1.1   *Three spears no two of which are parallel belong to one and only one cycle.*

Theorem 6.I.2   Let S be a spear belonging to a cycle c, and let T be a spear not in c and not parallel to S.   There exists one and only one cycle d which contains S and T and is tangent to c.

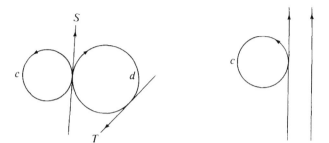

Figure A.6.4   **Figure A.6.5**

Theorem 6.I.3   If a spear T does not lie in a cycle c, there exists one and only one spear S in c which is parallel to T.

Theorem 6.I.4   There exist three spears in every cycle.   There exists a spear and a cycle not containing the spear.

As in inversive geometry, we write $\mathscr{C}(XYZW)$ for the statement that the spears X, Y, Z, W lie in a cycle.   It is a delicacy of the geometry of spears and cycles that Miquel's theorem is true.

Miquel's theorem   Let P, P', Q, Q', R, R', S, S' be different spears no two of which are parallel.   Then five of the relations $\mathscr{C}(PQRS)$, $\mathscr{C}(PP'QQ')$, $\mathscr{C}(RQS'P')$, $\mathscr{C}(RR'SS')$, $\mathscr{C}(SQ'R'P)$, $\mathscr{C}(P'Q'R'S')$ imply the sixth one (Fig. A.6.6).

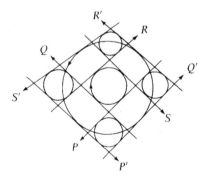

Figure A.6.6

If we are given (axiomatically) a system of "spears" and "cycles" with an incidence relation satisfying Theorems 6.I.1–6.I.4 (hereby two spears are called parallel if there is no cycle in which they both lie; two cycles are called

tangent if they have one and only one spear in common) we call this system a
Laguerre plane.

In an inversive plane we constructed a projective plane by deleting a point
R, calling all points $\neq R$ and the pencils of mutually tangent circles in R
"points," and taking as "lines" all circles through R as well as the system of
the above pencils. We apply a similar procedure to Laguerre planes:

Let R be a spear. As "points" we introduce:

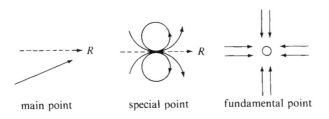

main point   special point   fundamental point

Figure A.6.7

1. All spears not parallel to R.
2. Any system of all cycles that are two-by-two tangent in R, together
   with R itself.
3. The system G of all parallel pencils of spears, where a parallel pencil of
   spears is defined as the set of all spears parallel to a spear S together
   with S itself.

The points of 1 are called **main points**; the points of 2, **special points**, and the
point of 3 is called the **fundamental point** (Fig. A.6.7).

As "lines" we define:

1. Any cycle c containing R. Its points are the spears $\neq R$ in it and the
   special point determined by R and c.

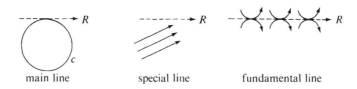

main line   special line   fundamental line

Figure A.6.8

2. A parallel pencil of spears not containing R. Its points are the spears
   of the pencil.
3. The set of all special points together with the fundamental point.

We call the lines of 1 **main lines**, the lines of 2 **special lines**, and the line of 3
the **fundamental line**.

Theorem 6.I.5 *The points and lines thus defined form a projective plane.*

Proof Two main points P, Q are joined by the cycle determined by P, Q, R (see Theorem 6.I.1) or by a special line. Two special points or a special point and a fundamental point are joined by a fundamental line. Theorem 6.I.2 guarantees that a main point and a special point are joined by a unique line. Finally, with the aid of Theorem 6.I.3, it is seen that every main point is situated on a unique special line which also contains a fundamental point.

Similarly, it can be shown that any two lines intersect. From Theorem 6.I.4 one deduces that every line contains at least three points, and that there exists a point and a line not containing the point. ◆

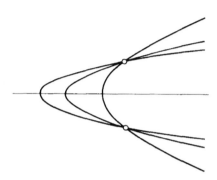

Figure A.6.9

We remark that with Miquel's theorem we can develop a theory analogous to the one developed for inversive planes in the preceding sections. The result is that in the projective plane of Theorem 6.I.5 all cycles appear as conics that are tangent to the fundamental line in the fundamental point. If we designate the fundamental line as the line at infinity, this is a system of all parabolas with parallel axes (see Fig. A.6.9 for a pencil of cycles).

Exercises

1. Prove Theorems 6.I.1–6.I.4 for spears and cycles as defined in the Euclidean plane.

2. *Blaschke model.* Let $\mathscr{Z}$ be a circular cylinder in Euclidean 3-space E^3 (given, for example, by the equation $x_1^2 + x_2^2 = 1$), and let the line g be a generator of $\mathscr{Z}$. We call all points on $\mathscr{Z}$ which are not on g spears, and we define cycles as plane sections of $\mathscr{Z}$ that do not contain a generator (nondegenerate conics). Two spears are called parallel if they are on the same generator of $\mathscr{Z}$. Show that this system of spears and cycles is a Laguerre plane.

3. In the Blaschke model of a Laguerre plane (Exercise 2) let A be a point on g, and let α be a plane of E^3 not passing through A and not containing

a generator of $\mathscr{Z}$. We extend E^3 to a projective space and denote the projective extension of α also by α. By projecting onto α from A, the cycles on $\mathscr{Z}$ are mapped onto conics of α as described at the end of Section 6.1. Discuss this in detail.

Appendix

4. Let the field $\mathscr{R}$ of real numbers be given and let ε be a symbol for which we set $\varepsilon^2 = 0$. Forming all linear polynomials $a + b\varepsilon$, where a, b are in $\mathscr{R}$, we obtain a system $\mathscr{L}$ of numbers called *dual numbers*. Show that $\mathscr{L}$ is a commutative ring (not a field, since $\varepsilon \cdot \varepsilon = 0$; for the definition of a ring see Appendix 4.II, Exercise 3).

5. Given the ring of dual numbers (see Exercise 4), define cross-ratios in the usual sense. Consider the numbers of $\mathscr{L}$ as spears and define cycles in strict analogy to Section 6.3 (ε instead of $\sqrt{d}$). Using the same methods as in Section 6.3, show that the cycles are conics as discussed in Exercise 3.

6. Prove Miquel's theorem for spears and cycles as defined in Exercise 5, following the method used in the proof of Theorem 6.4.3.

Suggestions for Further Reading

Chapter 1. This chapter arose out of the elaborate work of Bachmann and lecture notes by R. Lingenberg (unpublished). See also Hessenberg and Diller. In Weyl the idea of symmetry is beautifully treated from the viewpoint of both mathematics and art; this is also done in Fejes-Toth. Compare Bourbaki, Courant-Robbins, Hilbert and Cohn-Vossen, and Schaaf as well. For further reading on Sections 1.6 and Appendix 1.I, see Lanczos. For an elaborate book on finite geometries (Section 1.7, Exercise 5), see Dembowski. More about groups, discussed in Sections 1.8–1.10, can be found in Hall. See Jeger for more ideas about transformation geometry.

Chapter 2. Compare the classical treatment by Hilbert; see also Guggenheimer and Borsuk and Smielew. For an alternative treatment using metric postulates (in the sense of measure), see MacLane. On angles (Section 2.6), see Zassenhaus. About Section 2.8, compare Oppenheim. For more on ordinary Euclidean and hyperbolic planes (Sections 2.11 and 2.12), see Coxeter (1968), Eves, and Keedy and Nelson. Concerning Section 2.16, see Bachmann, Behnke and Fladt. More on finite and finitely generated groups of symmetries (Appendixes I and II for Chapter 2) is found in Fejes-Toth and Yale.

Chapter 3. Another elegant way of introducing coordinates (although somewhat more abstract) is presented in Artin, Chapter 2. A similar approach is given in Hartshorne. Concerning our treatment, compare Bachmann; the generalization to skew fields is due to Lingenberg (see Lingenberg). About the contribution of Descartes to coordinate geometry, see Boyer. For more on Section 3.4, see Klein (1939 and 1956) and Rédei. Further contributions to non-Desarguesian planes (Section 3.12) are in Levenberg. About quaternions (Section 3.13, Exercise 6), see Room and Du Val. For Sections 3.15 and 3.16, see Laugwitz and Moise. You will find historic notes about construction problems in boldface type.

Chapter 4. Among the numerous books on advanced analytic geometry and vector spaces, we recommend in particular, Artzy, Borsuk, Gruenberg, Weir, Halmos, and Kuiper. Concerning metric postulates for space geometry, see Brossard. For sections 4.13 and 4.14, see Pickert. About rotations and quaternions (Section 4.14), see Du Val. For further references on sections 4.15 and 4.16, see Coxeter (1961). See also Fejes-Toth. Applications to crystallography can be found in Yale.

Chapter 5. There are many classical treatments of purely synthetic projective geometry; we do not list them here. Bumcrot, Fishback, Hartshorne, Seidenberg, and Tuller combine the analytic and synthetic approaches. For a careful logical analysis of axioms in projective geometry, see Heyting. A treatment of classical projective geometry emphasizing projective metrics is given in Busemann and Kelly. For Section 5.2, see also Gans. Section 5.13: About quadrics in n-dimensional space, see Amir-Moéz and Fass. For more on projective geometry of arbitrary dimensions, see Baer.

Chapter 6. An elaborate work on plane inversive geometry (over $\mathscr{C}$) is that of Schwerdtfeger. It emphasizes nicely the group-theoretic standpoint but does not treat axioms of inversive geometry. See also Allen, Benz, Ewald, and Scherk. An elementary treatment of inversive geometry can be found in Pedoe. Concerning Laguerre's Geometry (Appendix 6.I), see Benz and Mäurer.

Chapter 7. The Fundamental theorem is due to F. Bachmann; see Bachmann, and also Hessenberg and Diller. For another treatment of Euclidean and non-Euclidean geometries using bilinear forms, see Busemann and Kelly. For the algebraic background of Chapter 7, compare Baer.

About the future of mathematics, in particular that of geometry, read the interesting article by Weil.

Bibliography

Allen, E. F. "An extended inversive geometry," *AMM*, 60 (1953): 233–237.

Amir-Moéz, A. R., and A. L. Fass. "Quadrics in R_n," *AMM*, 67 (1960): 632–636.

Artin, E. *Geometric Algebra.* New York: Interscience, 1957.

Artzy, R. *Linear Geometry.* Reading, Mass.: Addison-Wesley, 1965.

Bachmann, F. *Aufbau der Geometrie aus dem Spiegelungsbegriff* (Grundlehren der Mathematischen Wissenschaften 96). Berlin: Springer, 1959.

Bachmann, F., H. Behnke, and K. Fladt. *Grundzüge der Mathematik*, Band IIA: *Geometrie.* Göttingen: Vandenhoek & Rupprecht, 1967.

Baer, R. *Linear Algebra and Projective Geometry.* New York: Academic Press, 1952.

Barry, E. H. *Introduction to Geometrical Transformations.* Boston: Prindle, Weber & Schmidt Inc., 1966.

Benz, W. Über Möbiusebenen. Ein Bericht. *Jahresbericht DMV*, 63 (1960): 1–27.

Benz, W., and H. Mäurer. Über die Grundlagen der Laguerre-Geometrie. Ein Bericht. *Jahresbericht DMV*, 67 (1964): 14–42.

Blattner, John W. *Projective plane geometry.* San Francisco: Holden Day, 1968.

Bold, B. *Famous Problems of Mathematics; A History of Constructions with Straight Edge & Compass.* Van Nostrand, 1968.

Borsuk, Karol. *Multidimensional analytic geometry*, 1969. Warszawa: Polish Scient. Publ. Paris: Hermann, 1969.

Borsuk, K., and W. Smielew. *Foundations of Geometry.* Amsterdam: North-Holland, 1960.

Bourbaki, N. "The architecture of mathematics," *AMM*, 57 (1950): 221–232.

Boyer, C. B. "Descartes and the geometrization of algebra," *AMM*, 66 (1959): 390–393.

Brossard, R. "Metric postulates for space geometry," *AMM*, 74 (1967): 777–788.

Bumcrot, R. J. *Modern Projective Geometry.* New York: Holt, 1969.

Busemann, H., and P. J. Kelly. *Projective Geometry and Projective Metrics.* New York: Academic Press, 1953.

Courant, R., and H. E. Robbins. *What Is Mathematics?* New York: Oxford University Press, 1953.

Coxeter, H. S. M. *Introduction to Geometry.* New York: Wiley, 1961.

Coxeter, H. S. M. *Non-Euclidean Geometry.* Toronto, On.: University of Toronto Press, 1968.

Coxeter, H. S. M., and S. L. Greitzer. *Geometry Revisited.* New York: Random House, 1967.

Dembowski, P. *Finite Geometries.* Ergebnisse der Math. 44. Berlin: Springer, 1968.

Du Val, P. *Homographies, Quaternions, and Rotations.* New York: Oxford University Press, 1964.

Eberlein, W. F. "The spin model of Euclidean 3-space," *AMM,* 69 (1962): 587–598.

Eves, H. *A Survey of Geometry I, II.* Boston: Allyn and Bacon, 1963.

Ewald, G. "Ein Schließungssatz für Inzidenz und Orthogonalität in Möbiusebenen," *Math. Ann.,* 142 (1961): 1–21.

Fejes-Toth. *Regular Figures.* Elmsford, N.Y.: Pergamon, 1964.

Fishback, W. T. *Projective and Euclidean Geometry,* 2d ed. New York: John Wiley, 1969.

Gans, D. "Models of Projective and Euclidean Space," *AMM,* 65 (1958): 749–756.

Gruenberg, K. W., and A. J. Weir. *Linear Geometry.* Princeton: Van Nostrand, 1967.

Guggenheimer, H. W. *Plane Geometry and Its Groups.* San Francisco: Holden-Day, 1967.

Hall, M. *The Theory of Groups.* New York: Macmillan, 1959.

Halmos, P. R. *Finite-Dimensional Vector Spaces,* 2d. ed. Princeton, N.J.: Van Nostrand, 1958.

Hartshorne, R. *Foundations of Projective Geometry.* New York: Benjamin, 1967.

Hessenberg, G., and J. Diller. *Grundlagen der Geometrie.* Berlin: Walter de Gruyter u. Co., 1967.

Heyting, A. *Axiomatic Projective Geometry.* Groningen: Noordhoff, 1963.

Hilbert, D. *The Foundations of Geometry.* Chicago: Open Court, 1902.

Hilbert, D., and S. Cohn-Vossen. *Geometry and the Imagination.* New York: Chelsea, 1952.

Jeger, M. *Transformation Geometry.* New York: Wiley, 1966.

Keedy, M. L., and C. W. Nelson. *Geometry. A Modern Introduction.* Reading, Mass.: Addison-Wesley, 1965.

Klein, F. *Elementary Mathematics from an Advanced Standpoint: Vol. II, Geometry.* New York: Dover, 1939.

Klein, F. *Famous Problems of Elementary Geometry.* New York: Dover, 1956.

Kuiper, N. C. *Linear Algebra and Geometry.* Amsterdam: North-Holland, 1963.

Lanczos, C. *Albert Einstein and the Cosmic World Order.* New York: (Interscience) Wiley, 1965.

Laugwitz, D. "Eine Elementare Methode für Unmöglichkeitsbeweise bei Konstruktionen mit Zirkel und Lineal." *Elemente der Mathematik,* 17 (1962): 54–58.

Leisenring, K. "Area in non-Euclidean geometry," *AMM,* 58 (1951): 315–322.

Levenberg, K. "A class of non-Desarguesian plane geometries," *AMM,* 57 (1950): 381–387.

Lingenberg, R. *Grundlagen der Geometrie I.* Mannheim: BI, 1969.

MacLane, S. "Metric postulates for plane geometry," *AMM,* 66 (1959): 543–555.

Moise, E. *Elementary Geometry from an Advanced Standpoint.* Reading, Mass.:
Addison-Wesley, 1963.

Oppenheim, A. "The Erdös inequality and other inequalities for a triangle,"
AMM, 68 (1961): 226–230.

Pedoe, D. *Circles.* London: Pergamon Press, 1957.

Pickert, G. *Analytische Geometrie.* Leipzig: Geest & Portig.

Rédei, L. *Foundations of Euclidean and non-Euclidean Geometries According to
F. Klein.* Oxford: Pergamon Press, 1968.

Room, T. G. *A Background to Geometry.* New York: Cambridge University
Press, 1968.

Schaaf, W. L. "Art and mathematics: a brief guide to source materials,"
AMM, 58 (1951): 167–177.

Scherk, P. "Some concepts of conformal geometry," *AMM,* 67 (1960): 1–30.

Schwerdtfeger, H. *Geometry of Complex Numbers.* Toronto: University of
Toronto Press, 1962.

Seidenberg, A. *Lectures in Projective Geometry.* New York: Van Nostrand,
1962.

Tuller, A. *A Modern Introduction to Geometries.* Toronto: Van Nostrand,
1967.

Weil, A. "The future of mathematics," *AMM,* 57 (1950): 295–306.

Weyl, H. *Symmetry.* Princeton, N.J.: Princeton University Press, 1952.

Yale, P. B. *Geometry and Symmetry.* San Francisco: Holden-Day, 1968.

Zassenhaus, H. "What is an angle?" *AMM,* 61 (1954): 369–378.

Index of Axioms

Index of Axioms

Index

Index

Made in the USA
Middletown, DE
16 March 2017